Design Process for Additive Manufacturing

Design Process for Additive Manufacturing

Guest Editor

Paweł Turek

MDPI

Basel • Beijing • Wuhan • Barcelona • Belgrade • Novi Sad • Cluj • Manchester

Guest Editor
Paweł Turek
Department of
Manufacturing Techniques
and Automation
Rzeszów University of
Technology
Rzeszów
Poland

Editorial Office
MDPI AG
Grosspeteranlage 5
4052 Basel, Switzerland

This is a reprint of the Special Issue, published open access by the journal *Designs* (ISSN 2411-9660), freely accessible at: https://www.mdpi.com/journal/designs/special_issues/8A27O796W2.

For citation purposes, cite each article independently as indicated on the article page online and as indicated below:

Lastname, A.A.; Lastname, B.B. Article Title. *Journal Name* **Year**, *Volume Number*, Page Range.

ISBN 978-3-7258-5287-1 (Hbk)
ISBN 978-3-7258-5288-8 (PDF)
https://doi.org/10.3390/books978-3-7258-5288-8

Contents

designs

Editorial

Design Process for Additive Manufacturing

Paweł Turek

Faculty of Mechanical Engineering and Aeronautics, Rzeszów University of Technology, 35-959 Rzeszów, Poland; pturek@prz.edu.pl

1. Introduction

Additive Manufacturing (AM) techniques are rapidly emerging as leading technologies for the creation of complex models [1,2]. The AM process involves sequentially depositing material in layers until the complete model is achieved. The time required to produce finished models using additive techniques can vary significantly, ranging from several hours to several days, depending on factors such as the manufacturing technology used, the object's dimensions, and the complexity of its design [3]. Currently, there is a diverse range of AM techniques available in the market. In partnership with the International Organization for Standardization (ISO) and the American Society for Testing and Materials (ASTM), several standards, such as ISO/ASTM 52900 [4] and ISO/ASTM 52910 [5], have been established. These standards provide comprehensive descriptions of the additive techniques presently in use.

AM techniques are widely applied, including in the aerospace [6], automotive [7], and medical industries [8]. Since functional models are often produced using additive technologies, they must meet the design specifications. Unfortunately, there is still a lack of standardized design guidelines concerning areas such as mechanical strength assessments, geometrical tolerancing, and surface roughness. This lack of standards complicates the commercialization of finished products manufactured using these methods. The challenges in commercializing manufactured products are affected by the methods and parameters used during the design and AM phases. In the case of Computer-Aided Design (CAD) and Reverse Engineering (RE) design for AM, mechanical strength, geometric accuracy, and surface roughness are affected by the topology and geometry of a 3D-CAD model [9], CAD/RE software algorithms [10], chord and angular tolerance [11], mesh size, and scanning quality and data processing parameters in RE process [12]. The properties of a finished part produced through AM are influenced by a variety of factors. The mechanical strength of 3D-printed parts depends significantly on the type of material [13], the density and pattern of the internal infill in a model [14], number of shells [14], print direction [15], and proper nozzle [16] and bed temperatures [17], along with appropriate material cooling. The parameters that have the greatest impact on geometric accuracy and surface roughness include layer height [18], print speed [18], printer calibration [19], material shrinkage [19], print temperature and cooling [20], and print direction [21].

To achieve optimal mechanical strength, geometric accuracy, and surface roughness in the finished parts, a comprehensive approach is necessary. Conducting detailed scientific research is essential to optimize both the design process (CAD/RE) and the AM process. These two processes should not be treated separately; instead, they must be synchronized appropriately to ensure the highest quality, precision, and functionality of 3D prints.

Designs **2025**, *9*, 109 https://doi.org/10.3390/designs9050109

2. An Overview of Published Articles

When designing a 3D CAD model for AM, it is crucial to ensure that the geometric parameters align with the most favorable operating conditions, such as tightness, accuracy, component connections, wear, and deformation, among others. This task can be challenging, as each AM technology has its own technical limitations, often causing the final manufactured model to differ significantly from the designer's original intentions. Therefore, optimizing the geometry during the CAD modeling stage is essential to make it suitable for AM. Hence, the article by Metzger et al. (contribution 1) presents a new, user-friendly calculation model for the design of thin-walled, slit-shaped reactor modules created using AM. The simplified calculation model proved reliable, and a design graph was created based on the findings. Engineers can use this graph to design safe and robust additively manufactured reactor modules efficiently. The study highlights the potential of AM for creating high-performance process equipment with minimal material usage.

Geometry optimization is also undertaken in relation to the developed CAD models of prostheses. The authors of contribution 2 conducted a Finite Element Analysis (FEA) to compare two forelimb prosthetic designs for dogs: a solid model and a perforated model. Both models, made from ABS plastic, were tested under a simulated static load equivalent to 60% of a dog's body weight. The analysis revealed that both models remained within the material's elastic limit, indicating they would not experience permanent deformation under the applied loads. As a result of these findings, the perforated model was determined to be the more suitable option for developing canine prosthetics. Future work will involve analyzing the designs under dynamic loads, such as those encountered during walking and running.

Currently, research is also being conducted on optimizing geometry for AM using Artificial Intelligence (AI). The article by Zichar et al. (contribution 3) explores how AI tools, such as ChatGPT and Gemini, can assist or even replace students in code-based 3D modeling tasks. The study found that students are not yet able to delegate their work to AI tools fully. While AI can generate code for simple models, it often requires modifications and corrections to meet specific requirements. Furthermore, the article highlights that the increasing popularity of AI necessitates a reevaluation of assessment methods by educators. Instead of fearing that AI will take over students' work, teachers should focus on understanding how these tools can support, rather than replace, the learning process.

In the early stages of design, engineers often lack the necessary knowledge and tools to make informed decisions about manufacturing technologies. This can result in inefficient and costly choices. Although AM provides many advantages, it is not always the best solution for every application. The article by Salmi et al. (contribution 4) describes a hybrid Multi-Criteria Decision-Making (MCDM) method that aids engineers in selecting between AM and conventional manufacturing processes (such as machining) for specific components. This method analyzes several parameters to deliver a thorough assessment, including

- Part geometry: shape complexity and wall thickness;
- Material requirements: type of material and its properties;
- Production considerations: required accuracy, tolerances, and the number of parts to be produced;
- Economic factors: material and process costs, along with production time;
- Environmental factors: energy consumption and material waste.

This decision-making model is intended for use during the preliminary design phase.

When creating a 3D CAD model for AM, traditional CAD modeling techniques are commonly used. However, challenges often arise when there is a lack of technological or material documentation for a product. This situation is particularly prevalent when designing models of anatomical structures, museum artifacts, or other complex geometric shapes, where solid or surface design may not be feasible. The RE process can address these issues thanks to advancements in coordinate measuring systems, data processing software, and modern manufacturing techniques. Although this design process is frequently utilized for developing 3D CAD models intended for AM, it can sometimes result in geometric mapping errors during the design phase. Therefore, it is essential to establish procedures at the geometry design stage of the RE process to minimize these errors. The reconstruction process can be applied to different types of geometries. The article by Turek et al. titled (contribution 5) explores the errors that occur during RE and AM for models with simple geometries, including axisymmetric and primitive shapes. The research findings indicated that, overall, 95% of the points representing reconstruction errors are within the maximum deviation range of ±0.6 mm to ±1 mm. The highest errors in CAD modeling were attributed to the auto-surfacing method; overall, 95% of the points are within the average range of ±0.9 mm. In contrast, the smallest errors, averaging ±0.6 mm, occurred with the detect primitives method. Overall, on average, 95% of the points representing the surface of a model made using the additive manufacturing technology fall within the deviation range ±0.2 mm. The findings provide crucial insights for designers utilizing RE and AM techniques to create functional model replicas.

AM may also be used for models used in medicine. The article by Simarmata et al. (contribution 6) discusses a methodology for designing and producing personalized orthopedic insoles for patients with flat feet. The authors used an integrated design approach that combines several methods to create customized insoles that effectively reduce foot pressure. The analysis revealed that insoles 3D printed with a 20% auxetic infill made from TPU filament best aligned with patient preferences for both functionality and comfort. Measurements of pressure distribution confirmed that the insoles with the auxetic structure reduced maximum pressure by 25.4% compared to not using insoles, demonstrating their efficacy in distributing pressure across the foot. The article by Turek et al. (contribution 7) focuses on enhancing the accuracy of AM for medical applications, particularly in the reconstruction of the zygomatic bone. The authors explored the use of Masked Stereolithography (mSLA) technology as a production method for these models due to its cost-effectiveness and high precision. They tested two printing modes, standard and ultralight, on an Anycubic Photon M3 Premium printer to determine which mode yielded superior geometric accuracy. The models were then validated using a structured light scanner. The results indicated that the ultralight mode offered better surface accuracy, which is crucial for planning precise surgical procedures. Notably, over 70% of the models' surfaces fell within a deviation range of ±0.3 mm. This study demonstrates that when configured correctly, mSLA technology can effectively produce highly accurate surgical templates and implant-forming tools. This advancement significantly supports craniomaxillofacial reconstruction procedures.

The RE process and 3D printing can be used to create terrain models. The article by Chlost et al. (contribution 8) outlines a methodology for creating a 3D terrain model from point cloud data. This study aims to reduce mesh surface errors and incorporate a smoothing factor. The initial surface was generated from a square grid and subsequently converted to an input format suitable for a CAD environment. To minimize surface defects, a bilinear interpolation algorithm was applied. Accuracy analyses of the terrain mapping were conducted on three samples with different geometries using two options in Siemens NX software: Uniform Density (UD) and Variable Density (VD). The results indicated that changing the smoothing factor from 0% to 15% did not significantly affect accuracy. However, a marked increase in inaccuracy was noted at a smoothing factor of 20%. The developed methodology ensures high-accuracy mapping of digital data, which can be leveraged in manufacturing processes like 3D printing. Based on the results, three prints were produced using the Fused Deposition Modeling (FDM) method, each representing one of the analyzed terrain geometries.

The AM process is quite complex, and various 3D printing parameters, such as temperature, speed, and model orientation, directly influence the microstructure and mechanical properties of the material. Additionally, optimizing both the 3D printer design and the materials used in AM can impact issues like porosity, layer delamination, and internal stresses. These factors can significantly reduce the strength and dimensional accuracy of the final product. The article by Bradshaw et al. (contribution 9) is a review that examines how the unique layer-by-layer process of 3D Concrete Printing (3DPC) affects the long-term durability of the material. It outlines key challenges and recent advancements in mix design and testing methods. The review emphasizes that the durability of 3DPC can be significantly enhanced by optimizing both the mix design and the printing process. Additionally, it stresses the need for standardized testing protocols to accurately assess durability, particularly for issues such as freeze–thaw resistance, chloride ingress, and carbonation, which are particularly sensitive to the printing parameters and curing conditions. The article by Woods et al. (contribution 10) presents a novel design for a 3D-bioprinting printhead that enhances the precision and stability of the printing process. Many current bioprinters face challenges in accurately dispensing bio-inks, which are temperature-sensitive and contain living cells. Variations in the dispensing process or temperature can result in cell damage and decreased print quality. The authors propose a new printhead design that combines two essential control systems: pneumatic and thermal control. By integrating these systems into a modified extrusion bioprinter, they enable accurate and reliable dispensing of bio-inks, which is critical for tissue engineering. This study highlights how engineering innovations in hardware can advance bioprinting technology, facilitating the creation of more precise and functional tissue structures. Researching the implementation of 4D printing is currently essential. Four-dimensiaonal printing is an extension of traditional 3D printing. It involves the fabrication of objects using innovative materials that are capable of changing their shape, properties, or functionality over time in response to an external stimulus. The article by Jin et al. (contribution 11) defines the main areas of application:

- Biomedicine: smart implants, self-degrading tissue scaffolds, and drug delivery systems;
- Robotics: soft robots that can change shape and move without complex motors;
- Aerospace: components that can alter their shape in response to environmental conditions, such as aircraft wings;
- Civil engineering: self-healing structures.

The article concludes that 4D printing is still in its early stages, and the most significant challenges remain high costs, the limited stability of printed structures, and the need for further research into new materials.

3. Conclusions

The articles in the Special Issue discussed essential aspects of AM. They emphasize that this process, which involves the layer-by-layer deposition of material, requires optimization during both the design and manufacturing stages to achieve the desired properties, such as strength, accuracy, and surface quality. The quality of a finished 3D-printed part is influenced by a combination of complex factors that depend on the specific technology and machine settings used. These factors can be categorized into two main groups:

- Design parameters (CAD/RE): these include the model's topology, geometry, wall thickness, and the presence of features like sharp angles or hollow spaces. The algorithms used for triangulation and mesh optimization, as well as the chord and angular tolerances from the CAD to STL export process, also directly affect the final part's surface quality. In RE, the quality of the 3D scanner and data processing parameters are critical for accuracy;

- Manufacturing parameters: these factors directly impact the part's mechanical strength and geometric accuracy. They include the type of material used (e.g., PLA, ABS, nylon), infill density and pattern, the number of perimeters or "shells," print direction, and temperature and cooling settings. For geometric accuracy and surface roughness, the key parameters are layer height, print speed, printer calibration, and material shrinkage. The articles note that a smaller layer height improves precision but increases print time.

The articles also highlight several studies that demonstrate the importance of this optimization:

- Engineering and design: research on a simplified design method for thin-walled reactor modules allowed for the creation of safe and robust components with minimal material use. In another example, an FEA of a perforated prosthetic model for dogs was found to be more suitable than a solid model, highlighting the benefits of geometry optimization;
- Medical applications: studies show that AM is used to create personalized orthopedic insoles for flat-footed patients, where an auxetic infill reduced maximum foot pressure by 25.4%. Another study focused on using mSLA for surgical templates in zygomatic bone reconstruction, finding that an "Ultralight" printing mode provided the superior surface accuracy essential for surgical planning.
- Reverse engineering: reverse engineering is useful for creating models with complex shapes, which can lead to geometric mapping errors. One publication found that the highest errors occurred with the "auto-surfacing" method in CAD modeling.

The articles conclude that while standards like ISO/ASTM 52900 exist, there is still a lack of standardized design guidelines for critical areas like mechanical strength and dimensional tolerancing. Therefore, a comprehensive approach that links both design and manufacturing processes is essential to ensure high-quality and functional 3D prints.

Conflicts of Interest: The author declares no conflicts of interest.

List of Contributions:

1. Metzger, D.F.; Klahn, C.; Dittmeyer, R. A Simplified Design Method for the Mechanical Stability of Slit-Shaped Additively Manufactured Reactor Modules. *Designs* **2024**, *8*, 41. https://doi.org/10.3390/designs8030041.
2. Sarpong, J.; Khanafer, K.; Sheikh, M. 3D-Printed Prosthetic Solutions for Dogs: Integrating Computational Design and Additive Manufacturing. *Designs* **2025**, *9*, 107. https://doi.org/10.3390/designs9050107.
3. Zichar, M.; Papp, I. Contribution of Artificial Intelligence (AI) to Code-Based 3D Modeling Tasks. *Designs* **2024**, *8*, 104. https://doi.org/10.3390/designs8050104.
4. Salmi, A.; Vecchi, G.; Atzeni, E.; Iuliano, L. Hybrid Multi-Criteria Decision Making for Additive or Conventional Process Selection in the Preliminary Design Phase. *Designs* **2024**, *8*, 110. https://doi.org/10.3390/designs8060110.
5. Turek, P.; Bielarski, P.; Czapla, A.; Futoma, H.; Hajder, T.; Misiura, J. Assessment of Accuracy in Geometry Reconstruction, CAD Modeling, and MEX Additive Manufacturing for Models Characterized by Axisymmetry and Primitive Geometries. *Designs* **2025**, *9*, 101. https://doi.org/10.3390/designs9050101.
6. Simarmata, T.P.; Martawidjaja, M.; Harito, C.; Tobing, C.C.L. Three-Dimensional Printed Auxetic Insole Orthotics for Flat Foot Patients with Quality Function Development/Theory of Inventive Problem Solving/Analytical Hierarchy Process Methods. *Designs* **2025**, *9*, 15. https://doi.org/10.3390/designs9010015.

7. Turek, P.; Kubik, P.; Ruszała, D.; Dudek, N.; Misiura, J. Guidelines for Design and Additive Manufacturing Specify the Use of Surgical Templates with Improved Accuracy Using the Masked Stereolithography Technique in the Zygomatic Bone Region. *Designs* **2025**, *9*, 33. https://doi.org/10.3390/designs9020033.
8. Chlost, M.; Bazan, A. Comparison of Methods for Reconstructing Irregular Surfaces from Point Clouds of Digital Terrain Models in Developing a Computer-Aided Design Model for Rapid Prototyping Technology. *Designs* **2025**, *9*, 81. https://doi.org/10.3390/designs9040081.
9. Bradshaw, J.; Si, W.; Khan, M.; McNally, C. Emerging Insights into the Durability of 3D-Printed Concrete: Recent Advances in Mix Design Parameters and Testing. *Designs* **2025**, *9*, 85. https://doi.org/10.3390/designs9040085.
10. Woods, P.; Smith, C.; Clark, S.; Habib, A. Integrating Pneumatic and Thermal Control in 3D Bioprinting for Improved Bio-Ink Handling. *Designs* **2024**, *8*, 83. https://doi.org/10.3390/designs8040083.
11. Jin, Y.; Liu, J. 4D Printing: Research Focuses and Prospects. *Designs* **2024**, *8*, 106. https://doi.org/10.3390/designs8060106.

References

1. Gao, W.; Zhang, Y.; Ramanujan, D.; Ramani, K.; Chen, Y.; Williams, C.B.; Zavattieri, P.D. The status, challenges, and future of additive manufacturing in engineering. *Comput. Aided Des.* **2015**, *69*, 65–89. [CrossRef]
2. Ngo, T.D.; Kashani, A.; Imbalzano, G.; Nguyen, K.T.; Hui, D. Additive manufacturing (3D printing): A review of materials, methods, applications and challenges. *Compos. Part B Eng.* **2018**, *143*, 172–196. [CrossRef]
3. Gardan, J. Additive manufacturing technologies: State of the art and trends. *Addit. Manuf. Handb.* **2016**, *54*, 3118–3132. [CrossRef]
4. *ISO/ASTM 52900:2021*; Additive Manufacturing—General Principles—Fundamentals and Vocabulary. ISO: Geneva, Switzerland, 2021.
5. *ISO/ASTM 52910:2018*; Additive Manufacturing—Design—Requirements, Guidelines and Recommendations. ISO: Geneva, Switzerland, 2018.
6. Chu, M.Q.; Wang, L.; Ding, H.Y.; Sun, Z.G. Additive manufacturing for aerospace application. *Appl. Mech. Mater.* **2015**, *798*, 457–461. [CrossRef]
7. Böckin, D.; Tillman, A.M. Environmental assessment of additive manufacturing in the automotive industry. *J. Clean. Prod.* **2019**, *226*, 977–987. [CrossRef]
8. Javaid, M.; Haleem, A. Additive manufacturing applications in medical cases: A literature based review. *Alex. J. Med.* **2018**, *54*, 411–422. [CrossRef]
9. Jamshidi, P.; Aristizabal, M.; Kong, W.; Villapun, V.; Cox, S.C.; Grover, L.M.; Attallah, M.M. Selective laser melting of Ti-6Al-4V: The impact of post-processing on the tensile, fatigue and biological properties for medical implant applications. *Materials* **2020**, *13*, 2813. [CrossRef] [PubMed]
10. Jiang, J.; Xu, X.; Stringer, J. Support structures for additive manufacturing: A review. *J. Manuf. Mater. Process.* **2018**, *2*, 64. [CrossRef]
11. Baturynska, I. Statistical analysis of dimensional accuracy in additive manufacturing considering STL model properties. *Int. J. Adv. Manuf. Technol.* **2018**, *97*, 2835–2849. [CrossRef]
12. Klimecka-Tatar, D.; Krynke, M. Reverse engineering tools-3D scanning-as support for precise quality control in automated special processes. *Procedia Comput. Sci.* **2025**, *253*, 1933–1942. [CrossRef]
13. Saidi, T.; Hasan, M.; Amalia, Z. Tensile strength of natural fiber in different types of matrix. *Aceh Int. J. Sci. Technol.* **2022**, *11*, 136–144. [CrossRef]
14. Mazlan, M.A.; Anas, M.A.; Nor Izmin, N.A.; Abdullah, A.H. Effects of infill density, wall perimeter and layer height in fabricating 3D printing products. *Materials* **2023**, *16*, 695. [CrossRef] [PubMed]
15. Yao, T.; Deng, Z.; Zhang, K.; Li, S. A method to predict the ultimate tensile strength of 3D printing polylactic acid (PLA) materials with different printing orientations. *Compos. Part B Eng.* **2019**, *163*, 393–402. [CrossRef]
16. Triyono, J.; Sukanto, H.; Saputra, R.M.; Smaradhana, D.F. The effect of nozzle hole diameter of 3D printing on porosity and tensile strength parts using polylactic acid material. *Open Eng.* **2020**, *10*, 762–768. [CrossRef]
17. Grasso, M.; Azzouz, L.; Ruiz-Hincapie, P.; Zarrelli, M.; Ren, G. Effect of temperature on the mechanical properties of 3D-printed PLA tensile specimens. *Rapid Prototyp. J.* **2018**, *24*, 1337–1346. [CrossRef]
18. Buj-Corral, I.; Bagheri, A.; Sivatte-Adroer, M. Effect of printing parameters on dimensional error, surface roughness and porosity of FFF printed parts with grid structure. *Polymers* **2021**, *13*, 1213. [CrossRef] [PubMed]

19. Turek, P.; Bazan, A.; Kubik, P.; Chlost, M. Development of a Calibration Procedure of the Additive Masked Stereolithography Method for Improving the Accuracy of Model Manufacturing. *Appl. Sci.* **2025**, *15*, 7412. [CrossRef]
20. de Castro Gonçalves, C.; Malavolta, A.T.; Barbosa, G.F.; Shiki, S.B.; Parra, E.R. Influence of bed temperature on mechanical resistance, surface and dimensional quality of 3D printed parts by robotized fused pellet modeling. *Prog. Addit. Manuf.* **2025**, *10*, 3827–3841. [CrossRef]
21. Turek, P.; Bazan, A.; Bulicz, M. Effect of 3D Printing Orientation on the Accuracy and Surface Roughness of Polycarbonate Samples. *Machines* **2024**, *13*, 9. [CrossRef]

designs

MDPI

Review

Emerging Insights into the Durability of 3D-Printed Concrete: Recent Advances in Mix Design Parameters and Testing

James Bradshaw [1], Wen Si [1,2], Mehran Khan [1,2,*] and Ciaran McNally [1,2,*]

1 Centre for Critical Infrastructure, School of Civil Engineering, University College Dublin, D04 V1W8 Belfield, Ireland; wen.si@ucd.ie (W.S.)
2 Construct Innovate, School of Civil Engineering, University College Dublin, D04 V1W8 Belfield, Ireland
* Correspondence: khan.mehran@ucd.ie (M.K.); ciaran.mcnally@ucd.ie (C.M.)

Abstract

Although 3D-printed concrete (3DPC) offers advantages such as faster construction, reduced labour costs, and minimized material waste, concerns remain about its long-term durability. This review examines these challenges by assessing how the unique layer-by-layer manufacturing process of 3DPC influences key material properties and overall durability. The formation of interfacial porosity and anisotropic microstructures can compromise structural integrity over time, increasing susceptibility to environmental degradation. Increased porosity at layer interfaces and the presence of shrinkage-induced cracking, including both plastic and autogenous shrinkage, contribute to reduced durability. Studies on freeze–thaw performance indicate that 3DPC can achieve durability comparable to cast concrete when proper mix designs and air-entraining agents are used. Chemical resistance, particularly under sulfuric acid exposure, remains a challenge, but improvements have been observed with the inclusion of supplementary cementitious materials such as silica fume. In addition, tests for chloride ingress and carbonation reveal that permeability and resistance are highly sensitive to printing parameters, material composition, and curing conditions. Carbonation resistance, in particular, appears to be lower in 3DPC than in traditional concrete. This review highlights the need for further research and emphasizes that optimizing mix designs and printing processes is critical to improving the long-term performance of 3D-printed concrete structures.

Keywords: durability; 3D-printed concrete; chloride penetration; carbonation resistance; freeze–thaw resistance

1. Introduction

Three-dimensional (3D) printing, also known as additive manufacturing, is a technology that creates objects by sequentially depositing layers of material based on a digital model. In concrete manufacturing, this process involves extruding layers of concrete to form structures without the need for traditional formwork, vibration, or compaction [1]. The adoption of 3D printing in concrete construction presents numerous benefits, including significant reductions in construction time, labor costs, and material waste, thereby offering substantial social, economic, and environmental advantages [2–7]. For instance, 3D printing technology can reduce construction production times by approximately 50% to 70%, addressing critical housing shortages by enabling rapid provision of new homes [3,4]. An example of this is the social housing development at Grange Close in Dundalk, Ireland.

 https://doi.org/10.3390/designs9040085

Economically, 3D-printed construction can significantly cut labor costs by 50% to 80% compared to traditional methods [6,8]. Environmentally, this method can reduce construction waste by 30% to 60%, owing to increased precision and reduced use of materials, thereby also lowering associated carbon emissions [3,7].

Despite these benefits, 3D-printed concrete (3DPC) introduces specific challenges that can negatively affect its performance compared to conventionally cast concrete. The layer-by-layer deposition technique inherently leads to issues such as increased porosity, reduced interlayer bonding strength, and heightened susceptibility to shrinkage cracking [9–11]. Moreover, the directional nature of material deposition (anisotropy) can result in variations in mechanical properties, such as tensile strength and crack resistance, potentially compromising structural integrity under load and environmental exposure [12]. Specifically, the anisotropic properties observed in 3D-printed concrete significantly impact compressive and dynamic strength, varying according to printing direction due to differences in interlayer bonding and material alignment [13–15]. These drawbacks can collectively undermine the overall durability of 3D-printed concrete structures.

Durability is a critical parameter in evaluating the long-term performance and structural integrity of concrete structures and becomes even more essential for 3DPC due to the unique characteristics of the printing process. Issues such as increased porosity and weaker mechanical properties can accelerate deterioration mechanisms like freeze–thaw cycles, chloride ingress, carbonation, and chemical attacks, necessitating a thorough understanding of durability performance [11,12,16]. Figure 1 illustrates the key factors influencing the durability performance of 3D-printed concrete, which were selected as core criteria for evaluating long-term material performance in this study [17]. The diagram identifies four critical domains: (1) concrete mix proportion, which directly affects the mechanical properties, workability, and long-term stability of the printed structure; (2) printing process parameters, such as nozzle speed, layer height, and interlayer time, which significantly impact bond strength and structural integrity; (3) concrete porosity and permeability, which govern the ease with which moisture, gases, and aggressive substances penetrate the matrix, thus affecting durability; and (4) aggressive ions, type and concentration, including chlorides, sulfates, and other deleterious agents that can trigger chemical attacks, corrosion, and degradation over time. These criteria were selected based on their recurring significance across the reviewed literature and their interconnected influence on printed concrete structures' performance and service life.

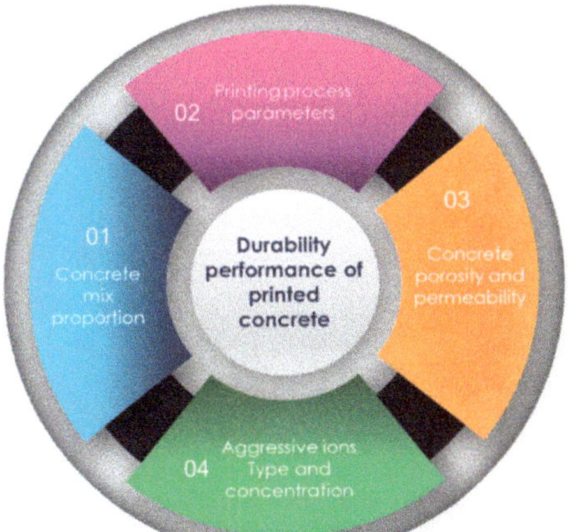

Figure 1. Durability performance of printed concrete [17].

This review aims to comprehensively address these durability challenges associated with 3D-printed concrete by synthesizing existing research and experimental evidence. It evaluates key deterioration mechanisms, including porosity, shrinkage, freeze–thaw resistance, sulfate/acid resistance, chloride penetration, and carbonation. Through quantitative summaries, comparative analyses, and critical discussions, this review identifies current knowledge gaps and highlights areas requiring further research and standardization. Ultimately, this comprehensive understanding is crucial to advancing 3DPC technology, ensuring its reliability and acceptance as a durable and sustainable construction material.

2. Durability Properties

2.1. Porosity

Porosity is one of the main factors that influences the durability of concrete [16]. Porosity is generally determined by the following characteristics: air/void pores, capillary pores, dead-end pores, and gel pores [16]. Conventionally cast concrete typically has a homogenous pore distribution, whereas 3DPC exhibits varying porosity, with the highest porosity found at the layer interfaces [17,18]. This can be explained by the construction process, with cast concrete being vibrated, reducing porosity, and 3DPC being used as printed [16]. The increased porosity at interlayer boundaries in 3DPC is worsened by anisotropic features from the directional layer-by-layer process. This can weaken mechanical strength and reduce durability [16,18], as shown in Figure 2.

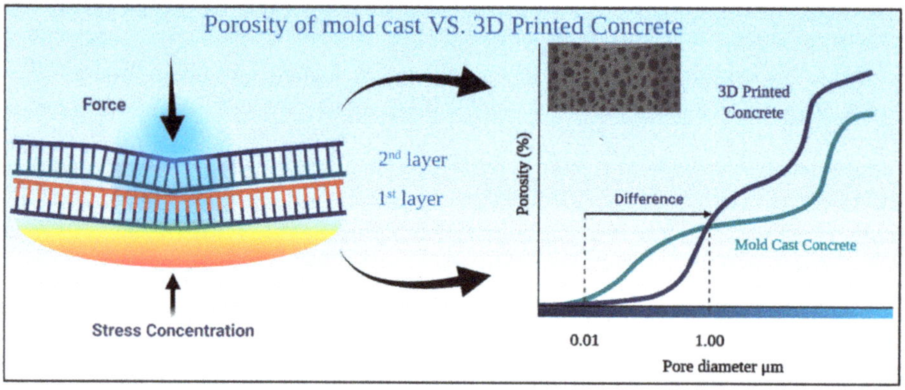

Figure 2. Porosity of conventionally cast versus 3DPC, reproduced with permission from [16].

Specifically, anisotropy introduces elongated and interconnected voids at layer interfaces. These pores align along the printing path. They are more open and better connected than the randomly distributed, isotropic pores in cast concrete. X-ray microcomputed tomography (X-CT) imaging has shown that interfacial zones in 3DPC contain pores with more irregular, crack-like shapes, and lower tortuosity, which enhance connectivity and allow fluids to migrate more easily along the printing direction [19,20]. This directional pore structure facilitates the transport of aggressive agents such as chlorides, sulfates, and CO_2, leading to faster deterioration. These characteristics stand in clear contrast to the isolated and more uniformly shaped pores in vibrated cast concrete. This contrast is further illustrated in Figure 3, which compares pore structure visualization between cast concrete and 3D-printed concrete using high-resolution 3D CT analysis [21]. Cast concrete displays a well-dispersed distribution of spherical pores with minimal connectivity, whereas the 3D-printed sample clearly shows a dense accumulation of elongated, interconnected pores along the interlayer zone, indicating higher directionality and permeability [22].

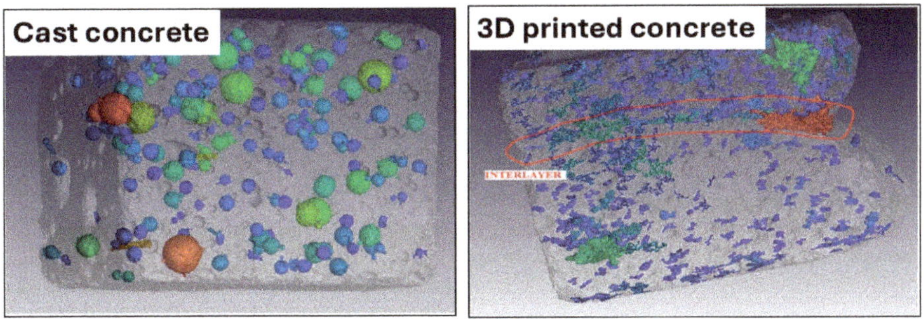

Figure 3. Micro CT investigation in void characteristics of conventionally cast concrete and 3D-printed concrete, reproduced with permission from [21].

These voids increase the permeability of the material and allow easier ingress of aggressive agents such as chlorides and sulfates. As a result, deterioration accelerates and overall durability decreases [16,20,23]. Despite the significance of this relationship, research explicitly examining how anisotropy-driven porosity variations impact long-term durability remains limited. Additives and admixtures such as nano-silica can be incorporated into 3DPC mixes to reduce porosity and subsequently improve durability [24].

Many different test procedures can study the porosity of concrete. Vacuum saturation and mercury intrusion porosimetry (MIP) are used to study the global porosity, and the interlayer porosity is studied by the spatial method, which uses optical or electronic microscopy and X-CT [25]. The advantage of the MIP test is its ability to characterize a wide range of pore sizes. However, there is a wide range of disadvantages, including damage to the microstructure caused by the injection of the mercury, cutting printed sections into small samples, and the impact of the drying procedure to remove pore water, which can heavily impact the pore structure through cracking and decomposition of hydrates [25]. The X-CT test is nondestructive, which can make it more viable.

Figure 4 below shows the process of the X-CT test. The equipment uses a high-resolution sample scan to generate a 3D image that shows the spatial distribution of porosity [26]. During scanning, X-rays penetrate the sample while rotating 360 degrees, and the transmitted signals are captured by a detector from multiple angles. These 2D projections are then computationally reconstructed into a 3D model using image reconstruction algorithms. In this study, slices were extracted from critical regions between printed layers and strips, allowing for localized pore segmentation and analysis. This enables a more accurate and detailed evaluation of the pore structure and interlayer heterogeneity, which are essential for understanding the performance and durability of 3D-printed concrete.

Table 1 demonstrates that porosity is highly dependent on both the mix composition and curing regime. Printed concrete generally exhibits higher porosity at layer interfaces compared to cast concrete. However, this disadvantage can be mitigated by incorporating supplementary cementitious materials (SCMs). For instance, the inclusion of 10–16% silica fume reduced porosity significantly, with values decreasing from 30.1% in the reference mix to as low as 23.2%. Similarly, a combination of metakaolin and nano-silica (C-MK-10%-NS-1%) yielded a 56.9% reduction in porosity compared to the control mix made with 100% OPC. Notably, porosity levels in 3DPC mixes ranged from a minimum of 0.09% to a maximum of over 40%, depending on the materials and testing methods used. These findings affirm that properly optimized mixes can lead to 3DPC having equal or even lower porosity than conventionally cast concrete, directly enhancing its long-term durability.

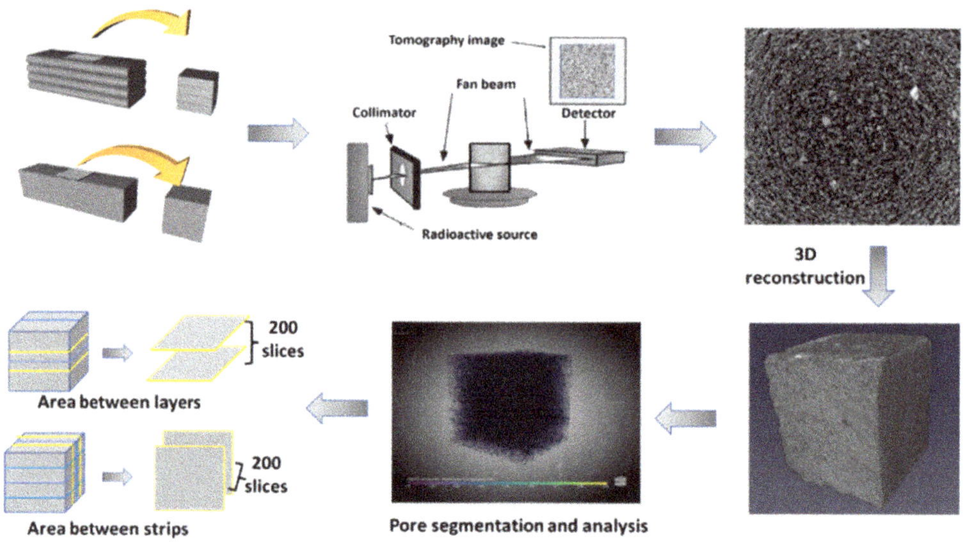

Figure 4. The flowchart of sample extraction and X-CT analysis, reproduced with permission from [27].

Table 1. Analysis of porosity in 3DPC literature: mix design parameters and testing.

Ref.	Mix	Binder Description	Mix Description	Test Method	Curing	Cast Porosity (%)	Printed Porosity (%)
[28]	A	100% OPC	VMA%/C 0.4, HRWR%/C 0.81	MIP	38 days @ 20 °C, 6 days @ 50 °C	13.58	13.11
	B	100% OPC	VMA%/C 0.47, HRWR%/C 0.95	MIP	38 days @ 20 °C, 6 days @ 50 °C	13.74	12.89
	C	100% OPC	VMA%/C 0.4, HRWR%/C 1.52	MIP	38 days @ 20 °C, 6 days @ 50 °C	11.23	11.67
[29]	M1	100% MK	-	XCT	28 days	4.48	2.98
				C + G	90 days	10	14.5
	M2	95% MK 5% Slag	-	XCT	28 days	4.07	1.81
				C + G	90 days	8	10.9
[26]	Control	100% OPC	100% SS	XCT	-	-	0.11
	AT50	90% OPC 10% SF	50% SS 50% AT	XCT	-	-	0.09
	AT100	90% OPC 10% SF	100% AT	XCT	-	-	0.3

Table 1. *Cont.*

Ref.	Mix	Binder Description	Mix Description	Test Method	Curing	Cast Porosity (%)	Printed Porosity (%)
[30]	REF	50% OPC 50% GGBS	S/B 1	VS	28 days @RH = 60	-	19.75
					28 days @RH > 95	-	9
	M1	100% OPC	S/B 1.5	VS	28 days @RH = 60	-	4.25
					28 days @RH > 95	-	4.5
	M2	75% OPC 25% GGBS	S/B 1.5	VS	28 days @RH = 60	-	12
					28 days @RH > 95	-	8.5
[31]	P	44.6% OPC 29.1% FA 26.3% LP	S/B 1 W/B 0.3	XCT	-	-	1.1
	H	61.7%OPC 32.7% FA 5.6% SF	S/B 1 W/B 0.2	XCT	-	-	1.8
[27]	REF	100% OPC	-	XCT	28 days @ 20 °C RH > 95%	38.9	30.1
	H012	100% OPC	0.6 g HPMC	XCT	28 days @ 20 °C RH > 95%	39.4	34.2
	H02	100% OPC	1 g HPMC	XCT	28 days @ 20 °C RH > 95%	41.2	36.4
	H03	100% OPC	1.5 g HPMC	XCT	28 days @ 20 °C RH > 95%	45.1	40.6
	S6	94% OPC 6% SF	-	XCT	28 days @ 20 °C RH > 95%	36.2	27.7
	S10	90% OPC 10% SF	-	XCT	28 days @ 20 °C RH > 95%	35.1	25.7
	S16	84% OPC 16% SF	-	XCT	28 days @ 20 °C RH > 95%	33.8	23.2
	S012 + S6	94% OPC 6% SF	0.6 g HPMC	XCT	28 days @ 20 °C RH > 95%	38.5	34
[32]	Cast	70% OPC 20% FA 10% SF	S/B 1.5 W/B 0.25	MIP	28 days	8.16	-
				XCT	-	8.07	-
	1Layer	70% OPC 20% FA 10% SF	S/B 1.5 W/B 0.25	MIP	28 days	-	7.86
				XCT	-	-	5.1
	3 Layer	70% OPC 20% FA 10% SF	S/B 1.5 W/B 0.25	MIP	28 days	-	6.7
				XCT	-	-	5.09

Table 1. *Cont.*

Ref.	Mix	Binder Description	Mix Description	Test Method	Curing	Cast Porosity (%)	Printed Porosity (%)
[33,34]	Cast	96% OPC 2% NC 2% SF	0.26% HRWR	XCT	-	4.89	-
	T1	96% OPC 2% NC 2% SF	0.26% HRWR	XCT	-	-	2.52
	T2	96% OPC 2% NC 2% SF	0.26% HRWR	XCT	-	-	2.73
	T3	96% OPC 2% NC 2% SF	0.26% HRWR	XCT	-	-	1.65
[35]	Cast	100% OPC	SS/B 1.5 W/B 0.32	XCT	28 days	14.6	-
	U_{T5}	100% OPC	SS/B 1.5 W/B 0.32	XCT	28 days	-	17.6
	L_{T14}	100% OPC	SS/B 1.5 W/B 0.32	XCT	28 days	-	12.4
[23]	Cast	83% OPC 17%SF	S/B 1.78 W/B 0.47	XCT	Ambient	2.67	-
					Sulfuric Acid	2.24	-
	3DP-6-X	83% OPC 17%SF	S/B 1.78 W/B 0.47	XCT	Ambient	-	1.26
					Sulfuric Acid	-	1.1
	3DP-12-X	83% OPC 17%SF	S/B 1.78 W/B 0.47	XCT	Ambient	-	2.31
					Sulfuric Acid	-	1.97
[36]	Mold	70% OPC 20% FA 10% SF	S/B 1.4 W/B 0.45	VS	28 days	6.05	-
	Lab	70% OPC 20% FA 10% SF	S/B 1.4 W/B 0.45	VS	28 days	-	7.025
	R-Site	70% OPC 20% FA 10% SF	S/B 1.4 W/B 0.45	VS	3 h @ 6.3 m/s airflow, 28 days	-	6.15
	U-Site	70% OPC 20% FA 10% SF	S/B 1.4 W/B 0.45	VS	3 h @ 6.3 m/s airflow, 28 days	-	6.5
[32,37],	M	70% OPC 20% FA 10% SF	S/B 1 W/B 0.35	XCT	90 days @ 20 °C RH > 95%	2.15	1.83
[38]	Cast/BM	55% OPC 30% MK 12% $CaCO_3$ 3%G	PVA%/B 0.225 S/B 1.5 W/B 0.4	VS	90 days @ 20 °C, 7 days @ 50 °C	7.5	14.9
	OL2	55% OPC 30% MK 12% $CaCO_3$ 3%G	PVA%/B 0.225 S/B 1.5 W/B 0.4	VS	90 days @ 20 °C, 7 days @ 50 °C	-	9
	OL4	55% OPC 30% MK 12% $CaCO_3$ 3%G	PVA%/B 0.225 S/B 1.5 W/B 0.4	VS	90 days @ 20 °C, 7 days @ 50 °C	-	8.2
	Z′	55% OPC 30% MK 12% $CaCO_3$ 3%G	PVA%/B 0.225 S/B 1.5 W/B 0.4	VS	90 days @ 20 °C, 7 days @ 50 °C	-	8.9

Table 1. *Cont.*

Ref.	Mix	Binder Description	Mix Description	Test Method	Curing	Cast Porosity (%)	Printed Porosity (%)
[39]	CM	100% OPC	S/B 1.47 W/B 0.35	MIP	28 days	-	10.2
	C-NS-1%	99% OPC 1% NS	S/B 1.47 W/B 0.35	MIP	28 days	-	7.6
	C-NS-2%	98% OPC 2% NS	S/B 1.47 W/B 0.35	MIP	28 days	-	8.5
	C-MK-5%	95% OPC 5% NS	S/B 1.47 W/B 0.35	MIP	28 days	-	7.45
	C-MK-10%	90% OPC 10% MK	S/B 1.47 W/B 0.35	MIP	28 days	-	6
	C-MK-10%-NS-1%	89% OPC 10% MK 1% NS	S/B 1.47 W/B 0.35	MIP	28 days	-	4.4
	C-MK-10%-NS-2%	88% OPC 10% MK 2%NS	S/B 1.47 W/B 0.35	MIP	28 days	-	5.45

Note: Ordinary Portland Cement (OPC), Viscosity-Modifying Agent (VMA), High-Range Water Reducer (HRWR), Mercury Intrusion Porosity (MIP), Metakaolin (MK), X-ray Computed Topography (XCT), Capillary + Gel (C + G) Porosity, Silica Fume (SF), Silica Sand (SS), Antimony Tailings (AT), Ground Granulated Blast-Furnace Slag (GGBS), Sand/Binder (S/B), Relative Humidity (RH), Vacuum Saturation (VS), Water/Binder (W/B), Fly Ash (FA), Limestone Powder (LP), (HPMC), Calcium Carbonate ($CaCO_3$), Gypsum (G), Polyvinyl Alcohol (PVA) Fibers.

The results found in papers, including porosity testing, are displayed in Figures 5 and 6. The bar chart shows the percentage decrease in porosity of the optimal mix compared to the reference mix. All the papers showed varying levels of improvement in terms of reducing porosity, excluding Bekaert et al. [30]. Bekaert et al. [30] compared printed mixes with different percentages of ground granulated blast-furnace slag (GGBS) under different curing conditions. The paper found that M2 (75% OPC and 25% GGBS) had a 180% increase in porosity compared to M1 (100% OP) [30]. Shafiq et al. [39] found the greatest decrease in porosity, with a 56.9% decrease when comparing CM (100% OPC) and C-MK-10%-NS-1%. The study found that adding metakaolin (MK) and nano silica (NS) can greatly improve the porosity of a printed sample. Du et al. [31] also looked at reducing the amount of ordinary Portland cement (OPC) used in the binder by supplementing it with fly ash (FA), limestone powder (LP), and silica fume (SF). The results found a 39% decrease in porosity for P (44.6% OPC 29.1% FA 26.3% LP) compared to H (61.7%OPC 32.7% FA 5.6% SF). This shows that reducing the portion of OPC decreases the porosity [31]. Liu et al. [27] also found that reducing OPC decreases the porosity. Comparing a reference mix (100% OPC) with S16 (84% OPC, 16% SF) showed that S16 had a 22.9% decrease in porosity [27]. The literature shows that decreasing the proportion of OPC used in the binder and replacing it with nano-clays and silica decreased the porosity of 3DPC.

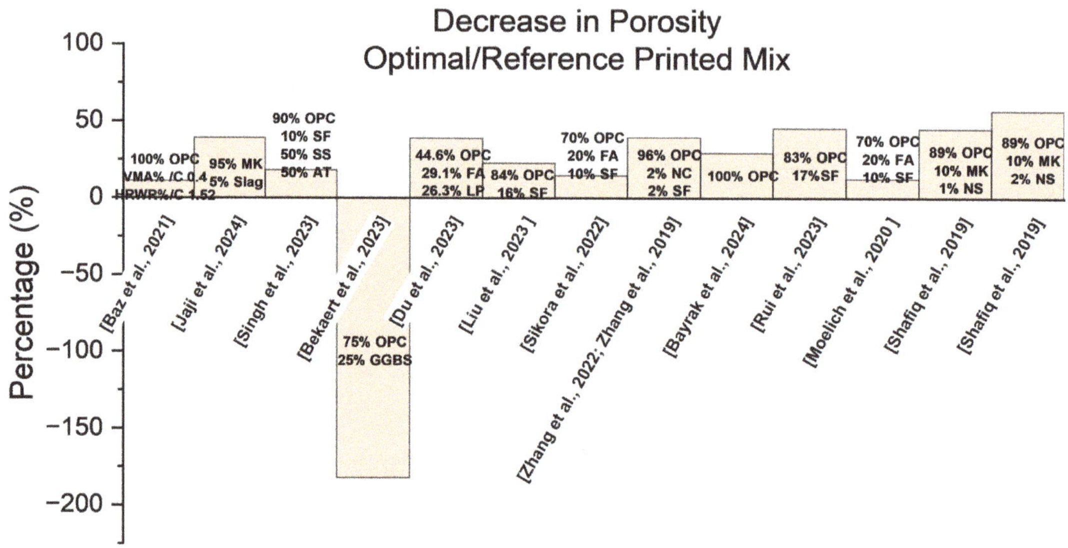

Figure 5. Decrease in porosity: optimal vs. reference printed mix [23,26–36,39].

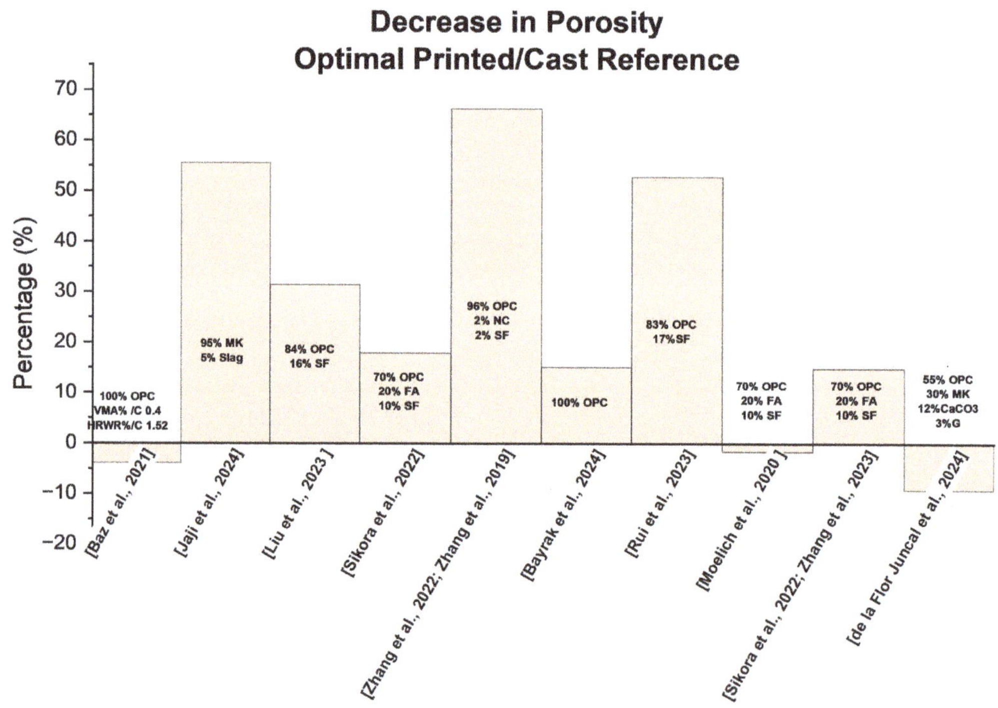

Figure 6. Decrease in porosity: optimal printed vs. cast reference [23,27–29,32–38].

According to the literature, 3DPC has a lower porosity than cast concrete. De la Flor Juncal et al. [38] found this exception when comparing the overlap between adjacent printed layers of concrete. The study used a concrete mix of 55% OPC, 30% MK, 12% $CaCO_3$, and 3% G. It found that an overlap of 4mm (Figure 7) had the lowest porosity of the printed samples; however, when compared to the cast sample, there was a 9.3% increase in porosity.

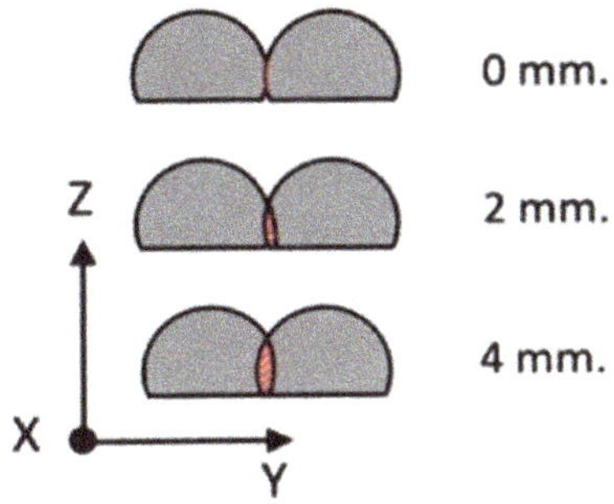

Figure 7. A section of the sample print layer overlapping [39].

2.2. Shrinkage

Overall, 3DPC is susceptible to three different forms of shrinkage: plastic, autogenous, and drying (Figure 8). This is due to the lack of formwork that helps prevent evaporation [16,29,40]. Plastic shrinkage is the moisture loss and contraction before setting [16,40]. It is caused by rapid moisture loss through evaporation. This evaporation rate increases due to the lack of formwork [40]. Plastic shrinkage can greatly reduce the durability of concrete as it can cause cracking and slippage between the printed layers [40]. Autogenous shrinkage refers to the continued reaction caused by cement hydration, leading to decreased volume. The process is known as self-desiccation. It occurs after the plastic shrinkage has ceased [16,40]. Autogenous shrinkage is the most dominant type of shrinkage in terms of 3DPC due to the faster setting time of 3DPC compared to cast concrete [16]. Autogenous shrinkage causes printed sections to deform before hardening and then crack once hardened, which may cause a large loss in durability [16]. Drying shrinkage is caused by water loss after hardening concrete [16,40].

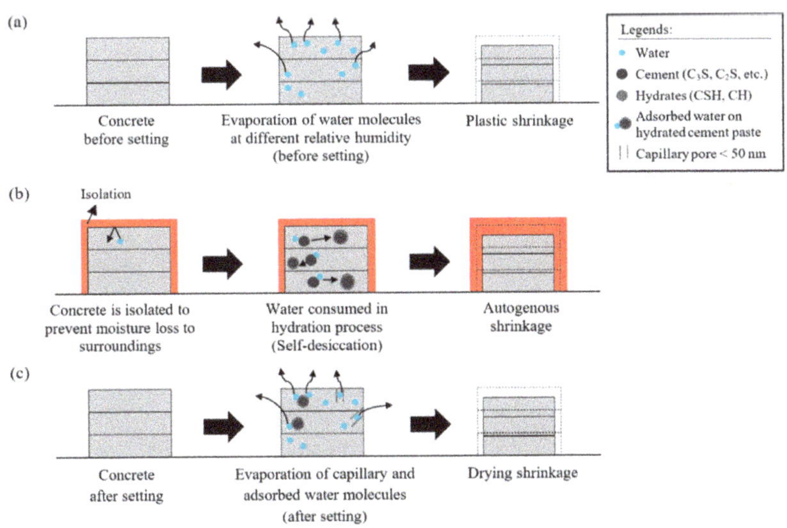

Figure 8. A schematic representation of (**a**) plastic shrinkage, (**b**) autogenous shrinkage, and (**c**) drying shrinkage, reproduced with permission from [40].

When the same mix is used, 3DPC experiences a similar drying shrinkage level to cast concrete [16]. However, 3DPC often uses admixtures such as geopolymers, which have been found to experience a higher level of drying shrinkage due to the network of micropores formed [41]. The increased cracking rate associated with the shrinkage of 3DPC greatly impacts its durability. There are two main test methods commonly used to assess shrinkage behavior in concrete. ASTM C1581-04, *Standard Test Method for Determining Age at Cracking and Induced Tensile Stress Characteristics of Mortar and Concrete under Restrained Shrinkage*, is known as the restrained ring test [42]. ASTM C1581 places concrete around a steel ring, as shown in Figure 9. As the specimen shrinks, it induces compressive stress in the steel and tensile stress in the concrete. Cracking is detected based on a measurable reduction in strain on the inner steel ring. This test provides a quantitative way to determine the age at which cracking occurs under restrained conditions and is suitable for evaluating autogenous and drying shrinkage. ASTM C1579-21, *Standard Test Method for Evaluating Plastic Shrinkage Cracking of Restrained Fiber Reinforced Concrete (Using a Steel Form Insert)*, focuses on plastic shrinkage behavior [43]. In this test, specimens are subjected to controlled drying conditions with set temperature, humidity, and wind speed. The extent and severity of surface cracking are observed and recorded. This method is particularly useful for comparing the effects of fiber reinforcement or admixtures on plastic shrinkage performance. Both ASTM standards are comparative in nature and do not define acceptance limits but offer a consistent framework to evaluate shrinkage tendencies under specific conditions [42,43].

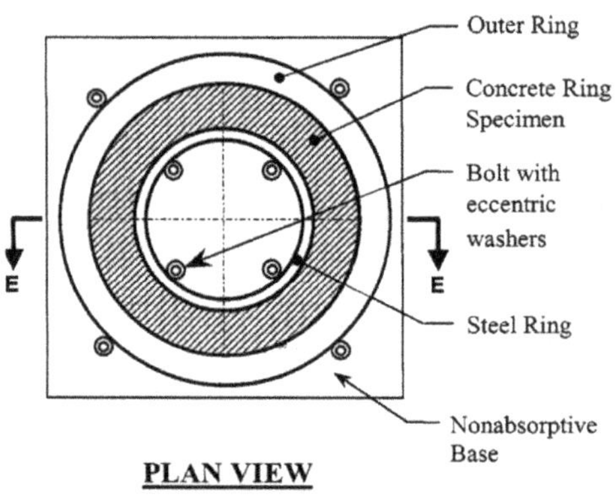

Figure 9. A plan view of the steel ring—ASTM C1581 [42].

Several papers were reviewed, and testing was carried out based on ASTM C1579 [43]. These papers had similar test conditions in terms of temperature of 25 °C and RH of 60%. However, the wind speed varied greatly, with 6.3 m/s [36] and 3.6 m/s [44]. Zhou et al. [45] also examine the drying shrinkage using a test designed for a Chinese standard JC/T C603-2004 [46]. The test used the same RH of 60% and a similar temperature of 20 °C as the previous papers [45]. Jaji et al. [29] also examined drying shrinkage. However, it is not accelerated using wind. Markers are placed along the different faces of the sample, and the change in length due to drying shrinkage is measured after 7, 14, 28, and 90 d [29]. Papachristoforou et al. [47] found the cracking age of samples due to restrained shrinkage in line with ASTM C1581. Table 2 outlines various shrinkage tests, highlighting how additives influence the shrinkage behavior of different mixes. Drying shrinkage ranged from 0.14% in covered samples to 2.98% in mixes made with 100% metakaolin. Restrained shrinkage results showed delayed cracking, with fly ash additions improving cracking resistance by up to 57% compared to control samples. In contrast, high contents of bauxite tailings increased shrinkage by over 47%, due to enhanced surface area and capillary pressure. Uncovered samples subjected to plastic shrinkage showed 61% greater deformation compared to covered ones. These results emphasize the need for effective shrinkage control measures, especially fly ash and slag additions, to mitigate early-age cracking and ensure dimensional stability in 3DPC structures.

Table 2. Analysis of shrinkage in 3DPC literature: mix design parameters and testing.

Ref.	Mix	Binder Description	Mix Description	Test Method	Curing	Shrinkage (%)	Time to Cracking (Days)
[29]	M1	100% MK	-	ASTM C596—Drying Shrinkage	90 days	2.98	-
	M2	95% MK 5% Slag	-	ASTM C596—Drying Shrinkage	90 days	2.86	-
[44]	L1-200	70% OPC 20%FA 10% SF	Layers 1, Length 200 mm	ASTM C1579—Plastic Shrinkage	-	-	-
	L3-200	70% OPC 20%FA 10% SF	Layers 3, Length 200 mm	ASTM C1579—Plastic Shrinkage	-	-	-
	L1-300	70% OPC 20% FA 10% SF	Layers 1, Length 300 mm	ASTM C1579—Plastic Shrinkage	-	-	-
	L3-300	70% OPC 20% FA 10% SF	Layers 3, Length 300 mm	ASTM C1579—Plastic Shrinkage	-	-	-
[47]	CL2	90% OPC 10% SF	50% SS 50% LF	ASTM C1581—Drying Shrinkage	Restrained	-	4.88
	FL2	70% OPC 20%FA 10% SF	50% SS 50% LF	ASTM C1581—Drying Shrinkage	Restrained	-	7.64
	F2	70% OPC 20% FA 10% SF	100% SS	ASTM C1581—Drying Shrinkage	Restrained	-	6.79
	LL2	70% OPC 20% Slag 10% SF	50% SS 50% LF	ASTM C1581—Drying Shrinkage	Restrained	-	5.95
	L2	70% OPC 20% Slag 10% SF	100% SS	ASTM C1581—Drying Shrinkage	Restrained	-	4.95

Table 2. *Cont.*

Ref.	Mix	Binder Description	Mix Description	Test Method	Curing	Shrinkage (%)	Time to Cracking (Days)
[45]	T0	100% OPC	100% N	JC/T C603-2004—Drying Shrinkage	90 days	0.15	-
	T15	100% OPC	85% N 15% BT	JC/T C603-2004—Drying Shrinkage	90 days	0.185	-
	T25	100% OPC	75% N 25% BT	JC/T C603-2004—Drying Shrinkage	90 days	0.205	-
	T35	100% OPC	65% N 35% BT	JC/T C603-2004—Drying Shrinkage	90 days	0.22	-
	T45	100% OPC	55% N 45% BT	JC/T C603-2004—Drying Shrinkage	90 days	0.285	-
[36]	Mold	70% OPC 20% FA 10% SF	S/B 1.4, W/B 0.45	ASTM C1581—Drying Shrinkage	28 days	-	-
	Lab	70% OPC 20% FA 10% SF	S/B 1.4, W/B 0.45	ASTM C1581 -Drying Shrinkage	28 days	-	-
	R-Site	70% OPC 20% FA 10% SF	S/B 1.4, W/B 0.45	ASTM C1581—Drying Shrinkage	3 h @6.3 m/s wind, 28 days	-	-
	U-Site	70% OPC 20% FA 10% SF	S/B 1.4, W/B 0.45	ASTM C1581—Drying Shrinkage	3 h @6.3 m/s wind, 28 days	-	-
[48–50]	M	90% OPC 10% SF	S/B 1.8, W/B 0.46	Plastic Shrinkage	210 min covered @25 °C, @5 m/s wind	0.14	-
					210 min uncovered @25 °C, @5 m/s wind	0.36	-

Note: Metakaolin (MK), 1 Layer (L1), 3 Layers (L3), Ordinary Portland Cement (OPC), Fly Ash (FA), Silica Fume (SF), Silica Sand (SS), Limestone Filler (LF), Natural (N) Sand, Bauxite Tailings (BT), Sand/Binder (S/B), Water/Band (W/B), Restrained on-Site (R-Site), Unrestrained on-Site (U-Site).

Figure 10 illustrates the influence of various experimental parameters on shrinkage reduction in concrete. Zhou et al. [45] investigated the impact of bauxite tailings (BT) on the durability of 3DPC and found that a mix containing 45% BT exhibited the highest shrinkage, whereas 15% BT achieved the best performance among all substituted mixes. Notably, the control mix (with no BT) demonstrated the lowest shrinkage overall. The inclusion of 15% BT increased shrinkage by 23.3%, which was attributed to the increased surface area introduced by the tailings. Moelich et al. [36] examined the effect of early-age shrinkage restraint in 3DPC by comparing unrestrained samples to those mechanically restrained with rods. Results showed that restraining shrinkage at early ages reduced the interlayer bond strength and facilitated increased chloride penetration. Jaji et al. [29] explored the influence of 5% slag addition in a metakaolin-based mix. The study revealed that drying shrinkage was more pronounced in the build-up (vertical) direction than in the horizontal direction for both the control and slag-modified mixes. However, the slag-enhanced mix experienced reduced drying shrinkage compared to the control. Papachristoforou et al. [47] further investigated the effects of slag, fly ash, and limestone additions on shrinkage in a mix containing 10% silica fume. In line with Jaji et al. [29], the inclusion of slag improved shrinkage resistance, with cracking delayed by 22% relative to the control. Among all additives studied, fly ash yielded the most significant improvement, extending the cracking age by 57%. In terms of surface exposure, Figure 10 also indicates that uncovered samples experienced a 61% increase in shrinkage compared to covered specimens. Moelich et al. [44] and Markin et al. [48] both emphasized the high susceptibility of 3DPC to plastic shrinkage. This vulnerability is primarily due to the absence of formwork, which exposes the concrete surface, accelerates evaporation, and intensifies water loss. The resulting capillary pressure buildup contributes to elevated plastic shrinkage levels. Additionally, interlayer slip was observed, primarily caused by the reduced cross-sectional area and increased porosity in printed layers [36,44]. As shown in Figure 11, the presence of a slip plane can undermine structural performance. Combined with early-age cracking, interlayer slip underscores the importance of shrinkage control in enhancing the durability of 3DPC.

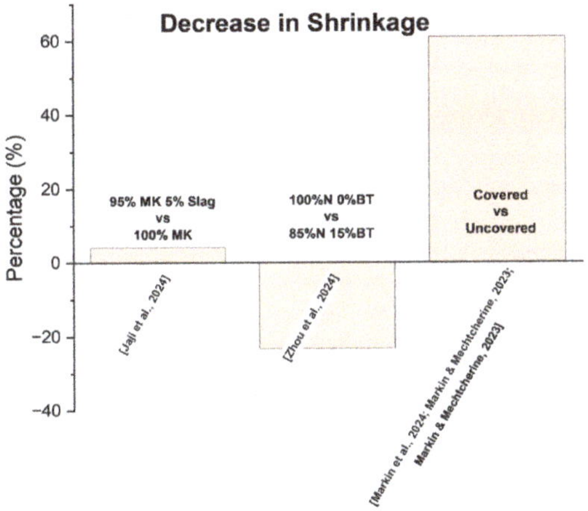

Figure 10. Decrease in shrinkage [29,45,48–50].

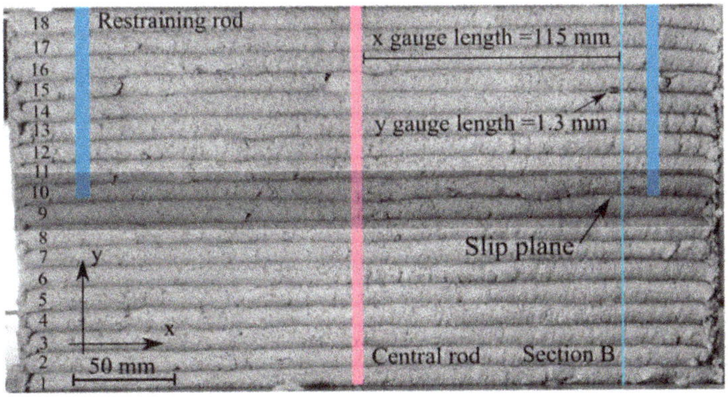

Figure 11. The restrained shrinkage specimen with induced slip plane, reproduced with permission from [36].

2.3. *Freeze–Thaw Resistance*

Freeze–thaw resistance in concrete is the capacity to resist the cyclic temperature changes that can cause degradation [18]. Establishing the freeze–thaw resistance of 3DPC contributes to its validity as a construction method, as it can be used in regions with temperatures that fluctuate above and below freezing. Freeze–thaw refers to water freezing and expanding by approximately 9% in volume in concrete pores, causing internal pressures and cracking [16,51], as shown in Figure 12. Freeze–thaw can also occur due to water penetrating the concrete through cracks and joints. This can particularly impact 3DPC. Mohan et al. [52] state that higher porosity at the layer interface can make 3DPC less freeze–thaw-resistant.

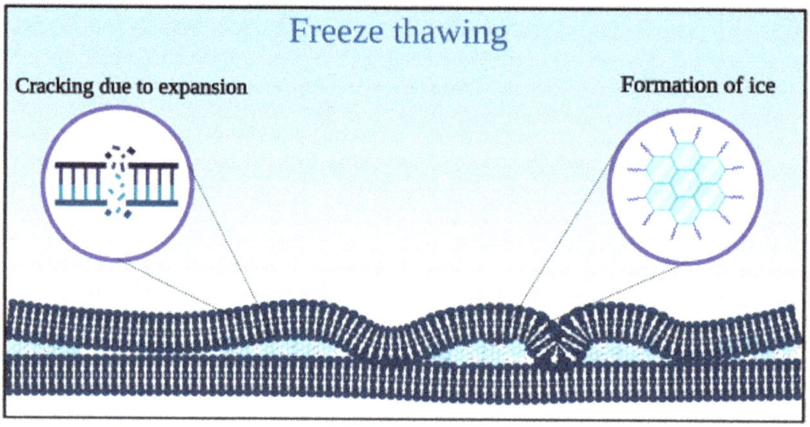

Figure 12. F/T process of 3DPC concrete layers, reproduced with permission from [16].

The freeze–thaw resistance of a sample is determined by applying F/T (freeze–thaw) cycles. A standard test is ASTM C666/C666M—*Standard Test Method for Resistance of Concrete to Rapid Freezing and Thawing* [53]. The test has two procedures: Procedure A, Rapid Freezing and Thawing in Water, and Procedure B, Rapid Freezing in Air and Thawing in Water. The test states that samples must be lowered from 4 to −18 °C and raised from −18 to 4 °C between 2 and 5 h. The test allows for a temperature variation of ±2 °C at the low temperature of −18 °C and the higher temperature of 4 °C. Samples should be between 75 and 125 mm in width, depth, and diameter and between 275 and 405 mm in length. Lastly, the sample should be left in the apparatus for 36 cycles.

Some of the studies from the papers reviewed are designed in line with the ASTM C666/C666M standard [31,32,53]. Sikora et al. [32] specified that sample measurements were taken after 25 and 50 cycles. Figure 13 shows the breakdown of the cycle. Du et al. [31] took measurements every 25 cycles up to 300 cycles and specified that the cycles lasted between 2 and 4 h with the temperature changing from −18 to 5 °C. A study by Dong et al. [54] also looks at a rapid freezing and thawing test. However, it is designed in line with a Chinese standard GB/T5008 [55]. Similar to the previous study, the measurements are taken every 25 cycles, up to 400 cycles [54]. An alternative test was used by Assaad et al. [51] that closely simulates the natural freeze–thaw cycle, freezing taking 12 h at −20 ± 4 °C and thawing taking a further 12 h at 20 ± 3 °C, with measurement being taken after 56 and 112 cycles. GivKashi et al. [56] state that the study samples used 24 h freezing and thawing cycles with a temperature range of −18 to 24 °C. It was also stated that samples were measured after 50 cycles [56]. Table 3 provides a comprehensive comparison of freeze–thaw performance across various 3DPC mixes. The compressive strength loss under cyclic freezing ranged from just 0.4% in air-entrained mixes to 28.7% in control mixes without additives. The addition of air-entraining agents (AEA) and styrene–butadiene rubber (SBR) significantly reduced strength loss, in some cases by more than 75%. For example, a mix that lost 24.2% of compressive strength without AEA saw this reduced to only 5.2% with AEA. In terms of mass loss, printed samples generally outperformed cast samples, with printed variants showing losses as low as 0.2%, while cast mixes exhibited up to 1.4% mass reduction. These findings confirm that with appropriate use of admixtures and SCMs, 3DPC can achieve or exceed the freeze–thaw durability of cast concrete, making it a reliable option for cold-region applications.

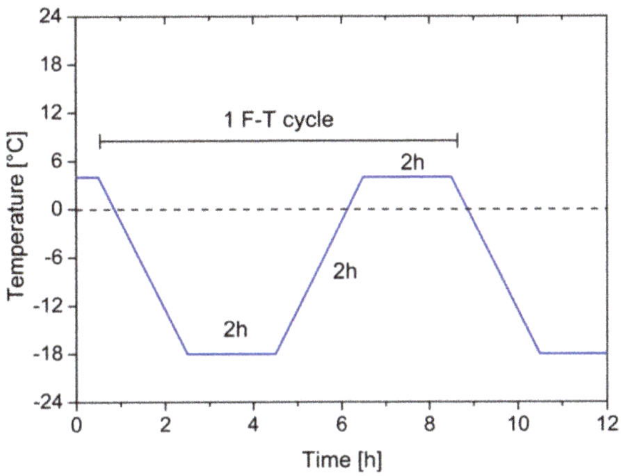

Figure 13. Breakdown of the F/T cycle [32].

Table 3. Analysis of freeze–thaw in 3DPC literature: mix design parameters and testing.

Ref.	Mix	Binder Description	Mix Description	Test Method	No. F/T Cycles	f′c Loss (%)	fr Loss (%)	Mass Loss (%)
[51]	550-0.45	92% OPC 8%SF	HRWR%/B 2.1	12 h @ −20 ± 4 °C 12 h @20 ± 3 °C	112	24.2	33.8	-
	550-0.45 Air	92% OPC 8%SF	HRWR%/B 1.9, AEA%/B 0.08	12 h @ −20 ± 4 °C 12 h @20 ± 3 °C	112	8.5	13.4	-
	550-0.45-SBR	92% OPC 8%SF	HRWR%/B 1.8, SBR%/B 15	12 h @ −20 ± 4 °C 12 h @ 20 ± 3 °C	112	15.6	9.4	-
	650-0.45	92% OPC 8%SF	HRWR%/B 0.95	12 h @ −20 ± 4 °C 12 h @ 20 ± 3 °C	112	17.2	22	-
	650-0.45-Air	92% OPC 8%SF	HRWR%/B 0.9, AEA%/B 0.1	12 h @ −20 ± 4 °C 12 h @ 20 ± 3 °C	112	5.5	9.4	-
	650-0.45-SBR	92% OPC 8%SF	HRWR%/B 0.8, SBR%/B 15	12 h @ −20 ± 4 °C 12 h @20 ± 3 °C	112	10.1	9.4	-
	750-0.35	92% OPC 8%SF	HRWR%/B 1.25	12 h @ −20 ± 4 °C 12 h @ 20 ± 3 °C	112	16.6	21	-
	750-0.35-Air	92% OPC 8%SF	HRWR%/B 1.15, AEA%/B 0.15	12 h @ −20 ± 4 °C 12 h @ 20 ± 3 °C	112	5.2	7.9	-
	750-0.35-SBR	92% OPC 8%SF	HRWR%/B 1, SBR%/B 15	12 h @ −20 ± 4 °C 12 h @ 20 ± 3 °C	112	10.2	10.6	-

Table 3. *Cont.*

Ref.	Mix	Binder Description	Mix Description	Test Method	No. F/T Cycles	f′c Loss (%)	fr Loss (%)	Mass Loss (%)
[56]	C	100% OPC	SP%/B 1.1	12 h F/T Cycles @ −18 to 24 °C	50	5.7	-	-
	A0.08	100% OPC	SP%/B 1.1, AEA%/B 0.08	12 h F/T Cycles @ −18 to 24 °C	50	4.3	-	-
	A0.10	100% OPC	SP%/B 1.1, AEA%/B 0.1	24 h F/T Cycles @ −18 to 24 °C	50	3.1	-	-
	A0.12	100% OPC	SP%/B 1.1, AEA%/B 0.12	24 h F/T Cycles @ −18 to 24 °C	50	0.4	-	-
[31]	P-O	44.6% OPC 29.1% FA 26.3% LP	S/B 1 W/B 0.3, Cast	2–4 h F/T Cycles @ −18 to 5 °C	300	28.7	48.3	1.4
	P-Z	44.6% OPC 29.1% FA 26.3% LP	S/B 1 W/B 0.2, Printed	2–4 h F/T Cycles @ −18 to 5 °C	300	19.5	41.3	0.2
	H-O	61.7%OPC 32.7% FA 5.6% SF	S/B 1 W/B 0.2, Cast	2–4 h F/T Cycles @ −18 to 5 °C	300	22.7	31.1	0.2
	H-Z	61.7%OPC 32.7% FA 5.6% SF	S/B 1 W/B 0.2, Printed	2–4 h F/T Cycles @ −18 to 5 °C	300	5.9	19.1	0
[32]	Cast	70% OPC 20% FA 10% SF	S/B 1.5 W/B 0.25	ASTM C666	50	2.7	11	-
	1 Layer	70% OPC 20% FA 10% SF	S/B 1.5 W/B 0.25	ASTM C666	50	5.1	9	-
	3 Layer	70% OPC 20% FA 10% SF	S/B 1.5 W/B 0.25	ASTM C666	50	1.3	21	-
[54]	JZ	93% OPC 7% SF	60% FS 40% AS	GB/T5008—Quick Freeze	400	21	-	3.02
	3D-X	93% OPC 7% SF	60% FS 40% AS	GB/T5008—Quick Freeze	400	22.4	-	3.12
	3D-Y	93% OPC 7% SF	60% FS 40% AS	GB/T5008—Quick Freeze	400	24	-	3.56

Note: Compressive Strength (f′c), Flexural Strength (fr), Ordinary Portland Cement (OPC), Silica Fume (SF), High-Range Water Reducer (HRWR), Air-Entraining Agent (AEA), Styrene–Butadiene Rubber (SBR), Superplasticizer (SP), Fly Ash (FA), Limestone Powder (LP), Sand/Binder (S/B), Water/Binder (W/B), Ferrochrome Sand (FS), Aeolian Sand (AS).

The results of the F/T test are normally determined by two factors: the change in mechanical strength and mass. All the papers with F/T tests that were reviewed included a compressive strength test [31,32,51,54,56]. Three papers looked at flexural strength before and after F/T tests [31,32,51]. The other two did not include flexural strength testing, with GivKashi et al. [56] focusing on compressive strength change and Dong et al. [54] including the mass loss as a measurement of the F/T resistance, as was also performed by Du et al. [31]. Equation (1) below shows how the F/T resistance is measured with the property varying from compressive strength to mass.

$$\Delta(Property),\ \% = \frac{Property\ of\ mix\ not\ exposed\ to \frac{F}{T} - Propert\ of\ mix\ after\ given \frac{F}{T}\ Cycle}{Propert\ of\ control\ mix\ not\ exposed\ to \frac{F}{T}} \times 100 \quad (1)$$

Sikora et al. [32] focused on the durability of cast vs. printed concrete specimens, using three different samples, cast, 1 layer of printed concrete (1 L), and 3 layers of printed concrete (3 L), ll of which used the mix shown in Table 3. The paper found cast concrete had higher initial strength properties than 3DPC, as seen across all the papers that compared cast and printed samples. They also found that the 3DPC behaved anisotropically [31,32,54].

However, Sikora et al. [32] and Dong et al. [54] found that printed samples had a slightly larger decrease in compressive strength and mass, whereas Du et al. [31] found the printed sample to have a smaller decrease in mass. The printed sample performs better with larger pore size and better pore connectivity [31].

Assaad et al. [51] and GivKashi et al. [56] compared 3D-printed concrete (3DPC) with and without air-entraining additives. Although the initial compressive strength of air-entrained mixes was lower than the control, the strength loss under the freeze–thaw cycle or acid attack was significantly smaller. This improvement is attributed to the modified pore structure, where uniformly distributed microbubbles help relieve internal pressure, reduce microcracking, and lower permeability. As a result, air-entrained 3DPC matched or even exceeded the durability performance typically expected from cast concrete with similar air-void systems. To optimize this balance, a controlled AEA dosage ($\leq$0.10%) and the incorporation of pozzolanic materials are recommended to mitigate strength loss while maintaining durability [51,56]. Du et al. [31] show evidence that a mix with silica fume performs better than one with limestone. This is seen by the initial strength characteristics and those after F/T cycles [31]. More research must be conducted to find if adding silica fume will improve the freeze–thaw resistance of 3DPC compared to traditional concrete mixes.

2.4. Chemical Resistance

For 3DPC to be considered a viable alternative to cast concrete, it must be ensured that it is effective in harsh environments. The harsh environments include acidic environments, sewage systems, maritime, and underground [40]. The chemical resistance of concrete is often challenged. Gu et al. [57] wrote that sewage systems create an environment with one of the most rapid rates of concrete degradation. There has been significantly less 3DPC used in these environments compared to traditional cast concrete. This lack of information requires extensive research into the chemical resistance of 3DPC to ensure that it is a feasible alternative to cast concrete.

2.4.1. Sulfuric Acid

Many papers state that Portland cement has little resistance to acid attacks due to its relatively high alkalinity [57,58]. This allows the degradation of concrete to occur due to many different processes. Sulfuric acid (H_2SO_4) is considered to be among the most harmful acids when it comes to the degradation of concrete. This is due to the combined effects of the sulfate and acid attacks [58], the result of which is shown in Figure 14. The sulfate attack is caused by the reaction between the sulfate and calcium hydroxide (CH) in the hydrated cement paste. The reaction produces gypsum. The gypsum causes a volume increase in the concrete. Further reactions cause larger volume increases. The large change in volume leads to internal pressure and, eventually, expansion. The expansion causes cracking and spalling, as well as a loss in mechanical strength and, eventually, complete failure [56,58]. GivKashi et al. [56] state that sulfuric acid causes damage to concrete through the following chemical reactions.

$$Ca(OH)_2 + H_2SO_4 \rightarrow CaSO_4 \cdot 2H_2O \quad (2)$$

(a) (b) (c)

Figure 14. Specimens after 495 days of exposure to 1% sulfuric acid: (**a**) OPCC_30; (**b**) OPCC_50; (**c**) AAC. For detailed explanations of the mix names, please refer to the original literature, reproduced with permission from [57].

Equation (2) shows the first reaction, which occurs between the sulfuric acid (H_2SO_4) and the calcium hydroxide (CH), producing calcium sulfate dihydrate, commonly known as gypsum ($CaSO_4{\cdot}2H_2O$). Gypsum can increase its volume by a factor of 2.2 [56]. The soluble gypsum then attempts to leech out of the concrete. In cases where that is not possible, the gypsum increases the internal pressure, leading to cracking and spalling.

$$CaSiO_2 \cdot 2H_2O + H_2SO_4 \rightarrow CaSO_4 + Si(OH)_4 + H_2O \quad (3)$$

Equation (3) shows a second reaction. The sulfuric acid (H_2SO_4) reacts with the hydrated calcium silicate ($CaSiO_2{\cdot}2H_2O$) to produce gypsum and silica hydroxide [56]. This reaction greatly impacts the mechanical properties of the concrete. In particular, it significantly reduces the compressive capacity. The fine network of the structure is broken down, causing the loss of cohesion and adhesion of the concrete. This, combined with the additional gypsum from the reaction, increases the degradation of the concrete.

$$3CaO \cdot Al_2O_3 \cdot 12H_2O + 3(CaSO_4 \cdot 2H_2O) + 14H_2O \rightarrow 3CaO \cdot Al_2O_3 \cdot 3CaSO_4 \cdot 32H_2O \quad (4)$$

Equation (4) shows the third reaction: the gypsum ($CaSO_4{\cdot}2H_2O$) produced from the first two reactions now reacts with calcium aluminate hydrate ($3CaO{\cdot}Al_2O_3{\cdot}12H_2O$), producing ettringite [56]. Ettringite has an increased volume by a factor of 7 in relation to the original compound [58]. This large expansion, combined with the structure's network breakdown, causes the concrete's failure.

The effects of a sulfuric acid attack are measured using a *Standard Test Method for Length Change of Hydraulic-Cement Mortars Exposed to a Sulfate Solution* (ASTM C109/C109M) [59]. The test measures the sulfate resistance of mortars using Portland cement, Portland cement with pozzolans or slags, and blended hydraulic types of cement. This makes the test suitable for comparing traditional cement mixes to those with nano silica admixtures. The standard solution for the test contains 352 moles of Na_2SO_4 per meter cubed (50 g/L). The test consists of 6 bars used for the acid attack test and up to 21 cubes used for mechanical property testing. The dimensions of the mortar bars used are 25 × 25 × 285 mm. Once molded, the bars must be placed in an oven at 35 ± 3 °C for 23½ h ± 30 min and removed from the molds. The samples are then tested in a curing tank of saturated lime water at 23 ± 2 °C. Once removed from the curing tank, the initial length is measured. The test should include 625–800 mL of solution per mortar bar. Once the samples are placed in the sulfate solution, such as Figure 15, the change in length is measured at 1, 2, 3, 4, 8, 13, and 15 weeks. The samples can be left in the solution, and further length changes are measured at 4, 6, 9, 12, 15, and 18 months.

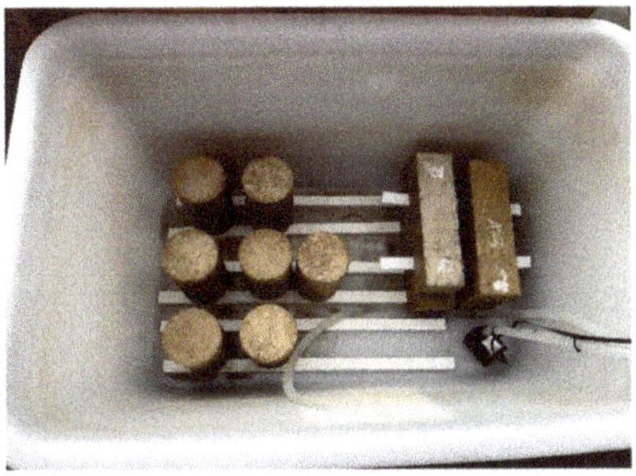

Figure 15. The specimen arrangement and test setup, reproduced with permission from [57].

Some of the studies [28,60,61] referred to a version of the ASTM C109/C109M test standard [59]. Baz et al. [28] and El Inaty et al. [61] both used a volume of solution four times the volume of the samples as recommended by ASTM C109/C109M. However, the studies used different concentrations of sulfuric acid. Baz et al. [28] had solution baths with 1 and 3% concentration compared to El Inaty et al.'s [61] 0.5% concentration. Three papers used a 5% solution [23,24,60]; the final paper did not state the concentration but had a pH of 2 [56]. Two papers used dry–wet cycles designed per GB/T 50082-2009 [23,60,62]. Rui et al. [23] used 1d cycles, with 15 h wet and 9 h dry cycles. Guo et al. [60] used 3, 7, 14, and 21 d cycles using 3:1 dry–wet ratio. The solution was replaced for tests that lasted longer than 30 d, with some replacement taking place every 30 d [56,60]. Others replaced the solution when mass or length measurement was taking place at 3, 7, 14, 21, 28, 42, 56, 70, 84, 98, 112, 126, and 140 days [28,61]. The sulfuric acid resistance of various 3DPC mixes, as shown in Table 4, reveals critical insights into durability under aggressive chemical environments. Mass loss in printed samples exposed to 3% sulfuric acid ranged up to 49.1%, though this was consistently lower than in cast samples in similar conditions. In one case, a 100% OPC printed mix lost 46.4% mass, while the cast equivalent lost 53.6%. The presence of silica fume dramatically improved acid resistance; a mix with 10% SF showed only 2.7% mass loss after 140 days. Compressive strength losses also varied, with printed samples sometimes outperforming cast ones despite being exposed to the same acid concentrations. These outputs underline the effectiveness of SCMs such as SF and the importance of microstructural control in resisting acid degradation in 3DPC.

Table 4. Analysis of sulfuric acid in 3DPC literature: mix design parameters and testing.

Ref.	Mix	Binder Description	Mix Description	Test Method	Procedure	Cast f′c Loss (%)	3DP f′c Loss (%)	Cast Mass Loss (%)	3DP Mass Loss (%)
[28]	A	100% OPC	VMA%/C 0.4, HRWR%/C 0.81	ASTM C1012/C1012M	56 d @ 1% Sulfuric Acid	-	-	10.9	10.9
					56 d @ 3% Sulfuric Acid	-	-	54.5	47.3
	B	100% OPC	VMA%/C 0.47, HRWR%/C 0.95	ASTM C1012/C1012M	56 d @ 1% Sulfuric Acid	-	-	10.7	9.1
					56 d @ 3% Sulfuric Acid	-	-	53.6	46.4
	C	100% OPC	VMA%/C 0.4, HRWR%/C 1.52	ASTM C1012/C1012M	56 d @ 1% Sulfuric Acid	-	-	19.1	14.5
					56 d @ 3% Sulfuric Acid	-	-	60.9	49.1
[56]	C	100%OPC	SP%/B 1.1	ASTM C349–14	90 d @ pH2 Sulfuric Acid	-	−5.64	-	-
	A0.08	100%OPC	SP%/B 1.1, AEA%/B 0.08	ASTM C349–14	90 d @ pH2 Sulfuric Acid	-	−6.53	-	-
	A0.10	100%OPC	SP%/B 1.1, AEA%/B 0.1	ASTM C349–14	90 d @ pH2 Sulfuric Acid	-	−10.61	-	-
	A0.12	100%OPC	SP%/B 1.1, AEA%/B 0.12	ASTM C349–14	90 d @ pH2 Sulfuric Acid	-	−13.44	-	-
[61]	A	90% OPC 10% SF	SP%/B 0.3	ASTM C1012/C1012M	140 d @ 0.5% Sulfuric Acid	7.5	8.75	3.7	3.5
	B	90% OPC 10% SF	SP%/B 0.6	ASTM C1012/C1012M	140 d @ 0.5% Sulfuric Acid	8.75	15	4	4.2
	C	90% OPC 10% SF	SP%/B 0.6	ASTM C1012/C1012M	140 d @ 0.5% Sulfuric Acid	4.75	5	3.2	2.7
[23]	Cast	83% OPC 17% SF	S/B 1.78 W/B 0.47	GB/T 50082-2009	150 d—15 h @ 5% Sulfuric Acid, 7 h drying	−42	-	−1.4	-
	3DP-6-Y	83% OPC 17% SF	S/B 1.78 W/B 0.47	GB/T 50082-2009	150 d—15 h @ 5% Sulfuric Acid, 7 h drying	-	−54.8	-	−1.8
	3DP-12-Y	83% OPC 17% SF	S/B 1.78 W/B 0.47	GB/T 50082-2009	150 d—15 h @ 5% Sulfuric Acid, 7 h drying	-	−43.5	-	−1.6
[24]	OPC	100% OPC	S/B 3.45 W/B 0.45	ASTM C267	30 d @ 5% Sulfuric Acid	34	-	1	-
	FA1GPC	70% FA1 30% SB	S/B 2.38 W/B 0.45	ASTM C267	30 d @ 5% Sulfuric Acid	32	-	−0.28	-
	FA1NS GPC	68% FA1 30% SB 2% NS	S/B 2.38 W/B 0.45	ASTM C267	30 d @ 5% Sulfuric Acid	19	-	−0.77	-
	FA2GPC	70% FA2 30% SB	S/B 2.38 W/B 0.45	ASTM C267	30 d @ 5% Sulfuric Acid	28	-	−0.39	-
	FA2NSG PC	68% FA2 30% SB 2% NS	S/B 2.38 W/B 0.45	ASTM C267	30 d @ 5% Sulfuric Acid	17	-	−0.74	-
[60]	M	80% OPC 20% FA	S/B 1.8 W/B 0.54	GB/T 50082-2009	231 d, 54 h @ 5% Sulfuric Acid, 18 h drying	-	-	4.03	-
					231 d, 126 h @ 5% Sulfuric Acid, 42 h drying	-	-	5.02	-
					231 d, 252 h @ 5% Sulfuric Acid, 84 h drying	-	-	3.28	-
					231 d, 378 h @ 5% Sulfuric Acid, 126 h drying	-	-	2.98	-

Note: Compressive Strength (f′c), Flexural Strength (fr), Ordinary Portland Cement (OPC), Viscosity-Modifying Agent (VMA), High-Ranger Water Reducer (HRWR), Superplasticizer (SP), Air-Entraining Agent (AEA), Silica Fume (SF), Sand/Binder (S/B), Water/Binder (W/B), Fly Ash Type 1 (FA1), Fly Ash Type 2 (FA2), Sodium Bisulfate (SB), Nano-Silica (NS), Fly Ash (FA).

The change in mass is the most common measurement recorded across the papers, followed by porosity, water porosity, and mechanical properties. Five of the six papers measured the loss or, in some cases, gain in mass [23,24,28,60,61]. Baz et al. [28] examined the microstructural analysis of 3DPC exposed to a sulfuric acid attack. The concrete mix used consisted of Portland cement, crushed limestone, and admixtures consisting of a water reducer and viscosity modifier. The results saw a decrease in mass across all samples over 53 days. El Inaty et al. [61] look at the long-term durability of 3DPC with concrete mixes consisting of Portland cement, sand, silica fume, and a superplasticizer. The results over 140 days showed that the mix with the largest silica fume content had the smallest loss in mass, implying that it had the greatest resistance to sulfuric acid. Rui et al. [23] look at how the anisotropy impacts the effects of sulfate attack on 3DPC, finding mass loss was the smallest when attacked in the x-direction, then z-direction, and the largest in the y direction. Loading directions are shown in Figure 16. It was also noted that the printed sample performed worse than the cast when the period between print layers was 6 min and better than the cast sample when the period between print layers was 12 min [23]. Two papers reported an increase in mass either throughout the testing [24,60]. This is likely due to the reactions that occur due to the sulfuric acid, cases where gypsum has not leached out of the samples, causing a reduction in mass.

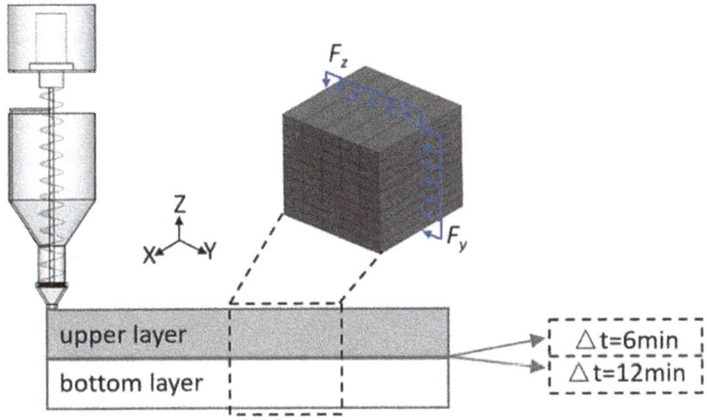

Figure 16. The experimental design and loading direction, reproduced with permission from [23].

In addition to mass change, sulfuric acid exposure also causes chemical degradation through gypsum and ettringite formation, which induce expansive stress and microcracking, particularly concentrated in near-surface zones. Microstructural investigations, such as those by Baz et al. [28] and Gu et al. [57], indicate the gradual decomposition of calcium hydroxide and C-S-H phases, leading to increased porosity and reduced cohesion of the cement matrix, as shown in the SEM images (Figure 17). XRD results [63] further reveal ettringite formation gradients that intensify with increasing acid concentration, confirming progressive sulfate ingress through the matrix (Figure 18). While such degradation mechanisms are well documented in conventional concrete, studies on 3DPC remain limited, particularly regarding the influence of anisotropic pore networks and interlayer interfaces on acid transport and mineral phase transformation.

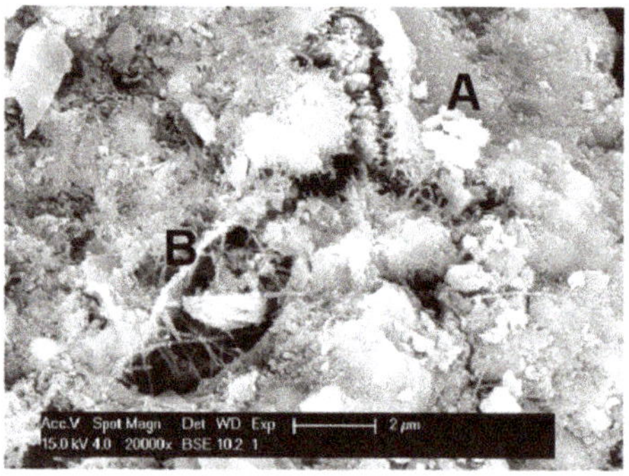

(a)

(b)

Figure 17. SEM images of OPC matrix before and after 495 days immersion, where A, B, and C indicate the areas selected for EDX analysis: (**a**) before; (**b**) after; reproduced with permission from [57].

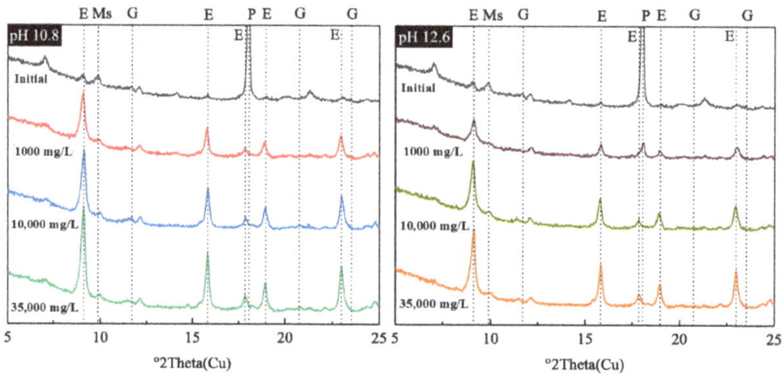

Figure 18. XRD pattern of PC + MK slices corroded under different sulfate concentrations for 150 days, reproduced with permission from [63].

2.4.2. Chloride

Chloride attack must be particularly considered for reinforced concrete structures, as it corrodes the reinforcement [40,64]. In comparison to cast concrete, 3DPC can have a higher porosity. This can lead to an increased rate of chloride ingress [25]. The combination of higher porosity with increased cracking rate makes 3DPC significantly more vulnerable to chloride attacks. This makes the resistance to chloride attacks one of the key durability criteria that must be assessed [25].

Two different tests measure chloride attack: the rapid chloride permeability (RCP) test and the chloride penetration depth (CPD) test [39]. ASTM C1202—17—*Standard Test Method for Electrical Indication of Concrete's Ability to Resist Chloride Ion Penetration*, a commonly used RCP test for evaluating concrete's resistance to chloride ion penetration, uses disk-shaped samples that are 50 mm thick and 100 mm in diameter [65]. The same dimensions are used for GB/T 50082-2009 (Figure 19) [62]. The electrical resistance of the sample is found by passing 60 V through the sample, which has one-half immersed in chloride solution and the other half in sodium hydroxide solution. The chloride resistance is then calculated using the electrical resistance [65]. The CPD test places a sample in the salt solution for an extended period, giving a closer description of the chloride resistance of the sample [39].

The RCP test can have a duration of 6 h, which leads to many studies opting for it over the lengthy CPD test [65].

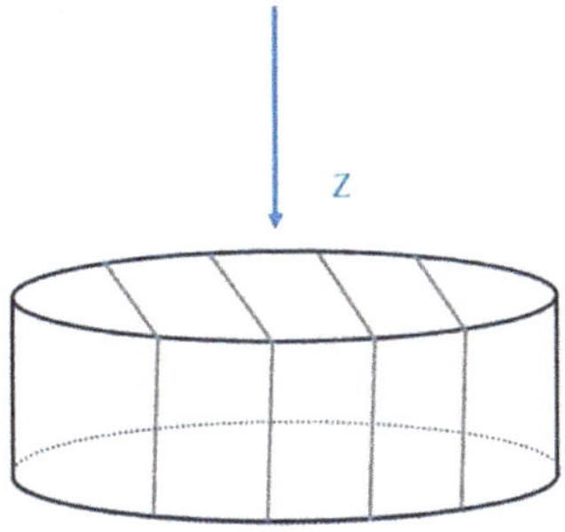

Figure 19. Test direction of chloride ion erosion of 3DPC, reproduced with permission from [45].

This review found that many papers used the RCP test when determining the chloride resistance of 3DPC. The majority of papers [30,31,36,39,66] took guidance from the ASTM C1202—17 [65], or the equivalent Chinese standard GB/T 50082-2009 [62]. Some of the studies used 60 V for the chloride migration inline from ASTM 1012-17 [31,39], while Moelich et al. used 10 V. There were varying concentrations used for the NaCl and NaOH solutions, with the NaCl varying from 3% [39] to 10% [30,45] and 1 M [66], and the NaOH varying between 0.1 M [66] and 0.3 M [30,39,45]. Shafiq et al. [39] ran the test for the shortest length at 6 h, and Du et al. [31] the longest at 96 h. Bran-Anleu et al. [66] and Bekaert et al. [30] tested the sample for 24 h, and the remaining papers did not specify a test duration [36,45]. As seen in Table 5, chloride migration coefficients (D-values) for 3DPC ranged from 311 to 2600 Coulombs (C), indicating a wide performance variation based on mix design and curing. High relative humidity curing (RH > 95%) consistently improved results, with one study showing a drop in permeability from 10.33 to 4.67 $\times 10^{-12}$ m^2/s. The integration of nano-silica and metakaolin reduced chloride permeability by up to 80%. The penetration depth was also substantially reduced in optimized mixes, from 12.36 mm to just 3.08 mm. These results demonstrate that 3DPC's chloride resistance can match or exceed traditional concrete when enhanced with SCMs and controlled curing.

Table 5. Analysis of chloride attack in 3DPC literature: mix design parameters and testing.

Ref.	Mix	Binder Description	Mix Description	Test Method	Procedure	Curing	D ($\times 10^{-12}$ m/s^2)	PD (mm)
[30]	REF	50% OPC 50% GGBS	S/B 1	RCM—NTBuild492	24 h @ 0.3 M NaOH 10% NaCl	28 d @ RH = 60	7.33	-
						28 d @ RH > 95	4.67	-
	M1	100% OPC	S/B 1.5	RCM—NTBuild492	24 h @ 0.3 M NaOH 10% NaCl	28 d @ RH = 60	10.33	-
						28 d @ RH > 95	8.33	-
	M2	75% OPC 25% GGBS	S/B 1.5	RCM—NTBuild492	24 h @ 0.3 M NaOH 10% NaCl	28 d @ RH = 60	20.33	-
						28 d @ RH > 95	10.33	-
[31]	P-O	44.6% OPC 29.1% FA 26.3% LP	S/B 1 W/B 0.3, Cast	RCM—NTBuild492, CPD	96 h	28 d	56.3	9.45
	P-Z	44.6% OPC 29.1% FA 26.3% LP	S/B 1 W/B 0.2, Printed	RCM—NTBuild492, CPD	96 h	28 d	78.8	12.36
	H-O	61.7%OPC 32.7% FA 5.6% SF	S/B 1 W/B 0.2, Cast	RCM—NTBuild492, CPD	96 h	28 d	21.3	3.82
	H-Z	61.7%OPC 32.7% FA 5.6% SF	S/B 1 W/B 0.2, Printed	RCM—NTBuild492, CPD	96 h	28 d	52.5	8.91

Table 5. *Cont.*

Ref.	Mix	Binder Description	Mix Description	Test Method	Procedure	Curing	D ($\times 10^{-12}$ m/s^2)	PD (mm)
[66]	P	48% OPC 26% MSS 26% FA	2, 13, 1440 min/layer	μXRF	24 h @ 0.1 M Na OH 1 M NaCl	28 d @ 20 °C RH = 65%	-	-
	Q	48% OPC 26% MSS 26% FA	2, 13, 1440 min/layer	μXRF	24 h @ 0.1 M NaOH 1 M NaCl	28 d @ 20 °C RH = 65%, 3 d @ 40 °C	-	-
[45]	T0	100% OPC	100% N	RCM-GB/T50082-2009 [62]	0.3 M NaOH 10% NaCl	28 d	5.8	-
	T15	100% OPC	85% N 15% BT	RCM—GB/T50082-2009	0.3 M NaOH 10% NaCl	28 d	5.2	-
	T25	100% OPC	75% N 25% BT	RCM—GB/T50082-2009	0.3 M NaOH 10% NaCl	28 d	4.33	-
	T35	100% OPC	65% N 35% BT	RCM—GB/T50082-2009	0.3 M NaOH 10% NaCl	28 d	4.2	-
	T45	100% OPC	55% N 45% BT	RCM—GB/T50082-2009	0.3 M NaOH 10% NaCl	28 d	4.6	-
[36,67]	Mold	70% OPC 20% FA 10% SF	S/B 1.4 W/B 0.45	CCI-UCT Manual	5 M NaCl	21 d @ ambient, 7 d @ 50 °C	0.325 mS/cm	-
	Lab	70% OPC 20% FA 10% SF	S/B 1.4 W/B 0.45	CCI-UCT Manual	5 M NaCl	21 d @ ambient, 7 d @ 50 °C	0.4 mS/cm	-
	R-Site	70% OPC 20% FA 10% SF	S/B 1.4 W/B 0.45	CCI-UCT Manual	5 M NaCl	21 d @ ambient, 7 d @ 50 °C	0.296 mS/cm	-
	U-Site	70% OPC 20% FA 10% SF	S/B 1.4 W/B 0.45	CCI-UCT Manual	5 M NaCl	21 d @ ambient, 7 d @ 50 °C	0.346 mS/cm	-
[38]	Cast	55% OPC 30% MK 12% $CaCO_3$ 3%G	PVA%/B 0.225 S/B 1.5 W/B 0.4	RCM	-	90 days @ 20 °C	284.5 kΩcm	
	BM	55% OPC 30% MK 12% $CaCO_3$ 3%G	PVA%/B 0.225 S/B 1.5 W/B 0.4	RCM	-	90 days @ 20 °C	165.5 kΩcm	
	OL2-Z	55% OPC 30% MK 12% $CaCO_3$ 3%G	PVA%/B 0.225 S/B 1.5 W/B 0.4	RCM	-	90 days @ 20 °C	99.6 kΩcm	
	OL4-Z	55% OPC 30% MK 12% $CaCO_3$ 3%G	PVA%/B 0.225 S/B 1.5 W/B 0.4	RCM	-	90 days @ 20 °C	113.3 kΩcm	
	Z′-Z	55% OPC 30% MK 12% $CaCO_3$ 3%G	PVA%/B 0.225 S/B 1.5 W/B 0.4	RCM	-	90 days @ 20 °C	177 kΩcm	
[39]	CM	100% OPC	S/B 1.47 W/B 0.35	RCM—ASTM C1202	3% NaCl	28 days	1575 C	11.83
	C-NS-1%	99% OPC 1% NS	S/B 1.47 W/B 0.35	RCM—ASTM C1202	3% NaCl	28 days	1650 C	-
	C-NS-2%	98% OPC 2% NS	S/B 1.47 W/B 0.35	RCM—ASTM C1202	3% NaCl	28 days	2600 C	-
	C-MK-5%	95% OPC 5% NS	S/B 1.47 W/B 0.35	RCM—ASTM C1202	3% NaCl	28 days	850 C	5.8
	C-MK-10%	90% OPC 10% MK	S/B 1.47 W/B 0.35	RCM—ASTM C1202	3% NaCl	28 days	350 C	9
	C-MK-10%-NS-1%	89% OPC 10% MK 1% NS	S/B 1.47 W/B 0.35	RCPT—ASTM C1202, CPD	3% NaCl	28 days	311 C	3.43
	C-MK-10%-NS-2%	88% OPC 10% MK 2%NS	S/B 1.47 W/B 0.35	RCPT—ASTM C1202, CPD	3% NaCl	28 days	318 C	3.08

Notes: Permeability Coefficient (D), Penetration Depth (PD), Ordinary Portland Cement (OPC), Ground Granulated Blast-Furnace Slag (GGBS), Sand/Binder (S/B), Rapid Chloride Migration (RCM), Relative Humidity (RH), Fly Ash (FA), Limestone Powder (LP), Water/Binder (W/B), Chloride Penetration Depth (CPD), Micro Silica Suspension (MSS), Micro-X-ray fluorescence (μXRF), Natural (N) Sand, Bauxite Tailings (BT), Silica Fume (SF), Metakaolin (MK), Chloride Conductivity Index (CCI), University of Cape Town (UCT), Calcium Carbonate ($CaCO_3$), Gypsum (G), Polyvinyl Alcohol (PVA) Fibers, Nano-Silica (NS), Rapid Chloride Permeability Test (RCPT).

The results of the RCP test are determined using Equation (5), which calculates D, the non-stationary chloride migration coefficient (m^2/s) [30,31,45]. U is the voltage passed through the test sample (V). T is the mean of the initial and final temperature of the test solution in (°C). L is the thickness of the sample (mm). X_d is the mean value of the chloride penetration depth (mm). t is the length of time that the sample was tested for (hours) [30,31,45].

$$D = \frac{0.0239 \times (275+T)}{(U-2)5}(X_d - 0.0238\sqrt{\frac{(273+T)LX_d}{U-2}}) \tag{5}$$

Zhou et al. [45] look at how adding bauxite tailings impacts the durability of 3DPC. The study compared printed samples with different concentrations of bauxite tailings. The results showed an optimal content of 35% bauxite tailings with a migration coefficient decrease from 5.8 to 4.2×10^{-12} m^2/s [45]. Bran-Anleu et al. [66] studied the impact of printing time intervals and curing conditions on chloride through cold joints using micro-XRF. The study found very little difference between samples that had 2 and 13 min between printing layers. However, the samples with 24 h printing intervals experienced a large increase in chloride ingress as the bond between layers was impacted [66]. Recent studies highlight the microstructural mechanisms by which layer interfaces accelerate chloride ingress. Interfaces formed during layer-by-layer deposition inherently possess increased porosity, interfacial voids, and microcracks, creating preferential pathways for chloride ions [68,69]. This effect is aggravated by drying–wetting cycles that cause dissolution, migration, and recrystallization of reaction products, further widening these pathways and accelerating deterioration processes [66,70,71]. Despite this critical relationship, explicit research into anisotropy-driven chloride penetration in 3DPC remains relatively limited.

A study by Bekaert et al. [30] investigated the service life of 3D-printed concrete (3DPC) by comparing a mix with ground granulated blast-furnace slag (GGBS) to one without it. The results showed that the mix containing GGBS had a lower chloride migration coefficient, indicating better resistance to chloride ingress. One of the most notable findings of the study was the effect of relative humidity on chloride permeability. Samples cured at 60 percent relative humidity exhibited a significant increase in chloride permeability, rising from 23.3% to 98.9%, compared to those cured at relative humidity above 95% [30]. Du et al. [31] examined the influence of pore structure on the durability of 3DPC, while Shafiq et al. [36] focused on the role of metakaolin and nano-silica. Both studies found that the addition of nano-silica improved the resistance of concrete to chloride attack, identifying it as a promising admixture for enhancing the durability of 3DPC [31,39]. Moelich et al. [36], in a study that included reference to the work of Du et al. [31], investigated the effect of early-age restraint on the durability of 3DPC. Their findings showed that printed samples had a lower chloride migration coefficient than cast specimens. However, Du et al. [31] reported that the chloride migration coefficient of printed samples was either similar to or slightly higher than that of cast concrete. Currently, there is a lack of comprehensive research comparing the durability of 3DPC and conventional cast concrete. Further studies are needed to better understand the durability characteristics of 3DPC and to support its development as a reliable and long-lasting construction material.

2.5. Carbonation

The carbonation process occurs due to air infiltration through the pores and cracks of concrete. The carbon dioxide in the air reacts with the calcium hydroxide and produces calcium carbonate [40]. Carbonation can cause the corrosion of steel reinforcements. Thus, the resistance to carbonation is paramount for the service life of a structure. This makes carbonation testing a key durability criterion to be tested [72]. The industry's understanding of the carbonation resistance of 3DPC is significantly lower than that of cast concrete. De la Flor Juncal et al. [38] stated that the durability of 3DPC structures should be further investigated and that they should be exposed to accelerated carbonation tests to obtain their resistance.

A standard test for accelerated carbonation is ISO-1920-12-2015—*Determination of the carbonation resistance of concrete—Accelerated carbonation method* [72]. The test exposes the concrete sample's vertical side to an increased carbon dioxide level. The storage room had a carbon dioxide concentration of 3 ± 0.5%, a temperature of 22 ± 2 °C, and a relative humidity of 65%. The test is used to compare the carbonation resistance of different concrete samples. Two different sizes are used for the concrete samples: a 100 mm cube for carbonation depth tested after one exposure period and a 400 mm long prism with a 100 × 100 mm face for a sample tested at multiple exposure periods. As previously mentioned, the samples' tops, bottoms, and ends are covered in paraffin wax, exposing the vertical faces. For samples tested at different lengths of exposures, 50 mm is removed from the end of the 400 mm sample. The sample is then resealed with paraffin wax. The carbonation depth is determined by coloration after 56, 63, and 70 d. The total time is 112 d, 28 d curing, 14 d conditioning, and 70 d testing [72].

Bekaert et al. [30] used 400 × 100 × 100 mm for cast concrete samples and printed a hollow wall 400 mm × 100 mm × layer width; they measured the carbonation depth at 0, 7, 14, 28, and 56 d exposure, and a high carbonation rate was observed. The other papers in the review used cubes [45,73]. Zhou et al. [45] used 100 mm cubes, and Sanchez et al. [73] used 25 mm printed hollow cubes. Sanchez [73] measured carbonation after 2 months. Zhou et al. [45] measured carbonation penetration in the z-direction (Figure 20) when the sample had reached the carbonation time in line with the Chinese standard GB/T 50082-2009. Table 6 evaluates the carbonation resistance of 3DPC using accelerated testing methods. Results show that carbonation depth ranged from 2.4 to 4.8 mm/$\sqrt{}$day, strongly influenced by curing conditions. Increasing curing humidity to above 95% halved the depth of carbonation compared to samples cured at 60% RH. Mixes containing 50% GGBS also showed superior carbonation resistance compared to mixes with higher OPC content. Although 3DPC tends to have higher carbonation rates than cast concrete due to its porous interfacial zones, these results affirm that carbonation resistance can be significantly improved through SCM inclusion and proper curing.

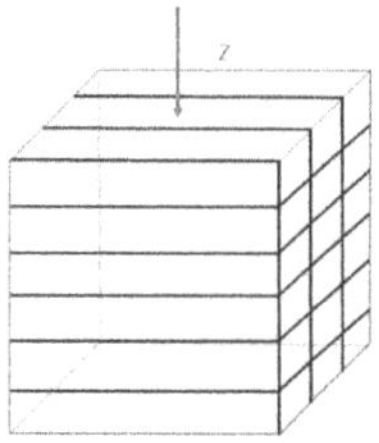

Figure 20. Test direction of 3DPC, reproduced with permission from [45].

Zhou et al. [45] compared the carbonation of samples with different concentrations of bauxite tailings. The results showed that the 35% bauxite tailing concentration sample had the largest carbonation resistance [45]. Bekaert et al. [30] studied the addition of GGBS (ground granulated blast-furnace slag) into the concrete mix, as well as having RH = 60% and RH > 95% (relative humidity). The results found that the 3DPC mix with GGBS had a high level of carbonation depth at RH = 60% and RH > 95%. The paper also noted that the carbonation depth was significantly higher at RH = 60% compared to RH > 95%. Sanchez et al. [73] performed the microstructural examination of carbonated

3DPC. The microstructural examination found zero carbonation penetration in the cast sample. However, significant carbonation was observed through the depth of the printed sample, with the highest level observed at the layer interfaces [73]. Both studies found that 3DPC had a higher level of carbonation penetration when compared to cast samples [30,73]. Sanchez et al. [73] used 15% limestone and 8% nano-silica as binder substitutions for the 3DPC mix. It is impossible to obtain the nano-silica's impact on the durability as the test compared a cast sample with the traditional mix to the 3DPC nano-silica mix. It is clear from the literature that printed samples have an increased carbonation rate. Further testing must be conducted to determine if adding nano-silica can reduce this rate.

Table 6. Analysis of carbonation in 3DPC literature: mix design parameters and testing.

Ref.	Mix	Binder Description	Mix Description	Test Method	Procedure	Curing	PD (mm)
[30]	REF	50% OPC 50% GGBS	S/B 1	AC	60 d @ RH = 70% 50–90% CO_2	28 d @ RH = 60	4.8/$\sqrt{day}$
						28 d @ RH > 95	2.4/$\sqrt{day}$
	M1	100% OPC	S/B 1.5	AC	60 d @ RH = 70% 50–90% CO_2	28 d @ RH = 60	7.4/$\sqrt{day}$
						28 d @ RH > 95	1/$\sqrt{day}$
	M2	75% OPC 25% GGBS	S/B 1.5	AC	60 d @ RH = 70% 50–90% CO_2	28 d @ RH = 60	12/$\sqrt{day}$
						28 d @ RH > 95	0.8/$\sqrt{day}$
[73]	C	100% OPC	W/B 0.6	-	60 d @ RH = 70% 50–90% CO_2	60 d	0.5
	3DP	77% OPC 15% FL 8% SF	S/B 1.9 W/B 0.30	-	60 d @ RH = 70% 50–90% CO_2	70 d	1
[45]	T0	100% OPC	100% N	GB/T 50082-2009	-	28 d	8.25
	T15	100% OPC	85% N 15% BT	GB/T 50082-2009	-	28 d	7.75
	T25	100% OPC	75% N 25% BT	GB/T 50082-2009	-	28 d	6.5
	T35	100% OPC	65% N 35% BT	GB/T 50082-2009	-	28 d	5.25
	T45	100% OPC	55% N 45% BT	GB/T 50082-2009	-	28 d	5.13

Note: Penetration Depth (PD), Ordinary Portland Cement (OPC), Ground Granulated Blast-Furnace Slag (GGBS), Sand/Binder (S/B), Accelerated Carbonation (AC), Relative Humidity (RH), Water/Binder (W/B), Filler Limestone (FL), Silica Fume (SF), Natural (N) Sand, Bauxite Tailings (BT).

Mechanistically, 3DPC is more prone to carbonation due to its distinct layered architecture, which introduces interfacial transition zones (ITZs) and lubrication layers with higher porosity and poor compaction. These features increase gas permeability, facilitating CO_2 ingress, especially along anisotropic paths parallel to the printing direction. Recent studies [73–75] using X-ray CT and SEM also confirm that interlayer voids and directional pore networks, which act as preferential carbonation pathways, are not present in cast concrete, as shown in Figure 21. Despite the growing body of data on carbonation resistance in traditional concrete, systematic evaluations of 3DPC under long-term carbonation exposure remain limited, particularly regarding the influence of print parameters on microstructure evolution.

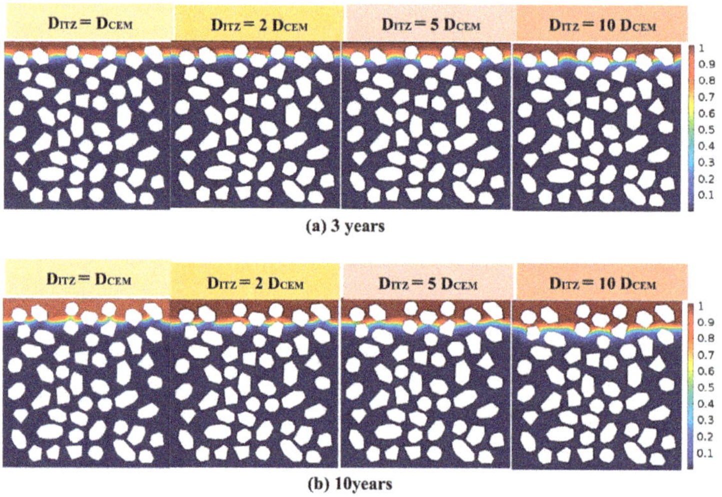

Figure 21. Relative carbonate content with different D_{ITZ}/D_{CEM} at 3 years (**a**) and 10 years (**b**), where D_{ITZ} is ITZ diffusivity of carbon dioxide and D_{CEM} is bulk paste diffusivity of carbon dioxide, reproduced with permission from [75].

3. Discussion and Future Work

The durability performance of 3DPC is governed by a complex interplay between material properties, printing process parameters, and environmental exposure conditions. This review identified key challenges, including elevated porosity at interlayer regions, increased susceptibility to shrinkage, and accelerated degradation under aggressive conditions such as freeze–thaw cycles, acid exposure, and chloride ingress. While the integration of SCMs and the refinement of printing parameters have shown considerable promise in improving durability, several research gaps persist.

Quantitative data summarized in six tables throughout this review underscore the multifaceted nature of 3DPC durability. Porosity in printed mixes ranged from as low as 0.09% to over 40%, with substantial reductions achieved through the use of SCMs such as silica fume and nano-silica. Shrinkage data revealed that plastic and drying shrinkage are more pronounced in 3DPC compared to cast concrete, but the use of fly ash and slag significantly mitigated cracking risk and improved dimensional stability, extending the crack initiation time by up to 57%.

Freeze–thaw resistance results indicated that the addition of air-entraining agents and styrene–butadiene rubber (SBR) emulsions led to marked improvements, with strength losses reduced by up to 79% compared to control mixes. In some cases, printed samples even outperformed their cast counterparts in terms of mass retention. Under sulfuric acid attack, 3DPC mixes containing silica fume showed enhanced chemical resistance, with mass losses reduced to as low as 2.7% after prolonged exposure, outperforming cast concrete in certain scenarios.

Chloride ingress resistance varied considerably, but the use of nano-silica and optimized curing methods (e.g., high relative humidity) reduced diffusion coefficients by up to 80%. Although 3DPC typically exhibited greater carbonation depth than cast concrete, performance improved significantly when GGBS was included in the binder and curing was conducted under RH > 95%, reducing carbonation depth by nearly 50%.

In addition to OPC-based systems, recent studies have demonstrated that geopolymer-based 3DPC systems can also benefit from similar durability strategies. The use of aluminosilicate precursors combined with optimized curing and SCM additions has shown potential in reducing porosity and enhancing chemical resistance. However, despite the presence of studies investigating the durability performance of geopolymer 3DPC in aspects such as acid resistance, carbonation, and chloride penetration, these systems are typically explored in isolation rather than through direct comparison with OPC-based 3DPC. Moreover, current literature lacks a systematic investigation of the differences and mechanisms between geopolymer and OPC-based 3DPC under various deterioration processes, including pore structure evolution, interlayer bonding performance, and degradation under environmental or chemical exposure.

In summary, while 3DPC presents inherent durability concerns due to its layered anisotropic structure and interfacial porosity, these challenges can be effectively addressed through comprehensive mix design optimization, targeted use of SCMs, appropriate curing strategies, and the incorporation of performance-enhancing admixtures. When properly engineered, 3DPC has the potential to meet or even exceed the durability standards of conventional cast concrete. To further improve the long-term performance of 3D-printed concrete (3DPC), future research should address the following key areas:

- Reducing porosity, particularly at interlayer regions, is essential for enhancing durability. Future studies should optimize the incorporation of nano-materials (e.g., nano-silica) and supplementary cementitious materials (SCMs) to improve pore structure uniformity and reduce overall permeability.
- Plastic and autogenous shrinkage remain critical durability concerns due to rapid moisture loss and the absence of formwork. Research should explore advanced curing techniques, such as controlled humidity environments and the use of internal curing agents or fiber reinforcements to limit early-age shrinkage cracking.
- While air-entraining agents and silica fume have shown positive effects, the long-term freeze–thaw durability of 3DPC under variable field conditions remains uncertain. Further investigation should focus on optimizing compaction quality, print parameters, and mix designs to ensure stable performance.
- Sulfuric acid poses a major threat to 3DPC durability. The development of acid-resistant binder systems using geopolymer formulations, hybrid cements, or surface treatments is necessary to withstand aggressive chemical environments.
- The anisotropic pore structure of 3DPC increases vulnerability to chloride penetration and steel reinforcement corrosion. Future work should design mixes with enhanced microstructural continuity and investigate admixtures such as corrosion inhibitors.
- Because 3DPC has shown higher susceptibility to carbonation than cast concrete, future experiments should employ accelerated carbonation testing across different mix designs, curing conditions, and print orientations to better understand and control carbonation depth.
- Currently, there is no dedicated durability standard for 3DPC. Adapting and validating existing ASTM, ISO, or EN test methods specifically for 3D-printed structures is crucial for establishing reliable long-term performance benchmarks.
- Most current assessments are based on accelerated laboratory testing. Long-term field monitoring of printed structures under varying environmental conditions and loading is needed to evaluate real-world performance.
- Parameters such as layer height, interlayer time gap, and print speed can significantly affect the formation of interfacial defects and long-term durability. Future studies should develop guidelines that balance durability with print efficiency.
- High-resolution synchrotron X-ray computed tomography (CT) is recommended for mapping CO_2 diffusion pathways in 3DPC, particularly across interlayer interfaces and anisotropic pore networks, enabling accurate correlation between microstructure and carbonation behavior.

- Future work should also focus on developing durability prediction models that account for anisotropic transport behavior, evolving interlayer bonding, and environmental influences over time. These models should incorporate the effects of creep, shrinkage, and temperature–humidity cycles to reflect realistic service conditions.
- Although several studies have reported promising durability results for geopolymer 3DPC, future research should systematically compare geopolymer and OPC-based systems under identical exposure conditions. Investigating their differences in pore structure evolution, interlayer bonding, and degradation mechanisms will help clarify their respective advantages and limitations in long-term durability performance.

4. Conclusions

This review systematically assessed key durability challenges associated with 3D-printed concrete (3DPC), a construction technology offering benefits such as faster construction, reduced labor requirements, and minimal material waste. This review focused on essential mix design parameters and testing methodologies addressing durability factors, including freeze–thaw resistance, chloride ingress, chemical attack, and carbonation. Key insights indicate significant impacts from porosity, shrinkage behaviour, microstructural development, and exposure conditions on the long-term durability of 3DPC. The primary conclusions from this study include:

- The durability of 3DPC significantly depends on optimized mix design, precise printing parameters, effective interlayer bond strength, controlled porosity, and environmental exposure conditions. The anisotropic behaviour and high interlayer porosity necessitate careful material and process optimization to achieve robust performance.
- The high porosity at layer interfaces poses a durability risk, which can be effectively mitigated by incorporating supplementary cementitious materials (SCMs) such as nano-silica, metakaolin, silica fume, and fly ash, significantly improving porosity and overall durability.
- Shrinkage-induced cracking, particularly plastic and autogenous shrinkage due to the absence of formwork, remains a critical issue. Slag and fly ash admixtures are recommended to reduce shrinkage rates and delay cracking.
- Enhanced freeze–thaw resistance is achievable through admixtures like silica fume and air-entraining agents, significantly reducing compressive strength loss and mass loss compared to traditional concrete.
- Sulfuric acid resistance in 3DPC can be notably improved by adding silica fume and nano-silica, minimizing mass loss and structural degradation in aggressive chemical environments.
- Chloride ingress is effectively controlled using optimized mixes containing metakaolin and nano-silica, significantly reducing chloride permeability.
- Carbonation depth tends to be higher in 3DPC; however, improvements through the use of SCMs like GGBS and nano-silica combined with high-humidity curing can substantially enhance resistance.
- In conclusion, while 3DPC holds substantial promise for sustainable and efficient construction, its durability needs targeted improvements. To enhance durability, it is recommended to strategically incorporate SCMs like silica fume, metakaolin, fly ash, and nano-silica, optimize printing parameters, and implement rigorous curing protocols. Standardized testing methods such as ASTM C666 [53] for freeze–thaw cycles, ASTM C1202 [65] for chloride permeability, and ASTM C1581 [42] and C1579 [43] for shrinkage assessments are essential for consistent evaluation and improvement. Future research should focus on standardizing long-term durability testing protocols,

developing specialized admixtures tailored to durability needs, and exploring deeper into how print-process parameters affect microstructural evolution.

Author Contributions: Conceptualization, J.B. and M.K.; methodology, M.K.; software, J.B.; validation, J.B., W.S. and M.K.; formal analysis, W.S. and M.K.; investigation, W.S. and M.K.; resources, M.K.; data curation, J.B. and W.S.; writing-original draft preparation, J.B.; writing-review and editing, M.K. and C.M.; visualization, W.S. and C.M.; supervision, M.K. and C.M.; project administration, M.K. and C.M.; funding acquisition, M.K. and C.M. All authors have read and agreed to the published version of the manuscript.

Funding: This publication emanated from 2 separate projects. The first is funded by Construct Innovate Technology Centre and Harcourt Technologies Limited (HTL) (Grant Code: CISFC1-23_013). The second is funded by Ecocem Materials and the Science Foundation Ireland (SFI) Research Centre in Applied Geosciences hosted by UCD (iCRAG-Phase 2-Grant Code: 13/RC/2092_P2).

Data Availability Statement: Not applicable.

Conflicts of Interest: The authors declare no conflicts of interest.

References

1. Craveiro, F.; Craveiro, F.; Nazarian, S.; Bartolo, H.; Bartolo, P.J.; Duarte, J.P. An automated system for 3D printing functionally graded concrete-based materials. *Addit. Manuf.* **2020**, *33*, 101146. [CrossRef]
2. Adaloudis, M.; Roca, J.B. Sustainability tradeoffs in the adoption of 3D Concrete Printing in the construction industry. *J. Clean. Prod.* **2021**, *307*, 127201. [CrossRef]
3. Fonseca, M.; Matos, A.M. 3D Construction Printing Standing for Sustainability and Circularity: Material-Level Opportunities. *Materials* **2023**, *16*, 2458. [CrossRef]
4. Firoozi, A.A. 3D Printing in Civil Engineering: Pioneering Affordable Housing Solutions. *J. Civ. Eng. Urban.* **2024**, *14*, 63–75. [CrossRef]
5. El-Sayegh, S.; Romdhane, L.; Manjikian, S. A critical review of 3D printing in construction: Benefits, challenges, and risks. *Arch. Civ. Mech. Eng.* **2020**, *20*, 34. [CrossRef]
6. Sakin, M.; Kiroglu, Y.C. Kiroglu, 3D Printing of Buildings: Construction of the Sustainable Houses of the Future by BIM. *Energy Procedia* **2017**, *134*, 702–711. [CrossRef]
7. AlZahrani, A.A.; Alghamdi, A.A.; Basalah, A.A. Computational Optimization of 3D-Printed Concrete Walls for Improved Building Thermal Performance. *Buildings* **2022**, *12*, 2267. [CrossRef]
8. Hossain, M.A.; Zhumabekova, A.; Paul, S.C.; Kim, J.R. A review of 3D printing in construction and its impact on the labor market. *Sustainability* **2020**, *12*, 8492. [CrossRef]
9. Hanratty, N.; Khan, M.; McNally, C. The Role of Different Clay Types in Achieving Low-Carbon 3D Printed Concretes. *Buildings* **2024**, *14*, 2194. [CrossRef]
10. Zhang, J.; Wang, J.; Dong, S.; Yu, X.; Han, B. A review of the current progress and application of 3D printed concrete. *Compos. Part A Appl. Sci. Manuf.* **2019**, *125*, 105533. [CrossRef]
11. Şahin, H.G.; Mardani, A. Mechanical properties, durability performance and interlayer adhesion of 3DPC mixtures: A state-of-the-art review. *Struct. Concr.* **2023**, *24*, 5481–5505. [CrossRef]
12. Chen, J.; Liu, X.; Tian, Y.; Zhu, W.; Yan, C.; Shi, Y.; Kong, L.B.; Qi, H.J.; Zhou, K. 3D-Printed anisotropic polymer materials for functional applications. *Adv. Mater.* **2022**, *34*, 2102877. [CrossRef] [PubMed]
13. Liu, C.; Yue, S.; Zhou, C.; Sun, H.; Deng, S.; Gao, F.; Tan, Y. Anisotropic mechanical properties of extrusion-based 3D printed layered concrete. *J. Mater. Sci.* **2021**, *56*, 16851–16864. [CrossRef]
14. Mo, Y.; Xing, J.; Yue, S.; Zhang, Y.; Zhou, Q.; Liu, X. Dynamic properties of 3D printed cement mortar based on Split Hopkinson Pressure Bar testing. *Cem. Concr. Compos.* **2022**, *130*, 104520. [CrossRef]
15. Ma, G.; Li, Z.; Wang, L. Printable properties of cementitious material containing copper tailings for extrusion based 3D printing. *Constr. Build. Mater.* **2018**, *162*, 613–627. [CrossRef]
16. Nodehi, M.; Aguayo, F.; Nodehi, S.E.; Gholampour, A.; Ozbakkaloglu, T.; Gencel, O. Durability properties of 3D printed concrete (3DPC). *Autom. Constr.* **2022**, *142*, 104479. [CrossRef]
17. Rehman, A.U.; Kim, J.H. 3D Concrete Printing: A Systematic Review of Rheology, Mix Designs, Mechanical, Microstructural, and Durability Characteristics. *Materials* **2021**, *14*, 3800. [CrossRef]
18. Zaid, O.; El Ouni, M.H. Advancements in 3D printing of cementitious materials: A review of mineral additives, properties, and systematic developments. *Constr. Build. Mater.* **2024**, *427*, 136254. [CrossRef]

19. van den Heever, M.; du Plessis, A.; Kruger, J.; van Zijl, G. Evaluating the effects of porosity on the mechanical properties of extrusion-based 3D printed concrete. *Cem. Concr. Res.* **2022**, *153*, 106695. [CrossRef]
20. Yuan, H.; Dong, E.; Jia, Z.; Jia, L.; Quan, S.; Ma, M.; Yang, Y.; Feng, M.; Banthia, N.; Zhang, Y. The influence of pore structure and fiber orientation on anisotropic mechanical property of 3D printed ultra-high-performance concrete. *Constr. Build. Mater.* **2025**, *471*, 140760. [CrossRef]
21. Kruger, J.; du Plessis, A.; van Zijl, G. An investigation into the porosity of extrusion-based 3D printed concrete. *Addit. Manuf.* **2021**, *37*, 101740. [CrossRef]
22. Mohan, M.K.; Rahul, A.V.; Van Stappen, J.F.; Cnudde, V.; De Schutter, G.; Van Tittelboom, K. Assessment of pore structure characteristics and tortuosity of 3D printed concrete using mercury intrusion porosimetry and X-ray tomography. *Cem. Concr. Compos.* **2023**, *140*, 105104. [CrossRef]
23. Rui, A.; Wang, L.; Lin, W.; Ma, G. Experimental study on damage anisotropy of 3D-printed concrete exposed to sulfate attack. *Constr. Build. Mater.* **2023**, *407*, 133590. [CrossRef]
24. Çevik, A.; Alzeebaree, R.; Humur, G.; Niş, A.; Gülşan, M.E. Effect of nano-silica on the chemical durability and mechanical performance of fly ash based geopolymer concrete. *Ceram. Int.* **2018**, *44*, 12253–12264. [CrossRef]
25. Van Tittelboom, K.; Mohan, M.K.; Šavija, B.; Keita, E.; Ma, G.; Du, H.; Kruger, J.; Caneda-Martinez, L.; Wang, L.; Bekaert, M.; et al. On the micro- and meso-structure and durability of 3D printed concrete elements. *Cem. Concr. Res.* **2024**, *185*, 107649. [CrossRef]
26. Singh, A.; Wang, Y.; Zhou, Y.; Sun, J.; Xu, X.; Li, Y.; Liu, Z.; Chen, J.; Wang, X. Utilization of antimony tailings in fiber-reinforced 3D printed concrete: A sustainable approach for construction materials. *Constr. Build. Mater.* **2023**, *408*, 133689. [CrossRef]
27. Liu, C.; Zhang, Y.; Banthia, N. Unveiling pore formation and its influence on micromechanical property and stress distribution of 3D printed foam concrete modified with hydroxypropyl methylcellulose and silica fume. *Addit. Manuf.* **2023**, *71*, 103606. [CrossRef]
28. Baz, B.; Aouad, G.; Kleib, J.; Bulteel, D.; Remond, S. Durability assessment and microstructural analysis of 3D printed concrete exposed to sulfuric acid environments. *Constr. Build. Mater.* **2021**, *290*, 123220. [CrossRef]
29. Jaji, M.B.; van Zijl, G.P.; Babafemi, A.J. Durability and pore structure of metakaolin-based 3D printed geopolymer concrete. *Constr. Build. Mater.* **2024**, *422*, 135847. [CrossRef]
30. Bekaert, M.; Van Tittelboom, K.; De Schutter, G. The Effect of Curing Conditions on the Service Life of 3D Printed Concrete Formwork. *Materials* **2023**, *16*, 6972. [CrossRef]
31. Du, L.; Zhou, J.; Lai, J.; Wu, K.; Yin, X.; He, Y. Effect of pore structure on durability and mechanical performance of 3D printed concrete. *Constr. Build. Mater.* **2023**, *400*, 132581. [CrossRef]
32. Sikora, P.; Techman, M.; Federowicz, K.; El-Khayatt, A.M.; Saudi, H.A.; Abd Elrahman, M.; Hoffmann, M.; Stephan, D.; Chung, S.Y. Insight into the microstructural and durability characteristics of 3D printed concrete: Cast versus printed specimens. *Case Stud. Constr. Mater.* **2022**, *17*, e01320. [CrossRef]
33. Zhang, Y.; Qiao, H.; Qian, R.; Xue, C.; Feng, Q.; Su, L.; Zhang, Y.; Liu, G.; Du, H. Relationship between water transport behaviour and interlayer voids of 3D printed concrete. *Constr. Build. Mater.* **2022**, *326*, 126731. [CrossRef]
34. Zhang, Y.; Zhang, Y.; She, W.; Yang, L.; Liu, G.; Yang, Y. Rheological and harden properties of the high-thixotropy 3D printing concrete. *Constr. Build. Mater.* **2019**, *201*, 278–285. [CrossRef]
35. Bayrak, A.T.; Shaban, N.; Choubi, S.S.; Tuncer, E.; Yang, S.H.; Yılmaz, H.D.; Alkilani, A.Z.; Dal, H.; Unluer, C.; Dino, İ.G.; et al. Spatial variation of physical, mechanical, and thermophysical properties of 3D printed concrete across a full-scale wall. *Constr. Build. Mater.* **2024**, *431*, 136574. [CrossRef]
36. Moelich, G.M.; Kruger, P.J.; Combrinck, R. The effect of restrained early age shrinkage on the interlayer bond and durability of 3D printed concrete. *J. Build. Eng.* **2021**, *43*, 102857. [CrossRef]
37. Zhang, Y.; Zhang, Y.; Yang, L.; Liu, G.; Du, H. Influence of the pore feature on the water uptake in 3D printed concrete. *Mater. Lett.* **2023**, *333*, 133642. [CrossRef]
38. de la Flor Juncal, L.; Loporcaro, G.; Scott, A.; Clucas, D. Influence of printing parameters on the durability of 3D-printed limestone calcined clay cement mortar: Overlap between filaments and nozzle offset. *Mater. Struct.* **2024**, *57*, 191. [CrossRef]
39. Shafiq, N.; Kumar, R.; Zahid, M.; Tufail, R.F. Effects of Modified Metakaolin Using Nano-Silica on the Mechanical Properties and Durability of Concrete. *Materials* **2019**, *12*, 2291. [CrossRef]
40. Ler, K.-H.; Ma, C.K.; Chin, C.L.; Ibrahim, I.S.; Padil, K.H.; Ab Ghafar, M.A.I.; Lenya, A.A. Porosity and durability tests on 3D printing concrete: A review. *Constr. Build. Mater.* **2024**, *446*, 137973. [CrossRef]
41. Zhong, H.; Zhang, M. 3D printing geopolymers: A review. *Cem. Concr. Compos.* **2022**, *128*, 104455. [CrossRef]
42. *ASTM C1581-04*; Standard Test Method for Determining Age at Cracking and Induced Tensile Stress Characteristics of Mortar and Concrete Under Restrained Shrinkage. American Society for Testing and Materials: West Conshohocken, PA, USA, 2004.
43. *ASTM C1579-21*; Standard Test Method for Evaluating Plastic Shrinkage Cracking of Restrained Fiber Reinforced Concrete (Using a Steel Form Insert). American Society for Testing and Materials: West Conshohocken, PA, USA, 2021.

44. Moelich, G.M.; Kruger, J.; Combrinck, R. Plastic shrinkage cracking in 3D printed concrete. *Compos. Part B Eng.* **2020**, *200*, 108313. [CrossRef]
45. Zhou, L.; Gou, M.; Ji, J.; Hou, X.; Zhang, H. Durability and hardened properties of 3D printed concrete containing bauxite tailings. *Mater. Today Sustain.* **2024**, *25*, 100704. [CrossRef]
46. *JC/T 603-2004*; Standard Test Method for Drying Shinkage of Mortar. China Building Materials Industry Press: Beijing, China, 2004.
47. Papachristoforou, M.; Mitsopoulos, V.; Stefanidou, M. Use of by-products for partial replacement of 3D printed concrete constituents; rheology, strength and shrinkage performance. *Frat. Integrità Strutt.* **2019**, *13*, 526–536. [CrossRef]
48. Markin, S.; Combrinck, R.; Mechtcherine, V. Specifics of plastic shrinkage in 3D-printed concrete elements. *Cem. Concr. Res.* **2024**, *184*, 107512. [CrossRef]
49. Markin, S.; Mechtcherine, V. Quantification of plastic shrinkage and plastic shrinkage cracking of the 3D printable concretes using 2D digital image correlation. *Cem. Concr. Compos.* **2023**, *139*, 105050. [CrossRef]
50. Markin, S.; Mechtcherine, V. Methods for measuring plastic shrinkage and related cracking of 3D-printed concrete. *J. Phys. Conf. Ser.* **2023**, *2423*, 012036. [CrossRef]
51. Assaad, J.J.; Hamzeh, F.; Hamad, B. Qualitative assessment of interfacial bonding in 3D printing concrete exposed to frost attack. *Case Stud. Constr. Mater.* **2020**, *13*, e00357. [CrossRef]
52. Mohan, M.K.; Rahul, A.V.; De Schutter, G.; Van Tittelboom, K. Extrusion-based concrete 3D printing from a material perspective: A state-of-the-art review. *Cem. Concr. Compos.* **2021**, *115*, 103855. [CrossRef]
53. *ASTM C666/C666M*; Standard Test Method for Resistance of Concrete to Rapid Freezing and Thawing. American Society for Testing and Materials: West Conshohocken, PA, USA, 2015.
54. Dong, W.; Wang, J.; Hang, M.; Qu, S. Research on printing parameters and salt frost resistance of 3D printing concrete with ferrochrome slag and aeolian sand. *J. Build. Eng.* **2024**, *84*, 108508. [CrossRef]
55. *GB/T 5008-2001*; Method of Test for Flexural Strength of Concrete. China Architecture & Building Press: Beijing, China, 2001.
56. GivKashi, M.R.; Tohidloo, M. The effect of freeze-thaw cycles and sulfuric acid attack separately on the compressive strength and microstructure of 3D-printed air-entrained concrete. *Constr. Build. Mater.* **2024**, *440*, 137411. [CrossRef]
57. Gu, L.; Bennett, T.; Visintin, P. Sulphuric acid exposure of conventional concrete and alkali-activated concrete: Assessment of test methodologies. *Constr. Build. Mater.* **2019**, *197*, 681–692. [CrossRef]
58. Barbhuiya, S.; Kumala, D. Behaviour of a Sustainable Concrete in Acidic Environment. *Sustainability* **2017**, *9*, 1556. [CrossRef]
59. *ASTM C109/C109M-20*; Standard Test Method for Compressive Strength of Hydraulic Cement Mortars (Using 2-in. or [50-mm] Cube Specimens). American Society for Testing and Materials: West Conshohocken, PA, USA, 2020.
60. Guo, J.-J.; Liu, P.Q.; Wu, C.L.; Wang, K. Effect of Dry-Wet Cycle Periods on Properties of Concrete under Sulfate Attack. *Appl. Sci.* **2021**, *11*, 888. [CrossRef]
61. El Inaty, F.; Baz, B.; Aouad, G. Long-term durability assessment of 3D printed concrete. *J. Adhes. Sci. Technol.* **2022**, *37*, 1921–1936. [CrossRef]
62. *GB/T 50082-2009*; Standard for Test Methods of Long-Term Performance and Durability of Ordinary Concrete. China Architecture & Building Press: Beijing, China, 2009.
63. Hou, W.; Liu, J.; Liu, Z.; He, F.; Zhu, J.; Cui, Y.; Jinyang, W. Calcium transfer process of cement paste for ettringite formation under different sulfate concentrations. *Constr. Build. Mater.* **2022**, *348*, 128706. [CrossRef]
64. Shumuye, E.D.; Mehrpay, S.; Fang, G.; Li, W.; Wang, Z.; Uge, B.U.; Liu, C. Influence of novel hybrid nanoparticles as a function of admixture on responses of engineered geopolymer composites: A review. *J. Build. Eng.* **2024**, *86*, 108782. [CrossRef]
65. *ASTM C1202-17*; Standard Test Method for Electrical Indication of Concrete's Ability to Resist Chloride Ion Penetration. American Society for Testing and Materials: West Conshohocken, PA, USA, 2017.
66. Bran-Anleu, P.; Wangler, T.; Nerella, V.N.; Mechtcherine, V.; Trtik, P.; Flatt, R.J. Using micro-XRF to characterize chloride ingress through cold joints in 3D printed concrete. *Mater. Struct.* **2023**, *56*, 51. [CrossRef]
67. Alexander, M.G. *Durability Index Testing Procedure Manual*; University of Cape Town: Cape Town, South Africa, 2018.
68. Hájková, K.; Šmilauer, V.; Jendele, L.; Červenka, J. Prediction of reinforcement corrosion due to chloride ingress and its effects on serviceability. *Eng. Struct.* **2018**, *174*, 768–777. [CrossRef]
69. Liu, H.; Liu, C.; Bai, G.; Zhu, C. Study on the effect of chloride ion ingress on the pore structure of the attached mortar of recycled concrete coarse aggregate. *Constr. Build. Mater.* **2020**, *263*, 120123. [CrossRef]
70. Xia, J.; Chen, K.; Hu, S.; Chen, J.; Wu, R.; Jin, W. Experimental and numerical study on the microstructure and chloride ion transport behavior of concrete-to-concrete interface. *Constr. Build. Mater.* **2023**, *367*, 130317. [CrossRef]
71. Surehali, S.; Tripathi, A.; Nimbalkar, A.S.; Neithalath, N. Anisotropic chloride transport in 3D printed concrete and its dependence on layer height and interface types. *Addit. Manuf.* **2023**, *62*, 103405. [CrossRef]
72. *ISO-1920-12-2015*; Determination of the Carbonation Resistance of Concrete—Accelerated Carbonation Method. International Organization for Standardization: Geneva, Switzerland, 2015.

73. Sanchez, A.M.A.; Wangler, T.; Stefanoni, M.; Angst, U. Microstructural examination of carbonated 3D-printed concrete. *J. Microsc.* **2022**, *286*, 141–147. [CrossRef] [PubMed]
74. Sahana, C.M.; Soda, P.R.K.; Dwivedi, A.; Kumar, S.; Kumar, A.; Pundir, A.; Dixit, A.; Gupta, S. 3D printing with stabilized earth: Material development and effect of carbon sequestration on engineering performance. *Cem. Concr. Compos.* **2024**, *152*, 105653. [CrossRef]
75. Han, J.; Liu, W.; Wang, S.; Du, D.; Xu, F.; Li, W.; De Schutter, G. Effects of crack and ITZ and aggregate on carbonation penetration based on 3D micro X-ray CT microstructure evolution. *Constr. Build. Mater.* **2016**, *128*, 256–271. [CrossRef]

designs

MDPI

Article

A Simplified Design Method for the Mechanical Stability of Slit-Shaped Additively Manufactured Reactor Modules

David F. Metzger [1,*], Christoph Klahn [2] and Roland Dittmeyer [1]

[1] Institute for Micro Process Engineering (IMVT), Karlsruhe Institute of Technology (KIT), Herrmann-von-Helmholtz-Platz 1, 76344 Eggenstein-Leopoldshafen, Germany; roland.dittmeyer@kit.edu

[2] Institute for Mechanical Process Engineering and Mechanics (MVM), Karlsruhe Institute of Technology (KIT), B. 30.70, Strasse am Forum 8, 76131 Karlsruhe, Germany; christoph.klahn@kit.edu

* Correspondence: david.metzger2@kit.edu

Abstract: Equipment integrity is an essential aspect of process engineering. Design guidelines facilitate the design and production of safe-to-operate and economic devices. Thin-walled, slit-shaped modules form a subgroup of process engineering devices made via additive manufacturing (AM). Being subject to internal pressure, they have lacked design guidelines until now. We derived a user-centered calculation model for such modules with regular internal structures. It was validated with Finite Element Analysis (FEA) and practical pressure tests for which the modules were manufactured additively. The performance of the calculation could be confirmed, and a design graph was derived. Slit-shaped modules with appropriate internal structures can withstand high pressure at a minimum wall thickness, and they are efficiently fabricated. These structures, being pins, fins, lattice, or heat transfer enhancing fluid-guiding elements (FGEs), occupied approximately 10% of the modules' internal volume.

Keywords: laser-based powder bed fusion; pressure vessel; stainless steel 316L; periodic open cellular structures; fluid-guiding elements

1. Introduction

Process engineering requires devices containing hazardous materials at a high pressure and temperature. Processes like chemical reactions, adsorption, and heat transfer are often carried out at a high pressure [1]. In Fischer-Tropsch synthesis (FTS), being one example out of many, the product yield was increased using a high pressure of typically 20 to 30 bar [2], while reactants were flammable and toxic [3]. Thus, stability is a very important factor in safe operation.

Slit-shaped devices made with conventional manufacturing are common in micro-process engineering [4,5]. A metal additive manufacturing (AM) reactor consisting of slit-shaped modules was used successfully to carry out challenging FTS [6]. Utilizing function integration and design freedom, the AM reactor required fewer sealing surfaces and assembly steps compared to a conventional micro-reactor for FTS [7,8].

Slits can be extended in width and length, and they can be stacked to increase their capacity without a significant performance loss, making them important for engineers and researchers in the disciplines of chemical engineering and flow chemistry. Therefore, we investigated these particular modules in terms of mechanical integrity.

Finite element analysis (FEA) is one option to determine stability and optimize the part weight of almost arbitrary parts [9]. However, it requires the re-meshing of every design and computational effort. FEA interpretation depends on the designer, especially in the presence of singularities in the analysis [10].

For fast equipment development, a reliable and user-friendly calculation model for design with regard to stability is imperative. Such a model being accessible to the above-mentioned target group, who are usually not mechanical engineers, is beneficial. A model that could be handled by non-mechanical engineers and directly applied to typical slit modules encountered in micro-structured devices was not found in the literature.

For comparatively simple geometries, rectangle and circular disc formulas for stability calculation were found in a standard reference book [11]. A subset of slit-shaped modules investigated in this article can be abstracted to simple geometries and assessed using the aforementioned formulas.

The aim of this article was to perform this abstraction and validate it with FEA and experimental data. Comparing the experiment, formulas, and simulations of failure in AM parts helps expand knowledge and increase trust in this promising manufacturing method [12].

2. Materials and Methods

The investigated modules consisted of parallel planar walls forming a narrow slit with fluidic connectors at the bottom and top ends. For its material, stainless steel 316L was chosen due to its strength, corrosion resistance, and prevalence [13,14]. For the composition, refer to Table S1. The pressure at which a module is operated could be above the outside pressure, the outside being the surroundings or another module operated at a lower pressure. Planar walls tend to buckle when subjected to a pressure difference across the two sides. Under otherwise constant conditions, buckling is more likely when the smaller of the rectangular side's length increases or the wall thickness decreases [11]. Excess wall thickness must be avoided, but with every increase in the device's size, the walls become larger.

The following approaches were pursued to enhance mechanical stability: 1. the deformation of the walls in the direction of the y-axis and 2. the insertion of internal structures (Figure 1).

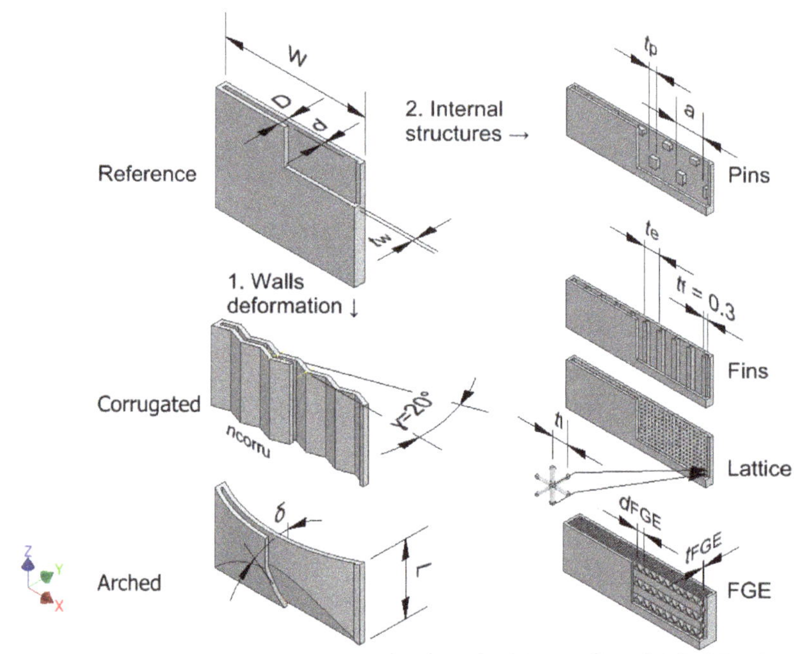

Figure 1. Designs for mechanical stability testing.

Walls bent in three-dimensional (3D) space are expected to be more stable than planar ones [15]. Evenly distributed structures connecting neighboring walls, preventing buckling, were introduced into the design, the structures being hexagonally arranged pins, evenly distributed fins, lattice, or fluid guiding elements (FGE). Topology optimization was not used, mainly due to two reasons. The first reason was that process engineering devices cannot be designed only by optimizing the mechanical load and weight. The flow pattern and accessibility of the catalyst and sensors are crucial. The second reason was that the chemical industry has strong regulatory requirements, especially for pressure vessels. Validation is facilitated using well-known features such as regular pins instead of topology-optimized shapes.

Internal structures have possible downsides: a reduction in the fluidic volume and the addition of material to the design. The volume of the internal structure per slit volume is called the volume fraction φ, and it was kept as low as possible. Nevertheless, there is a physical minimum for the volume fraction of internal structures. Based on a force equilibrium (Equation (1)), internal structures are expected to rupture below a certain volume fraction. This resulted in $\varphi_{\text{crit}} = 0.008$ for $p = 30\,\text{bar}$, a margin of error of $S = 1.5$, and a yield strength of $R_{\text{p0.2}} = 582\,\text{MPa}$ taken from [16] for vertical tensile samples.

$$p \cdot A \cdot (1 - \varphi) = S \cdot R_{\text{p0.2}} \cdot A \cdot \varphi \quad (1)$$

Internal structures have been identified by other persons as well as means to increase the stability of pressure vessels by adding a minimum mass [17–19]. The internal structures proposed in this article are unlikely to change the flow pattern significantly, as the flow is usually strongly laminar [20] in micro-structured devices.

Possible tubular portions are only minor and might occur close to fluid connections. They can be designed to protect against rupture according to well-established formulas, and they are not in the scope of this article [21].

2.1. Design and Fabrication

The empty planar module served as a reference and starting point for the design (Figure 1, top left).

2.1.1. Empty Modules

The first approach was bending the slit walls in three-dimensional (3D) space. The entire volume between the corrugated or arched walls was available for fluid ($\varphi = 0$). The deformation was effected in a way that still permitted multiple modules to be stacked seamlessly. The wall thickness of these samples was $t_w = 1\,\text{mm}$.

Three of the modules comprised corrugated walls. Corrugated-5 and Corrugated-10 had five and ten corrugations parallel to the z-axis, respectively. Corrugated-7′ had seven corrugations parallel to the x-axis. The angle between the corrugation and the xz-plane was always $\gamma = 20\,^\circ\text{C}$.

Arched-5, -15, -30, and Arched-45 were four modules with walls warped over the xz-plane. The digits denote the cutting angle between the wall and the xz-plane δ.

2.1.2. Modules with Internal Structures

Four categories of internal structures were used inside narrow planar slits in this work: pins, fins, lattice, and fluid guiding elementss(FGE) [22].

Connecting both sides of the slit directly, pins were characterized by their shape and arrangement. The cross-section of pins was a square with side length t_{p}, which was the characteristic parameter. To prevent horizontal overhangs, small angles were added at both slit walls. Pins were arranged in a hexagonal manner with spacing a.

To study mechanical stability, the size and spacing of internal pins varied over a wide range. The side length of quadratic pins was $t_{\text{p}} = 0.26\,\text{mm}$–$5\,\text{mm}$, while the spacing was $a = 1\,\text{mm}$–$20\,\text{mm}$. The wall thickness was $t_w = 0.6\,\text{mm}$. The width and length were $w = 40\,\text{mm}$ and $l = 60\,\text{mm}$, respectively.

The volume fraction was calculated from geometrical considerations, it and took values of $\varphi_{\text{p}} = 0.01$–$0.19$.

Fins connected both sides of the slit directly and separated the slit into parallel channels. The thickness of fins was $t_f = 0.28\,\text{mm} = \text{const}$. The characteristic parameter was the distance between two fins $t_e = 1\,\text{mm}–4\,\text{mm}$. Fillets were added where the fins transitioned into the walls. Again, the volume fraction was calculated from geometrical considerations, and it took values of $\varphi_f = 0.1–0.33$.

The lattice was of a body-centered cubic type with a side length of $t_l = 1\,\text{mm}$. Unlike the other internal structures, this one was not derived from CAD but from the printer's software. Under the scan strategy used, the strut diameter was $t_{strut} = 0.2\,\text{mm}$ as measured with a caliper. From this, a volume fraction of $\varphi_l = 0.186$ was calculated.

The geometry of a planar fluid guiding elements (FGE), consisting of one fluid guiding unit (FGU) repeated with certain distances in three dimensions, can be varied in multiple ways. However, the following simplifications were made: the ratio of the depth, width, and length of an FGU was fixed to 1:2:4 (Figure 2, top), and distances between repeating units were minimal, e.g., spacing of $a_{FGE} = l_{FGE}$ (Figure 2, bottom).

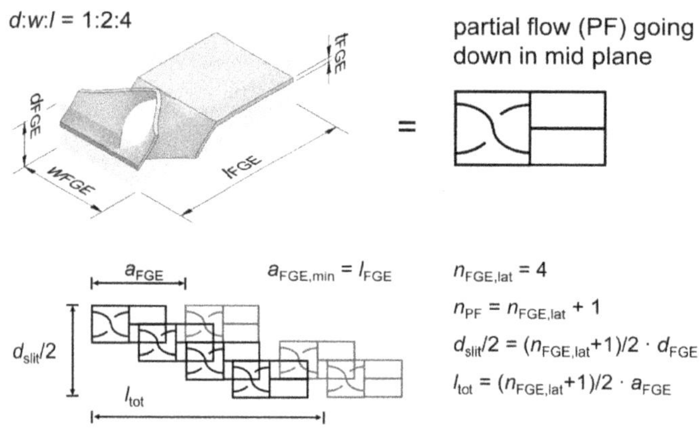

Figure 2. Definitions concerning fluid guiding elements (FGE).

In the case of FGE, assessing the volume fraction φ was not straightforward. An explicit formula for φ_{FGE}, based on d_{FGE} and the wall thickness t_{FGE}, was established and checked with fabricated samples (Equation (2)).

$$\varphi_{FGE} = \frac{5.03 \cdot d_{FGE}^2 \cdot t}{8 \cdot d_{FGE}^3} = 0.629 \frac{t}{d_{FGE}} \tag{2}$$

The numerator of Equation (2) was obtained with the help of computer aided design (CAD). Knowing it is an ideal, the thickness in CAD was set to $t_{FGE} = 0.25\,\text{mm}$. The mass of the fabricated FGE samples was assessed, and the wall thickness was measured as $0.3\,\text{mm}–0.4\,\text{mm}$. Based on this, φ could be calculated. The results from the explicit formula, as well as the experiments, are shown in Table 1. Especially at low values of d_{FGE}, the φ was much higher than expected, suggesting that this wall thickness was much higher than in CAD. This is probably due to the strong curvature of a small FGE.

Table 1. Calculated and measured values of φ_{FGE}.

d_{FGE}/mm	n. of FGU/-	φ_{FGE} meas./-	φ_{FGE} calc./-	
			$t_{FGE} = 0.3$ mm	$t_{FGE} = 0.4$ mm
1.3	32	0.299	0.145	0.194
2	18	0.149	0.094	0.126
3	8	0.101	0.063	0.084

FGE-equipped modules for stability testing had design parameters of d_{FGE} = 1.333 and 4 mm. They could only be manufactured with t_w = 0.8 mm. At the interface of the FGE and the wall, the spacing equaled l_{FGE} in the range of 5.33 mm–16 mm.

2.1.3. Connections and Fabrication

Internal threads enabled the slits to be attached to standard externally threaded connectors. The internal G1/8″-thread was designed individually according to [23] with 4 % higher diameter values to compensate for tolerances. It had clamping surfaces on the outside, and cones transitioned the rectangular cross-section of the slit into the circular cross-section of the thread section, resulting in a high wall thickness for these sections; t_w = 1 mm–2 mm (Figure 3a). The cones, which were not in the scope of the study, allowed for little overhangs, a flow-through with a smooth cross-section transition, and good depowdering [24].

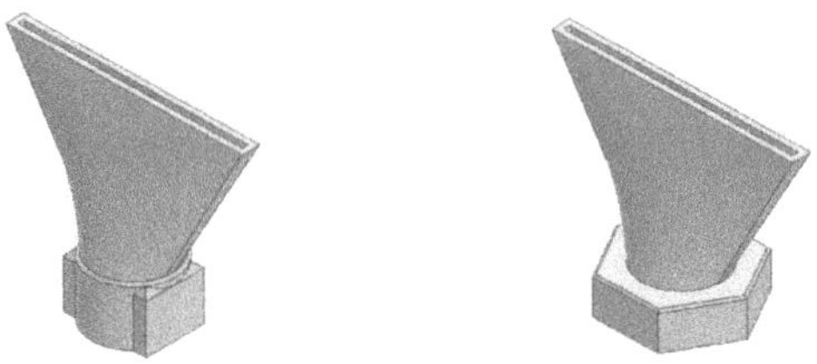

(**a**) (**b**)

Figure 3. Computer aided design (CAD) representation of module connectors. (**a**) Additive threaded connection: G1/8″. (**b**) Hybrid threaded connection: G1/8″ (possible at bottom end only).

The internal thread section at the lower end could be omitted when the part was manufactured on a nut that had been positioned inside the printer beforehand (Figure 3b). This approach combining conventional nuts and PBF-LB/M additive manufacturing is called hybrid manufacturing, and it and it greatly facilitated the production of modules for pressure testing [25].

The conventional substrate was 19 mm stainless steel nuts with a G1/8″ thread from Schrauben-Jäger AG (Karlsruhe, Germany, art.-n. 103806). The nuts were positioned with the help of parallel rests and fixed to the build plate with Loctite superglue by Henkel AG (Düsseldorf, Germany). Figure 4 depicts the manufacturing approach with an exemplary module. All modules tested are depicted in Supplementary Information (SI) Tables S2–S6.

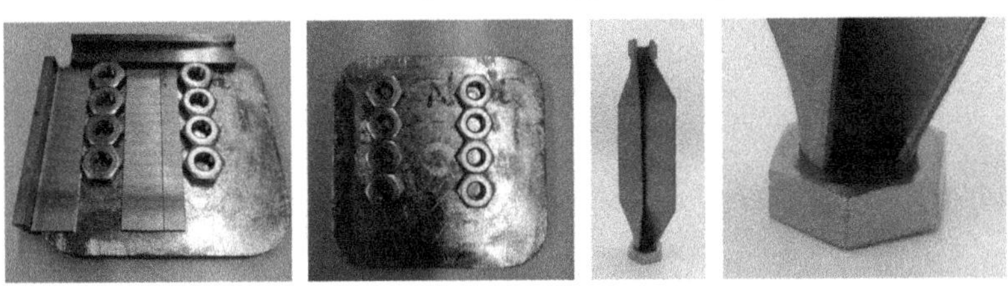

Figure 4. Hybrid manufacturing of modules on nuts with an exemplary empty module (three-quarter cut).

The standard tessellation language (STL) files obtained from the CAD software Inventor 2022 by Autodesk (San Rafael, CA, USA) were repaired in Autodesk® Netfabb® (NF). This was necessary since, under the unrevised version, the locally very complex designs (FGE) resulted in slicing errors in the pre-process software.

Modules were printed on a Realizer SLM125 (now LASERTEC SLM12) using DMG MORI (Bielefeld, Germany). The powders used, stainless steel 316L, were supplied by Carpenter Additive (Philadelphia, PA, USA) and SLM Solutions (Lübeck, Germany).

After printing, the threads at the ends were tapped and countersunk. The end surfaces were milled with a plain cutter on a Deckel (today DMG MORI) FP5 numerically controlled (NC) mill.

2.2. Experimental Method

The procedure described and shown in Figure 5 was applied to a clean and leak-tight part in order to determine the mechanical stability. The metallic modules were connected to a pressure test rig by Konstandin und Partner engineering GmbH (Pfinztal, Germany) with threaded connectors and metal-bonded polymer gaskets. The test pressure calculated with $1.43 \cdot PS$ (Ref. [26]) was rounded to $p_{\text{test}} = 30\,\text{bar}$. The external micrometer used had a measuring tip with $d = 5\,\text{mm}$. For modules with non-planar walls, parallel rests were employed.

(**a**) (**b**) (**c**)

Figure 5. Experimental procedure to test the mechanical stability. (**a**) Place marking in the middle of a module wall, measure the outside depth D_0, and attach to the test port of the pressure test rig. (**b**) Fill with water via a manual pump, close the opposite connector, and increase the pressure to $p_{\text{test}} = 30\,\text{bar}$. (**c**) After a minimum $t = 15\,\text{min}$, measure the outside depth, D_1, before decreasing to the ambient pressure.

2.3. Calculation Method

For modules equipped with pins, case 2j from Ref. [11] Table 11.2 (p. 457) was used in theoretical calculations. This case concerned an annular disc, fixed in the middle with guided outer edges, that was subject to a uniform load. The hexagonally arranged pins were abstracted to the annular disk in said reference as follows: The radius of the circle area equivalent to one quadratic pin was the inner radius in the reference. The radius of the circle area equivalent to one hexagonal repetition unit was the outer radius in the reference. The pressure was equal to the load per unit area q. The vertical deflection of the plate Δy at the outer radius was the most important: it was assumed as the deflection between pins with spacing a.

For empty modules and those equipped with fins, case 6a from Ref. [11] Table 11.4 (p. 506) was used in theoretical calculations. The formulas were evaluated with the help of Microsoft Excel (Redmond, WA, USA) and Matlab R2022b by Mathworks Inc. (Nattick, MA, USA).

Material properties can be found in Table 2. These properties, being standard values for wrought 316L from Ref. [27], have been used and confirmed by several authors for additive 316L [16,28].

Table 2. Material properties of 316L at room temperature used in calculation and simulation [27].

Young's Modulus E/MPa	Poisson's Ratio ν/-	Density ρ/kg m^{-3}
200,000	0.3	8000

2.4. Simulation Method

For finite element analysis (FEA), software ANSYS 2020 R2 Workbench and Mechanical by ANSYS Inc. (Canonsburg, PA, USA) were used. A custom material with the properties found in Table 2 was used. Only repetition units of the modules were subjected to FEA, also utilizing symmetries. The element size was 0.5 mm if not otherwise mentioned.

Mechanical strength was considered sufficient when the displacement was below $\Delta y \leq 0.01$ mm or $\Delta D = D_1 - D_0 \leq 0.02$ mm when referring to a module with external depth D. This quantity is accessible with all three investigation methods, while stress was not accessible for the experimental method chosen. The maximum value was chosen to limit the deformation of a slit with d = 1 mm to 2%, which is especially important to avoid crushing the catalyst particles present.

3. Results and Discussion

Almost all modules without internal structures (φ = 0) showed poor mechanical stability in experiments, the displacement being y = 0.2 mm–10 mm. The exception was the empty slit with highly arched walls Arched-45, which showed ΔD below measuring accuracy (Figure 6).

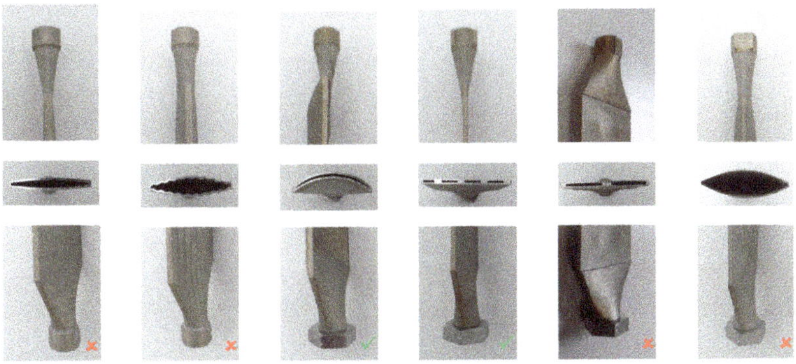

(a) (b) (c) (d) (e) (f)

Figure 6. Modules after pressure test. (**a–c**) t_w = 1 mm. (**d–f**) t_w = 0.6 mm. Green check marks indicate passing; red x marks indicate failing. (**a**) Empty planar. (**b**) Corrugated-10. (**c**) Arced-45. (**d**) Pins: t_p = 3 mm a = 9 mm. (**e**) Pins: t_p = 4 mm a = 20 mm. (**f**) Pins: t_p = 0.42 mm a = 5 mm.

Some pin-equipped modules showed a mode of failure referred to as "Telegraphing" in the experiments. This is known for fiber-reinforced composite materials [29]. When the distance between pins was too high $a \gtrsim 10$ mm, the walls curved outside between pins (red open circles in Figure 7). Below a certain volume fraction, $\varphi \leq \varphi_{crit} \approx 0.014$, a rupture of the pins occurred, and the structure bent outside as a whole (red open square). Apart from that, most modules with pins showed excellent mechanical stability in the investigated region of φ = 0.02–0.19 and t_p = 0.26 mm–5 mm (green full circles). Calculations predicted stable designs to be in the green area below the solid line in Figure 7, which concurs with experimental observations.

Modules with internal fins, lattice, and FGE all passed the pressure test.

In Figure 8, for one module experiment, the calculation and simulation are compared. The module with t_p = 4 mm and a = 20 mm deformed notably. The measured displacement of one side was Δy = 0.20 mm, while the calculated and simulated displacement had been Δy = 0.235 and 0.215 mm, respectively. Both the calculation and the simulation overestimated the experimental displacement a little and reflected the characteristic shape well.

The results from the simulation, experiment, and calculation concurred. Table 3 shows the results of finite element analysis (FEA) of rectangular units. Full-size images can be found in the SI (Figures S1–S3). Mesh size independence was confirmed for each simulation. The mesh size was reduced until the maximum deformation did not change anymore (0.5 mm for most cases). An example is provided in Figure S5. For a low width of up to 4 mm, displacement was not significant. This confirmed not only the experimental results of fin-equipped modules but also the fact that the modules did not fail at their narrow sides. Rectangular walls with w = 40 mm and t_w = 1 mm were displaced severely in the experiment, calculation, and simulation. Displacement was only limited below the accepted value at t_w = 5 mm.

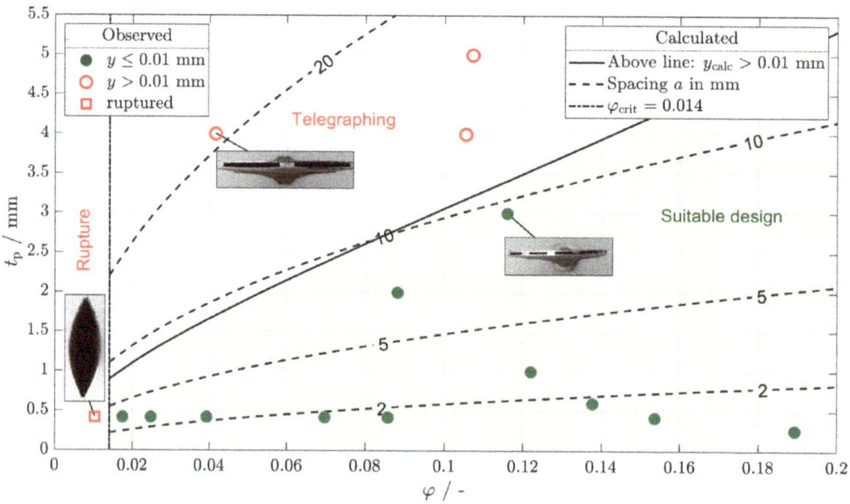

Figure 7. Results of mechanical stability calculation and experiment.

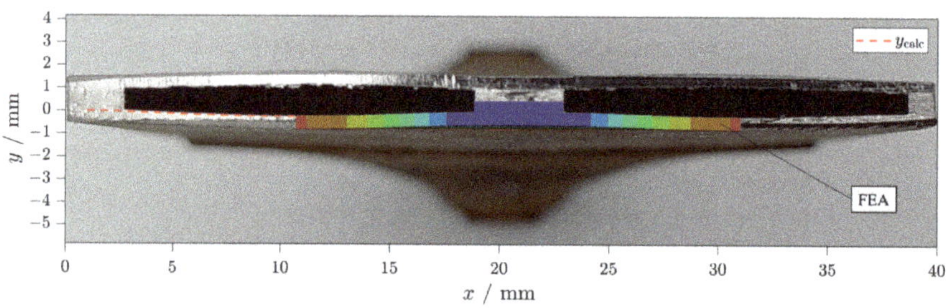

Figure 8. Failed pin-equipped module t_p = 4 mm a = 20 mm: photography of cross-cut in background, calculated deformation (dotted red line), and simulated deformation (multicolored area). Maximum deformation is visible at $x = 11$ mm.

In Table 4, pin-equipped modules are shown. Abstracting the hexagonally arranged quadratic pins led to an insignificant change in FEA displacement. The calculation and experiment yielded slightly higher values for displacement (first two columns). While the calculated and simulated deformation correctly predicted that there was no telegraphing in the module in column three, the volume fraction was close to its critical value. The simulation predicted that the stress inside the pin reached almost yield strength. Both the volume fraction and the simulation result explain the rupture of the pin-equipped module with φ = 0.01 (third column). All three methods agreed in revealing that a module, as shown in column four, did not deform significantly.

Table 3. Results concerning the stability of empty and fin-equipped modules: l = 60 mm.

Module		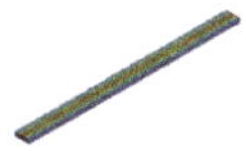	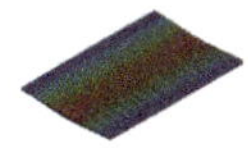	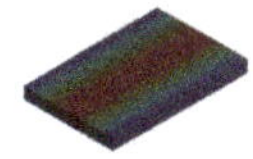
t_w/mm	1	1	1	5
w/mm	1	4	40	40
$y_{\text{max,FEA}}$/mm	-1.007×10^{-5}	-2.201×10^{-4}	-1.2373	-0.0105
$y_{\text{max,calc}}$/mm	-4.275×10^{-7}	-1.094×10^{-4}	-1.0074	-0.0081
$y_{\text{max,exp}}$/mm	0	0	≈ -5	-

Table 4. Results concerning the stability of pin-equipped modules. t_w = 0.6 mm.

Module	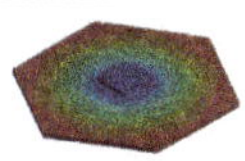	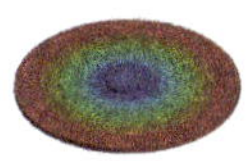	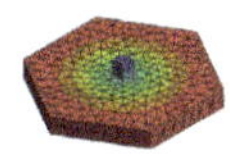	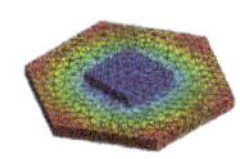
t_p/mm	4	-	0.42	3
d_i/mm	-	4.514	-	-
a/mm	20	-	5	9
d_a/mm	-	21.0	-	-
φ/-	0.0382		0.0081	0.1283
$y_{\text{max,FEA}}$/mm	-0.219	-0.215	-2.994×10^{-3} [a]	-5.451×10^{-3}
$y_{\text{max,calc}}$/mm	-0.235		-1.286×10^{-3}	-3.236×10^{-3}
$y_{\text{max,exp}}$/mm	-0.20		≈ -10	0

[a] Stress reached σ = 436 MPa inside pin, equaling 75% $R_{p0.2}$.

Table 5 features modules for which no calculation method was available. Finite element analysis (FEA) and experiments were in agreement that no significant deformation occurred for the investigated modules with an FGE and lattice.

Table 5. Results for stability of FGE-equipped (t_w = 0.8 mm, t_{FGE} = 0.4 mm) and lattice-equipped (t_w = 0.6 mm) modules.

Module			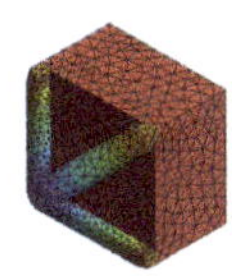
Dimensions/mm	d_{FGE} = 1.333	d_{FGE} = 4	d_{strut} = 0.2
d/mm	2.667	8	0.5
w/mm	2.667	8	1
l/mm	5.333	16	1
φ/-	0.1887	0.0629	0.2177
Element size/mm	0.2	0.2	0.1
$y_{\text{max,FEA}}$/mm	$-1.887 \times 10^{-4}/2$	$-3.985 \times 10^{-3}/2$	-4.820×10^{-4}
$y_{\text{max,exp}}$/mm	0	0	0

A combination of calculation and checking the volume fraction against its critical value was effective for the stability assessment. Having implemented the formula in a programmable script, it was possible to extract information on failure criteria based on all design parameters. In Figure 9, the design parameter wall thickness varied, while the load was constant, q = 3 MPa, referring to an operation pressure of p = 20 bar and a safety factor of S = 1.5.

When assuming that a design with $\varphi = 0.1$ is desired, the pin size can be determined by starting on the abscissa and moving upward until the solid line referring to the aimed-for wall thickness is reached. The pin size must not exceed the respective value on the ordinate. The selected pin size defines the spacing. At $t_w = 0.6$ mm, a pin size of $t_p = 2$ mm is selected, leading to a spacing of $a = 6.8$ mm.

Another design graph where the wall thickness is kept constant, while the load is varied, can be found in the SI (Figure S4).

The hybrid manufacturing connection was never a source of failure in our experience, which is consistent with the findings of [25].

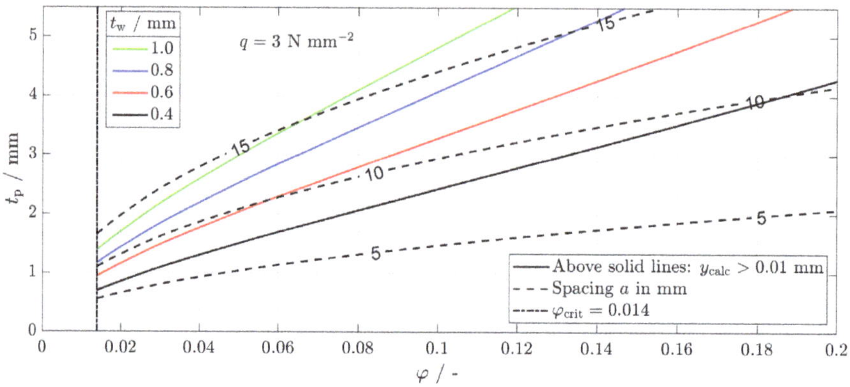

Figure 9. Calculated stability criterion for various thicknesses at constant load.

4. Conclusions and Outlook

The wall thicknesses of several millimeters necessary for slit modules at a reasonable size without internal structures to withstand $p = 30$ bar at room temperature are unacceptable.

The first approach (deformation of the walls) was not successful in increasing the mechanical stability of planar slits. Neither corrugated nor arched walls could, except for one instance with a comparatively high thickness, withstand the pressure difference occurring at $p = 30$ bar at room temperature.

Significant insights into the stability of micro-structured devices were gained. The second approach of introducing regular internal structures of a minimal mass and volume led to stable modules. Evenly spaced internal structures with a sufficiently high volume fraction and sufficiently low spacing were stable; the exact values depended on the material properties, wall thickness, and load.

For external pressure, no failure is expected since the internal structures are unlikely to buckle with the highest slenderness ratio being $\lambda_{\text{buckling}} = 3.33$ for $t_p = 0.26$ mm. This ensures both process safety and smooth operation without the compression of particles on the inside of devices, e.g., a catalyst.

To be on the safe side regarding a rupture, the volume fraction aimed for should be above the critical value. This is reasonable, as there is a chance of cracking formation starting from the points where internal structures meet the walls at steep angles, as well as from surface roughness [30].

The experiment, simulation, and calculation were in good agreement. The calculation is easy to use, especially when being implemented into tables in the form of a graph, as shown in this article.

Supplementary Materials: The following supporting information can be downloaded at https://www.mdpi.com/article/10.3390/designs8030041/s1: Table S1. Composition of metal powders.; Table S2. Modules without internal structures.; Table S3. Pin-equipped modules.; Table S4. Fin-equipped modules.; Table S5. FGE-equipped modules.; Table S6. Lattice-equipped module.; Figure S1. FEA results of empty and fin-equipped modules.; Figure S2. FEA results of pin-equipped modules.; Figure S3. FEA results of FGE-equipped and lattice-equipped modules.; Figure S4. Calculated stability criterion for various loads at constant thickness.; Figure S5. Results of mesh independence study.

Author Contributions: Conceptualization, D.F.M., C.K. and R.D.; methodology, D.F.M.; software, D.F.M.; validation, D.F.M.; formal analysis, D.F.M.; investigation, D.F.M.; resources, C.K. and R.D.; data curation, D.F.M.; writing—original draft preparation, D.F.M.; writing—review and editing, D.F.M., C.K. and R.D.; visualization, D.F.M. and C.K.; supervision, C.K. and R.D.; project administration, C.K. and R.D.; funding acquisition, R.D. All authors have read and agreed to the published version of the manuscript

Funding: We are grateful to the German Federal Ministry of Education and Research (BMBF) for funding this work through project Kopernikus P2X-Phase 2 (Funding ID: 03SFK2K0-2).

Data Availability Statement: Data can be made available upon request.

Acknowledgments: We sincerely thank the KIT-Publication Fund of the Karlsruhe Institute of Technology Library for its support and the funding bodies of the Large Scale Data Facility (https://www.scc.kit.edu/forschung/11843.php, accessed on 5 March 2024). We appreciate the help of our colleagues Manuel Hofheinz for fabrication and Cornelia Schorle for testing.

Conflicts of Interest: The authors declare no conflicts of interest.

Abbreviations

Symbol	Meaning	Unit
A	Area	mm^2
a	Spacing	mm
D	Depth external	mm
y	displacement	mm
d	Depth	mm
d	Diameter	mm
l	Length	mm
p	Pressure	Pa
q	Mechanical load	MPa
$R_{p0.2}$	Proof strength at 0.2 % strain	MPa
t	Dimension	mm
t	Thickness	mm
t	Time	s
w	Width	mm
Δy	Deflection	mm
crit	critical	
e	empty	
f	fins	
FGE	related to FGE	
l	lattice	
p	pins	
strut	strut	
w	wall	
3D	three-dimensional	
AM	additive manufacturing	
CAD	computer aided design	
FEA	finite element analysis	
FGE	fluid guiding element	
FGU	fluid guiding unit	
FTS	Fischer-Tropsch synthesis	

NC	numerically controlled
NF	Autodesk Netfabb
SI	Supplementary Information
STL	standard tessellation language

References

1. Jess, A.; Wasserscheid, P. *Chemical Technology: An Integral Textbook*; CourseSmart; Wiley-VCH: Weinheim, Germany, 2013.
2. de Klerk, A. *Fischer-Tropsch Refining*, 1st ed.; Wiley-VCH: Hoboken, NJ, USA; Weinheim, Germany, 2011. [CrossRef]
3. Deutsche Gesetzliche Unfallversicherung e.V.. GESTIS-Stoffdatenbank. 2023. Available online: https://gestis.dguv.de/ (accessed on 11 November 2023).
4. Herz, G.; Gallwitz, M.; Näke, R.; Megel, S.; Jahn, M.; Reichelt, E. Lab-Scale Demonstration of By-Product Utilization in an Integrated Power-to-Liquid Process Applying Co-Electrolysis. *Energy Technol.* **2023**, *11*, 2201057. [CrossRef]
5. Woo, Y.; Oh, D.B.; Park, J.E.; Han, S.J.; Lee, Y.J.; Park, M.J. CFD Modeling of a Multichannel Fischer–Tropsch Reactor Module with Microscale Cooling Channels: Effects of Mirrored Structure Cooling Layers. *Korean J. Chem. Eng.* **2023**, *40*, 2572-2580. [CrossRef]
6. Metzger, D.F.; Klahn, C.; Dittmeyer, R. Downsizing Sustainable Aviation Fuel Production with Additive Manufacturing—An Experimental Study on a 3D Printed Reactor for Fischer-Tropsch Synthesis. *Energies* **2023**, *16*, 6798. [CrossRef]
7. Almeida, L.C.; Sanz, O.; D'olhaberriague, J.; Yunes, S.; Montes, M. Microchannel Reactor for Fischer–Tropsch Synthesis: Adaptation of a Commercial Unit for Testing Microchannel Blocks. *Fuel* **2013**, *110*, 171–177. [CrossRef]
8. Cao, C.; Hu, J.; Li, S.; Wilcox, W.; Wang, Y. Intensified Fischer–Tropsch Synthesis Process with Microchannel Catalytic Reactors. *Catal. Today* **2009**, *140*, 149–156. [CrossRef]
9. Kroll, E.; Buchris, E. Weight Reduction of 3D-printed Cylindrical and Toroidal Pressure Vessels through Shape Modification. *Procedia Manuf.* **2018**, *21*, 133–140. [CrossRef]
10. Unnikrishnan Nair, S.; Somanath, S. (Eds.) *Introduction to Finite Element Analysis: A Textbook for Engineering Students*; Springer: Singapore, 2024. [CrossRef]
11. Young, W.C.; Budynas, R.G. *Roark's Formulas for Stress and Strain*, 7th ed.; McGraw-Hill: New York, NY, USA, 2007.
12. Vu, H.M.; Meiniger, S.; Ringel, B.; Hoche, H.; Oechsner, M.; Weigold, M.; Seidel, C. Investigation on Surface Characteristics of Wall Structures out of Stainless Steel 316L Manufactured by Laser Powder Bed Fusion. *Prog. Addit. Manuf.* **2024**. [CrossRef]
13. Deutsche Edelstahlwerke GmbH. *Werkstoffdatenblatt X2CrNiMo17-12-2 1.4404*; Technical Report; Deutsche Edelstahlwerke GmbH: Witten, Germany, 2015.
14. Ashby, M.F. *Materials Selection in Mechanical Design*, 4th ed.; Butterworth-Heinemann: Amsterdam, The Netherlands, 2011.
15. Zhuang, W.; Yang, C.; Wu, Z. Mechanical Stability of Hybrid Corrugated Sandwich Plates under Fluid-Structure-Thermal Coupling for Novel Thermal Protection Systems. *Appl. Sci.* **2020**, *10*, 2790. [CrossRef]
16. Charmi, A.; Falkenberg, R.; Ávila, L.; Mohr, G.; Sommer, K.; Ulbricht, A.; Sprengel, M.; Saliwan Neumann, R.; Skrotzki, B.; Evans, A. Mechanical Anisotropy of Additively Manufactured Stainless Steel 316L: An Experimental and Numerical Study. *Mater. Sci. Eng. A* **2021**, *799*, 140154. [CrossRef]
17. Meyer, J. Pressure Vessel. US20160238193A1, 18 August 2016.
18. Kotilar, I.K. Pressure Vessels, Design and Method of Manufacturing Using Additive Printing. US20160061381A1, 3 March 2016.
19. Kiener, C.; Boschert, S.; Küsters, Y.; Nicolai, A.; Otto, R. Hochdruck-geeignete AM-Konstruktionselemente Mit Hierarchisch-funktionalen Metallstrukturen. *Chem. Ing. Tech.* **2022**, *94*, 1040–1045. [CrossRef]
20. Kockmann, N. *Transport Phenomena in Micro Process Engineering*; SpringerLink Bücher; Springer: Berlin/Heidelberg, Germany, 2008. [CrossRef]
21. *DIN EN 13445-1:2018-12*; Unbefeuerte Druckbehälter—Teil1: Allgemeines. DIN e.V.: Berlin, Germany, 2018.
22. Hansjosten, E.; Wenka, A.; Hensel, A.; Benzinger, W.; Klumpp, M.; Dittmeyer, R. Custom-Designed 3D-printed Metallic Fluid Guiding Elements for Enhanced Heat Transfer at Low Pressure Drop. *Chem. Eng. Process.—Process Intensif.* **2018**, *130*, 119–126. [CrossRef]
23. *DIN ISO 228-1*; Rohrgewinde Für Nicht Im Gewinde Dichtende Verbindungen: Teil 1: Maße, Toleranzen Und Bezeichnung (ISO 228-1:2000). DIN e.V.: Berlin, Germany, 2003.
24. Pei, E.; Bernard, A.; Gu, D.; Klahn, C.; Monzón, M.; Petersen, M.; Sun, T., Eds. *Springer Handbook of Additive Manufacturing*; Springer Handbooks; Springer International Publishing: Cham, Switzerland, 2023. [CrossRef]
25. Ferchow, J.; Bühler, M.; Schlüssel, M.; Zumofen, L.; Klahn, C.; Hofmann, U.; Kirchheim, A.; Meboldt, M. Design and Validation of a Sheet Metal Clamping System for Additive Manufacturing and Post-Processing. *Int. J. Adv. Manuf. Technol.* **2022**, *119*, 7947–7967. [CrossRef]
26. BG RCI. *Druckprüfungen von Druckbehältern Und Rohrleitungen: Flüssigkeitsdruckprüfungen, Gasdruckprüfungen*; BG RCI: Heidelberg, Germany, 2012.
27. IMA Materialforschung und Anwendungstechnik GmbH. WIAM(R) Metallinfo. 2022. Available online: https://www.wiamonline.de/wiamdb/wiamlogin.php (accessed on 19 February 2021).
28. Hitzler, L.; Hirsch, J.; Heine, B.; Merkel, M.; Hall, W.; Öchsner, A. On the Anisotropic Mechanical Properties of Selective Laser-Melted Stainless Steel. *Materials* **2017**, *10*, 1136. [CrossRef] [PubMed]

29. Türk, D.A. Exploration and Validation of Integrated Lightweight Structures with AM and FRP Parts. Ph.D. Thesis, ETH Zürich, Zürich, Switzerland, 2017.
30. van Hooreweder, B.; Apers, Y.; Lietaert, K.; Kruth, J.P. Improving the Fatigue Performance of Porous Metallic Biomaterials Produced by Selective Laser Melting. *Acta Biomater.* **2017**, *47*, 193–202. [CrossRef] [PubMed]

designs

MDPI

Article

Integrating Pneumatic and Thermal Control in 3D Bioprinting for Improved Bio-Ink Handling

Perrin Woods [1], Carter Smith [2], Scott Clark [2] and Ahasan Habib [1,*]

[1] Department of Manufacturing and Mechanical Engineering Technology, Rochester Institute of Technology, Rochester, NY 14623, USA; pcw3635@g.rit.edu
[2] Department of Sustainable Product Design and Architecture, Keene State College, Keene, NH 03431, USA; carter.smith@keene.edu (C.S.); scott.clark@keene.edu (S.C.)
* Correspondence: mahmet@rit.edu

Abstract: The rapid advancement of 3D bioprinting has created a need for cost-effective and versatile 3D printers capable of handling bio-inks at various scales. This study introduces a novel framework for a specialized nozzle-holding device designed for an extrusion-based 3D bioprinter, specifically tailored to address the rigorous requirements of tissue engineering applications. The proposed system combines a pneumatically actuated plunger mechanism with an adaptive nozzle system, ensuring the safe inhibition and precise dispensing of bio-inks. Rigorous thermal management strategies are employed to maintain consistently low temperatures, thereby preserving bio-ink integrity without changing chemical stability. A key component of this design is a precision-milled aluminum block, which optimizes thermal characteristics while providing a protective barrier. Additionally, a 3D-printed extruder head bracket, fabricated using a high-precision resin printer, effectively mitigates potential thermal inconsistencies. The integration of these meticulously engineered components results in a modified extrusion-based 3D bioprinter with the potential to significantly advance tissue engineering methodologies. This study not only contributes to the advancement of bioprinting technology but also underscores the crucial role of innovative engineering in addressing tissue engineering challenges. The proposed bioprinter design lays a solid foundation for future research, aiming to develop more accurate, efficient, and reliable bioprinting solutions.

Keywords: 3D bioprinting; nozzle system; extruder; tissue engineering

1. Introduction

As of June 2024, the organ transplant waitlist in the United States includes over 103,223 (Kidney: 89,101; Liver: 9862; Heart: 3436) individuals, significantly outnumbering the approximately 46,000 (Kidney: 27,332; Liver: 10,660; Heart: 4545) transplants performed in 2023 [1]. This growing disparity results in an average of 17 daily fatalities due to organ shortages, underscoring the urgent need for alternative solutions. Recent advancements in tissue engineering, particularly in 3D bioprinting technology, offer promising alternatives [2–4]. Some organs such as the Kidneys, Liver, and Heart need more exploration to make them ready for transplantation, and a substantial amount of research [5–10] is ongoing to make this happen in the long run. Extrusion-based 3D bioprinting, the most prevalent technique, allows for the deposition of various materials containing multiple cell types and concentrations [11–13]. This method is categorized into pneumatic-, piston-, and screw-driven systems, with print quality determined by factors such as temperature, nozzle diameter, pressure, and speed. The quality of printed scaffolds is controlled by a series of process parameters such as temperature, nozzle diameter, extrusion pressure, movement speed, extrusion speed, path interval, etc. [14]. The market offers a range of commercial 3D bioprinters, varying in price from USD 5000 to USD 250,000, with capabilities suited for different applications [15]. Notable examples include the 3D Bioplotter [16,17], NovoGen [18], and BioX [19], among others from various global manufacturers. While these printers provide advanced functionalities, they may limit researchers in pursuing specific objectives due to their predetermined features.

While commercial 3D bioprinters offer sophisticated features, their closed-source nature and limited customization options can constrain research potential. These proprietary systems often confine users to materials designed and developed in-house, printing parameters, and hardware setups, impeding the investigation of innovative bio-inks, printing methods, or the integration of specialized components. For example, current commercial printers often lack in situ physical crosslinking capabilities, which can compromise shape fidelity. Even though some printers feature UV in situ photo-crosslinking capacity, it is like a "one-shoe-fits-all" approach where limited wavelengths are available to cure. In contrast, custom-built 3D bioprinters provide researchers with the flexibility to adapt the system to their specific requirements, enabling them to advance bioprinting technology.

Some efforts of designing and developing custom-made 3D printers have been reported [20]. To address material exploration and development for 3D printers, the Additive Manufacturing Autonomous Research System (AM ARES) was developed, which uses automated image analysis and Bayesian optimization to autonomously modulate print parameters [21]. However, this printer mostly focused on synthetic thermoplastic polymers, not hydrogel materials. A standard 3D printer was converted into an open-source extrusion-based 3D bioprinter, addressing the high costs associated with commercial bioprinting systems [22]. A FlashForge Finder 3D printer was converted into a bioprinter using our Replistruder 4 syringe pump and Duet3D Duet 2 WiFi, for less than USD 900. The modified bioprinter demonstrated a travel accuracy better than 35 μm in all axes and printed collagen scaffolds with less than 2% error. The UV-crosslinked system was not considered for both systems. Moreover, a dual crosslinked system was also not considered in [22] which will be one of the considerations of our proposition.

While there are some rules of thumb for bio-ink selection [23,24], many commercial 3D bioprinters face challenges in maintaining consistent bio-ink viscosity and printability throughout the printing process. This can lead to clogging, uneven material deposition, and reduced cell viability [25–27]. For example, Dubbin et al. demonstrated that certain bio-inks like GelMA and PEGDA can damage up to 10% of cells during the printing process, with even higher cell damage (>50%) occurring at the edges of bioprinted droplets during crosslinking [28]. Temperature control is also a crucial factor for maintaining bio-ink properties and cell viability [29]. Inconsistent temperature control can affect the rheological properties of bio-inks, leading to inconsistent print quality. For instance, collagen-based bio-inks are particularly sensitive to temperature changes, which can impact their gelation kinetics and final mechanical properties [30]. On the other hand, an improved bio-ink handling system can enable better precision over multi-material bioprinting, allowing for the creation of complex tissue constructs with varying mechanical and biological properties [31,32]. For example, in cartilage tissue engineering, the importance of precise bio-ink deposition for creating zonal organization similar to native cartilage was highlighted [33]. Additionally, enhanced bio-ink handling could improve the printing of vascularized tissues [34].

Having full access to hardware and software components, researchers can modify and optimize various aspects of the bioprinter, including extruder design, motion control systems, and printing algorithms. This level of customization facilitates the development and testing of novel bio-ink formulations, the incorporation of specialized sensors or imaging systems, and the investigation of innovative printing strategies tailored to specific applications [22,35,36]. Table 1 summarizes key findings from previous studies on bio-ink handling and thermal control. Developing custom bioprinters fosters a deeper understanding of the fundamental principles and mechanics of bioprinting, enabling researchers to more effectively identify and address limitations. This hands-on approach promotes interdisciplinary collaboration, allowing experts from various fields to contribute their knowledge and push the boundaries of bioprinting technology [15,20].

To address these limitations, our long-term goal is to develop a custom 3D bioprinter capable of depositing multiple materials with diverse crosslinking capabilities, including both physical and UV-cured methods. The proposed printer will allow for accommodation to various wavelengths based on printing requirements. Building on our previous research demonstrating multi-hydrogel extrusion [37], this article presents a conceptual framework for a specialized nozzle-holding device designed for extrusion-based 3D bioprinting in tissue engineering applications. The proposed nozzle-holding device will focus on pneumatic extrusion methods, chosen for their high cell survivability rates and prevalence in the market [38,39]. Moreover, being brand-independent, this customized nozzle-holding device can attach to various existing bioprinters that can increase the effectiveness of handling hydrogels having various viscosities. The engineering design, material selection, and manufacturing process for the proposed nozzle-holding device were accomplished based on diverse factors such as ease-of-manufacturability, better fitment, temperature control, and ease-of-attachment. Furthermore, we propose to develop a range of attachments for future consideration that are compatible with the nozzle-holding device presented in this article. These attachments will facilitate various crosslinking methods, including chemical crosslinking with extrusion flow, misting, and UV curing. The aim is to expand the applicability of our system to a diverse array of hydrogel materials, enhancing both printability and cellular functions such as viability, proliferation, and differentiation. This approach aims to overcome the limitations of commercial printers, offering greater flexibility and customization options for researchers in the field of tissue engineering.

Table 1. A summary of bio-ink challenges, thermal control, methodologies, and key findings.

Bio-Ink Handling Challenges	Thermal Control Issues	Methodologies	Key Findings
Inconsistent bio-ink viscosity, clogging	Not specified	Quantitative criteria for bio-ink benchmarking	Up to 10% cell damage during printing, >50% cell damage at edges during crosslinking [28]
Uneven material deposition, reduced cell viability	Not specified	Overview of bio-inks for 3D bioprinting	Highlighted need for bio-inks with consistent rheological properties [26]
Impact of pH and crosslinking on printability	Sensitivity of collagen-based bio-inks to temperature changes	Rheological analysis, photo-crosslinking with riboflavin	pH and riboflavin photo-crosslinking affect gelation kinetics and mechanical properties [40]
Precise bio-ink deposition for tissue engineering	Not specified	Bioprinting of cartilage tissue with zonal organization	Yield stress determines bioprintability, crucial for creating native-like cartilage [41]
Multi-material bioprinting challenges	Not specified	Creation of perfusable vascular networks	Precise control over bio-inks is essential for vascularized tissue engineering [42]
Cell compatibility in bio-inks	Not specified	Printing stem cells for skeletal regenerative medicine	Emphasized importance of bio-ink properties on cell viability and differentiation [43]

2. Development of Nozzle Holder

2.1. Conceptualization

This section highlights the conceptualization of the proposed nozzle holder that will substitute the standard filament extruder of regular 3D printers with a specialized extruder head developed in-house for bioprinters, as shown in Figure 1. The modified extruder head consists of two main parts: a milled aluminum block and an extruder head bracket. The aluminum block facilitates heat transfer and shields the heating element from damage, while the 3D-printed resin bracket ensures precise alignment between the hydrogel container (e.g., dispensing syringe) and the heated clamp of the extruder head. This manufacturing approach guarantees precision and high heat resistance, minimizing errors. A series of operations such as extruder holder design, cartridge holder redesign, extruder bracket design, pneumatic design, and finally the circuitry design and development will be discussed in the following sections. These improvements collectively contribute to the optimization of the system's performance and functionality.

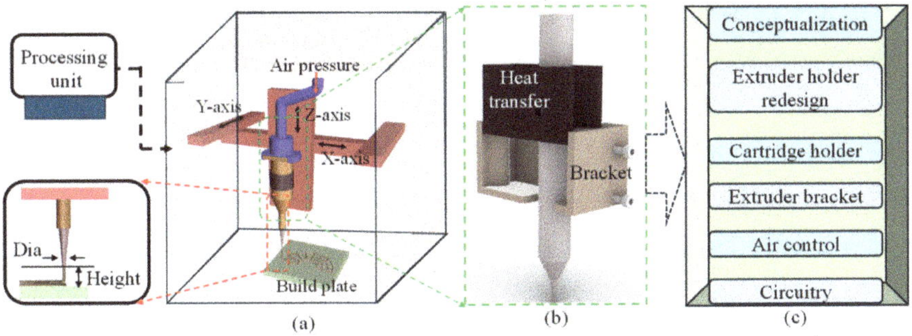

Figure 1. An overview of the nozzle holder design: (**a**) a schematic of the bioprinter with a processing unit and process parameters, (**b**) a concept proposition, and (**c**) a flow chart of related experiments to modify for the final design and development.

2.2. Replacement of Extruder

The development of the new extruder head involved multiple iterations, primarily focusing on securely attaching it to the 3D printer's extruder carriage while maintaining proximity to the original filament extrusion point. The initial attempts to incorporate a syringe hole between the carriage bolt points proved impractical due to spacing issues. A two-part design concept was then explored, starting with a dovetail design proof of concept, which provided valuable insights into necessary tolerances and design flaws, as shown in Figure 2. It had inaccurate overall dimensions and lacked a means of attachment to the 3D printer motion system. Further refinement followed, incorporating holes for carriage bolts in the extruder bracket and improving the fit between the bio clamp and the extruder bracket. The bio clamp itself underwent improvement to better secure the bio-cartridge, introducing slots to facilitate easier bolt removal. Ultimately, efforts were made to bring the extrusion point closer to the printer's original position and to streamline the manufacturing process, reducing the difficulty associated with the bio clamp. The initial and improved designs were printed using MakerBot Replicator by Ultimaker (Manhattan, NY, USA) with default settings to investigate the fitment with the printer and syringe, as shown in Figure 2.

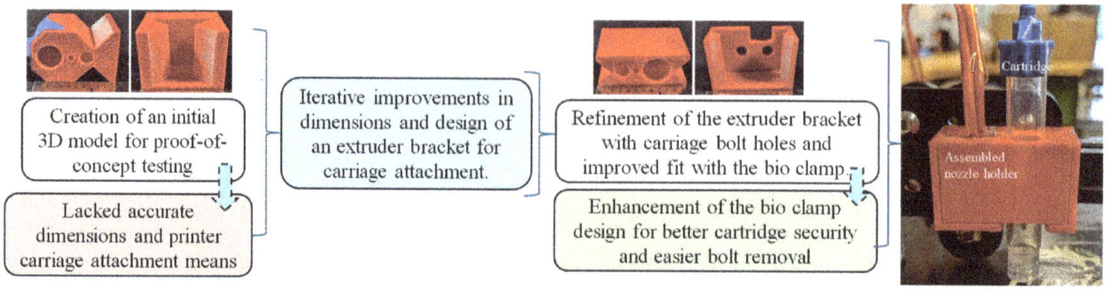

Figure 2. Extruder head replacement from concept development to enhanced design.

2.3. The Design of the Extruder

2.3.1. Engineering Drawing of Extruder Bracket

Three-dimensional modeling and simulation software, Solid Works v 2023 (Dassault Systèmes, Waltham, MA, USA), was used to develop engineering drawing. The final engineering drawing of the extruder bracket is shown in Figure 3.

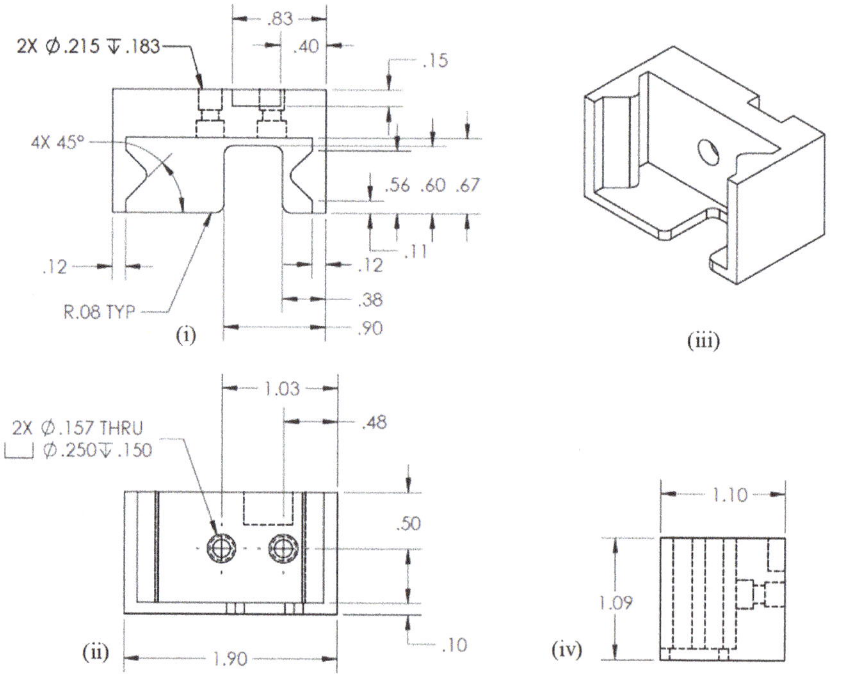

Figure 3. Engineering drawing for extruder bracket: **(i)** Top view, **(ii)** Front view, **(iii)** Isometric view, and **(iv)** Side view.

2.3.2. Engineering Drawing of Syringe Holder and Assembly

Similar 3D modeling and simulation software as mentioned in Section 2.3.1 was used to develop the engineering drawing. The final engineering drawings of the nozzle/syringe holder and overall assembly are shown in Figures 4a and 4b, respectively.

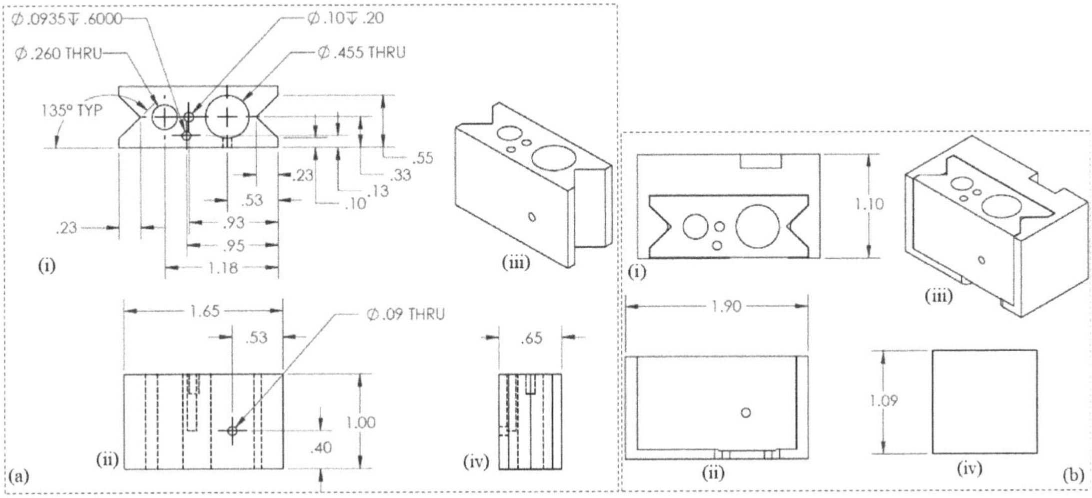

Figure 4. Engineering drawings for (**a**) nozzle holder: (**i**) Top view, (**ii**) Front view, (**iii**) Isometric view, and (**iv**) Side view. and (**b**) nozzle syringe assembly: (**i**) Top view, (**ii**) Front view, (**iii**) Isometric view, and (**iv**) Side view.

2.4. The Manufacturing of the Extruder

2.4.1. Manufacturing of Syringe Holder

At the core of this innovation lies a precisely engineered aluminum block, milled to accommodate the heating element, heat sensor, and syringe. General-purpose Aluminum 6061 alloy was used for this fabrication. This design choice reflects a calculated approach to optimize thermal properties crucial for bio-ink preservation. The selection of milled aluminum for this component serves two key purposes: first, enhancing heat transfer efficiency and second, providing a protective shield against potential heating element damage. A Bridgeport milling machine (Bridgeport, CT, USA) was utilized for the machining process to achieve the desired dimensions. Moreover, rigorous controls were implemented to maintain a consistently low temperature within the bioprinting environment, safeguarding the bio-ink's integrity. This integration of carefully engineered components resulted in an enhanced extrusion-based 3D bioprinter with the potential to transform tissue engineering practices.

Figure 5a,b illustrate the planning and fabrication processes, respectively. The extruder head bracket demonstrates remarkable precision in connecting the 3D printer's trunnion to the extruder head's heated clamp. The choice of this manufacturing method is based on its established accuracy and relatively high heat resistance, effectively minimizing potential temperature-related inconsistencies.

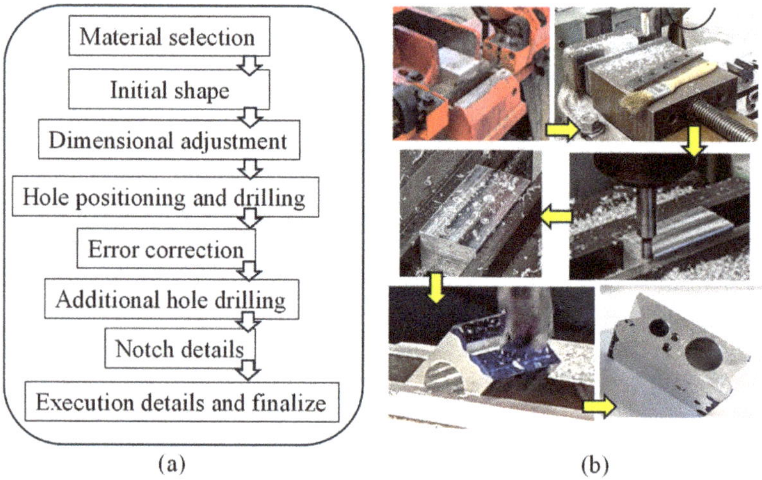

Figure 5. (**a**,**b**) Workflow to manufacture the nozzle holder from material selection to final product.

2.4.2. Manufacturing of Extruder Bracket and Assembly

A crucial element in the improved extruder head assembly is the 3D-printed extruder head bracket, produced using a freeform resin printer (Form 3, Formlab, Somerville, MA, USA), as shown in Figure 6a. Once the bracket and nozzle holder were manufactured, they were assembled to demonstrate the fitment. Figure 6b,d show the pneumatic connector, heating element, and temperature sensor attachment to the assembled extruder. This showcases an exceptional accuracy in the interconnecting parts between the 3D printer's trunnion and the heated clamp of the extruder head. The assembly's components were fastened together using screws and washers. To maintain an appropriate temperature for the cells (around 37 °C), the printer's original heating element and temperature sensor were reused.

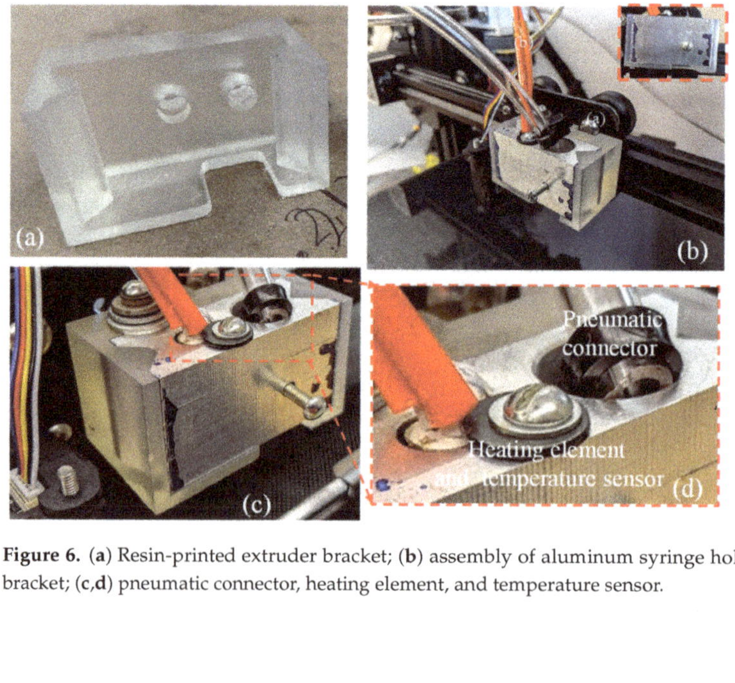

Figure 6. (**a**) Resin-printed extruder bracket; (**b**) assembly of aluminum syringe holder and resin bracket; (**c**,**d**) pneumatic connector, heating element, and temperature sensor.

2.4.3. Syringe Air Control for In-House Bio-Ink Printing Confirming Shape Fidelity and Viability

The decision to utilize pneumatic extrusion in this project underscores the importance of a well-designed air compressor system. This system is crucial for supplying the 3D bioprinter with the necessary compressed air at various pressure levels to extrude materials designed and developed in-house. The compressor setup was designed with portability, allowing for easy transportation alongside the 3D printer. Additionally, the required amount of air supply was confirmed to maintain a consistent pressure throughout the printing process. In our setup, the power source was passed through the pressure sensor switch. This switch was designed to deactivate the compressor when the pressure reaches 100 psi and reactivate it when it drops to a minimum amount governed by the material viscosity (e.g., pressure required for A_4C_4 [44] extrusion) to maintain a consistent flow. To complete the system loop, we connected the ground directly to the compressor ground. If the compressor does not turn on when plugged in, the user may need to switch the cables leading to the pressure switch. The occasional cable swaps can cause malfunctions that can be resolved by reconnecting them correctly. The specific connection method for the pneumatic tubing is not critical if it passes through the compressor safety system before reaching the pressure regulator. From the pressure regulator, the tubing extends to the 3D printer through the solenoids. Apart from this, the system is relatively straightforward, and if the current air tank is open, the setup should function smoothly. The entire connection setup is illustrated in Figure 7.

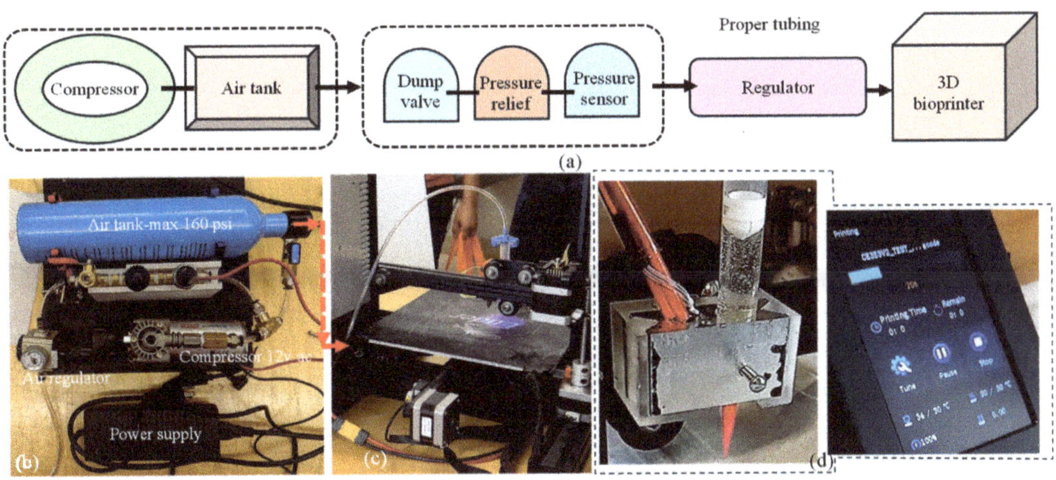

Figure 7. (**a**) A schematic of all connections, (**b**) all components used to build air control systems and assembled air control systems, (**c**) assembled air control system connected to the customized nozzle holder, and (**d**) nozzle holder attached to the Ender 3 printer and working with our customized G-code.

2.4.4. Circuitry Design and Implementation

This section provides a brief overview of the wiring required for signal interpretation in this article, focusing on the connection between the Arduino and the voltage divider. It explains the voltage divider's purpose and includes mathematical insights to facilitate potential adjustments or rewiring if necessary. The electrical path continues from the positive output to the terminal, entering the voltage divider's resistor setup. Current flows through this setup, around to resistor setup B, and then returns to a ground cable via the other terminal for Signal A. The Arduino reads the voltage at the midpoint between resistor setups A and B. This midpoint positioning is crucial because the Arduino's analog inputs are limited to 5 volts AC, while the signal output exceeds the standard 12 volts.

The voltage divider's primary function is to reduce the incoming 12 volts AC to below 5 volts AC. In this case, a 150-Ohm resistor setup is used for resistor A and a 100-Ohm resistor setup for resistor B. This configuration yields a 4.8-volt output, which is within the Arduino's operational range. Figure 8a illustrates the connection between the Arduino and the voltage divider, including a detailed view of the voltage divider's connection with the resistors. Figure 8b presents a comprehensive circuit diagram of the setup.

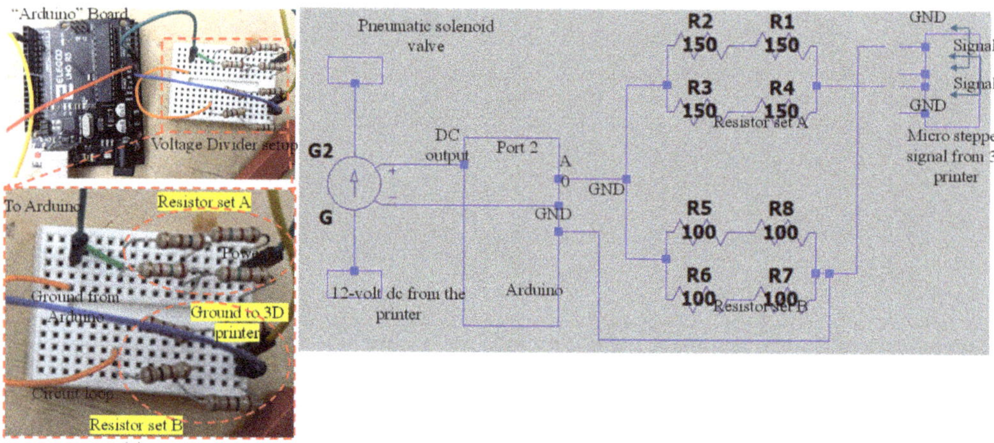

(a) (b)

Figure 8. (**a**) Connection between Arduino and voltage divider and detailed view of voltage divider connection with resistors and (**b**) circuit diagram generated using LTspice (Linear Technology, Milpitas, CA, USA).

3. Implementation and Application

3.1. Three-Dimensional Printing of Filaments and Construct with Proposed System

To demonstrate the implementation of our custom-made 3D printing system including a nozzle holder, circuitry design, air supply and control system, Arduino code, and G-code, our previously developed hybrid hydrogel composed with 4% alginate and 4% Carboxymethyl Cellulose [44] was used to extrude several filaments through a 410 µm nozzle, as shown in Figure 9a,b. A range of applied pressure from 10 to 15 psi along with 20–25 mm/s print speed was used. Toolpath was generated using in-house Arduino code integrated with slicer software. For this experimental test, we printed filaments at room temperature with the intention that the filament will maintain defined geometry after solidification. This system was able to extrude consistent and constant filaments successfully, as shown in Figure 9b. We did not observe significant difference in a set of three filaments we printed; to produce continuous filaments and consequently defined architectures of 3D-printed constructs, several process parameters need to be optimized such as extrusion pressure, nozzle diameter, printing speed and distance, and material composition and viscosities [45]. Extruded filaments showed $\leq$20% deviation compared to the nozzle diameter used. Maintaining this consistency is crucial to achieve defined porosities that affect cellular activities [46,47]. Having experience in solving a problem related to filament width and process parameters in our recently published article [26], our ongoing research will allow us to address these issues in the near future. Moreover, GelMA- and PEG-based photosensitive hybrid hydrogels are undergoing experiments to validate UV crosslinking. Related process parameters such as UV wavelengths (200–600 nm), the exposure time and sequence of LED on/off, the distance from LEDs to print beds, the choice of photosensitive polymers, and the type and amount of photo initiators will be optimized for continuous filaments and defined 3D constructs.

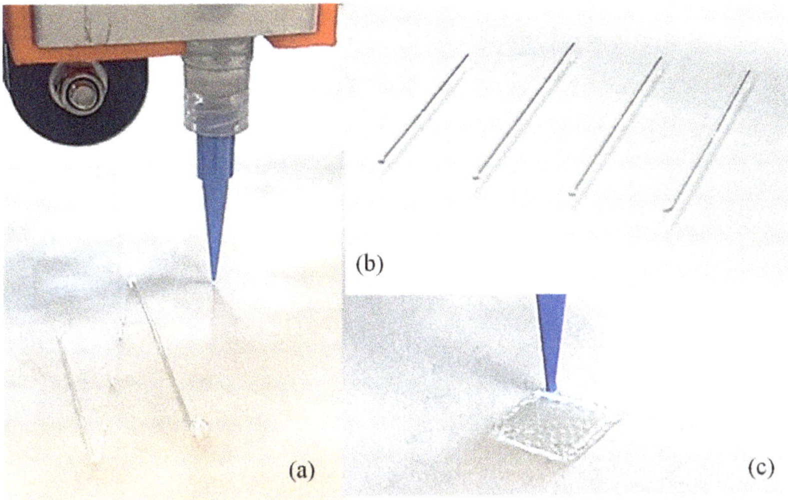

Figure 9. (**a**) Extruding a single filament using our custom-made nozzle-holding system, (**b**) a series of printed filaments to show consistency, and (**c**) a 3D-printed construct. Additionally, 4% alginate and 4% CMC were used to print the filament and scaffold.

3.2. Application for Various Bio-Inks

To further apply our developed nozzle system in future, we will utilize a series of novel bio-inks such as alginate–Carboxymethyl Cellulose [44], alginate–Carboxymethyl Cellulose–Montmorillonite nano-clay [45], alginate–Carboxymethyl Cellulose–Tempo-Mediated nano-Fibrillated Cellulose (TONFC) [48], and pre-crosslinked alginate–Carboxymethyl Cellulose [49], these were developed in-house to achieve shape fidelity for clinically relevant scaffolds (up to 5 cm) and higher cell viability ($\geq$80%). A range of applied pressure, e.g., 8–20 psi, was utilized for extruding these bio-inks. The process parameters (nozzle diameter, applied pressure, and print speed and distance) used for these bio-inks will be mimicked in the proposed system, and performance will be evaluated compared to the shape fidelity and cell viability of these bio-inks.

4. Attachment for Extrusion, Crosslinking, and UV Curing

4.1. Proposition of Attachments

To accommodate photo-crosslinking with ultraviolet (UV) light having user-defined wavelengths (with a range of 200–500 nm) and physical crosslinking (with the application of Ca^{2+}) methods, three semi-modular attachments were designed to be compatible with the proposed nozzle holder, as shown in Figure 10. While the range of 200–250 nm can be harmful for cells [50], this range can be useful for acellular crosslinking. This module design will allow the 3D bioprinter to be equipped with any combination of these attachments, ranging from a simple attachment to a complex one. The choice of combination will depend on the specific materials (photo-crosslinking or chemical crosslinking), scale of scaffolds, and cell viability requirements.

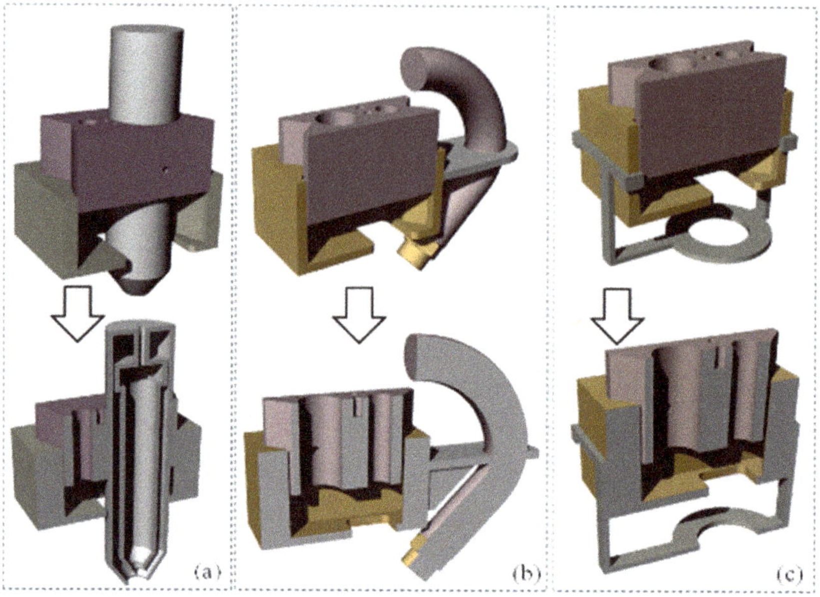

Figure 10. Proposed 3D models for next step extruders: (**a**) syringe mount, (**b**) hose mount, and (**c**) UV light mount.

The first attachment, shown in Figure 10a, is an extrusion-based crosslinking attachment. This attachment allows for the use of a pneumatic system and solenoid, similar to that powering the syringe, to extrude the crosslinker out of a vessel surrounding the syringe, such that it combines with the hydrogel as it exits the orifice of the syringe. This allows for direct control over the amount of crosslinker that the hydrogel is receiving. However, this attachment requires a modification of the nozzle holder to account for the larger diameter of the syringe-surrounding assembly, with the size of both the aluminum cartridge holder and the resin block needing to be moderately increased. Figure 10b depicts a spray-based crosslinking nozzle attachment. Its design and functionality are similar to the first attachment, providing an alternative option for crosslinker application during the printing process. While the design proposed in Figure 10b can offer a higher crosslinking rate and quicker solidification after releasing from the nozzle, this can a create a nozzle clog if the wait time is not controlled properly for printing subsequent layers [51,52]. This challenge can be resolved by spraying, where the released hydrogel will be partially crosslinked and leave more time to complete subsequent printing without nozzle clogs [53]. This attachment is intended to be affixed to the side of the nozzle holder and contain a small tube and misting nozzle, such that the crosslinker can be sprayed onto the printed hydrogel either while it is extruding or during additional curing passes. This misting nozzle is held close to the end of the cartridge so that the crosslinker is misted onto the extruded hydrogel in a predictable manner. This misting attachment will use the same pneumatic system as the primary nozzle system, with a separate control method based on instructions sent to the same microcontroller. Finally, Figure 10c showcases a UV curing attachment. This device is designed to accommodate a UV LED light strip around its ring, which can be connected to a microcontroller for power regulation. The attachment secures to the existing nozzle holder block by snapping onto its upper four corners. This component is specifically intended for photo-active hydrogels, and alternative attachments would be necessary for curing other hydrogel types.

4.2. The 3D Printing of Attachments and Future Recommendations

To expand the practical implications in various biomedical applications, such as tissue engineering, regenerative medicine, and drug delivery systems, we can update the scalability of our bioprinter. One such addition is a dual crosslinking feature including physical (with $CaCl_2$) and photo-crosslinking (with UV) [54]. While UV crosslinking can assist in holding the shape during printing, physical crosslinking will increase the overall shape fidelity after printing. Moreover, we will explore the pre-crosslinking technique to enhance scalability that we reported in our earlier research [49]. Controlling material preparation to be capable of dual or pre-crosslinking along with appropriate process parameter selection can allow users to create patient-specific implants or scaffolds for wound healing, which could significantly improve patient outcomes. Furthermore, based on the dimensions of the printer used to attach to our system, the scale can vary. Finally, we will consider the implications for commercial viability, including partnerships with healthcare providers and the integration of this technology into existing manufacturing processes, thereby showcasing its transformative potential in the field of biomanufacturing.

As proof of concept, we successfully 3D printed models of a syringe mount, hose mount, and UV light mount, as illustrated in Figure 11a, Figure 11b, and Figure 11c, respectively. Our future work will focus on several key areas of improvement and expansion:

- Enhancing attachment design for better stability and user-friendliness.
- Investigating alternative materials to improve durability and compatibility with various 3D printers.
- Developing a modular Multi-Attachment System for easy switching between curing methods.
- Creating plugins for popular 3D printing software to streamline printing and curing processes.
- Expanding microcontroller functionality to include real-time monitoring, feedback mechanisms, and smart device connectivity.
- Researching alternative UV light sources to optimize curing for specific hydrogel formulations.
- Collecting and incorporating user feedback from diverse fields to iteratively improve the UV curing attachment's design and functionality.

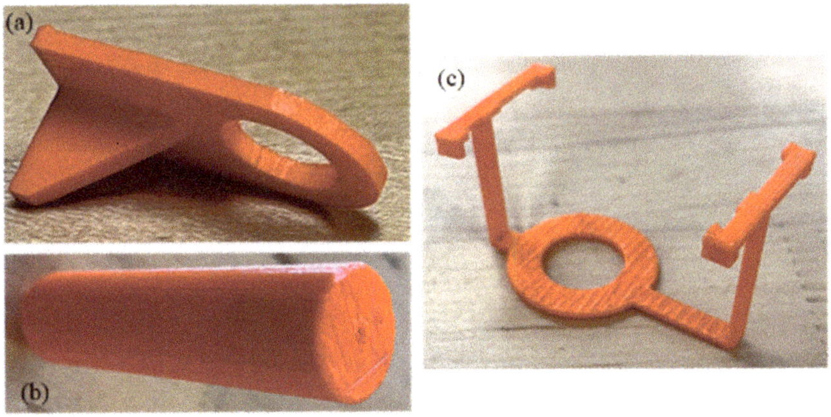

Figure 11. Three-dimensional-printed models as proof of concept: (**a**) hose mount, (**b**) syringe mount, and (**c**) UV light mount.

These future developments aim to enhance the versatility, efficiency, and user experience of our 3D printing system for hydrogel-based applications.

5. Conclusions

In conclusion, this study not only signifies a significant advancement in bioprinting technology but also underscores the critical role of innovative engineering in overcoming the unique challenges inherent in tissue engineering. The adaptive nozzle system ensures the secure containment of the bio-ink, while the pneumatically driven plunger mechanism facilitates precise dispensing. The careful selection of materials and manufacturing techniques in the extruder head components highlights a commitment to safeguarding the bio-ink's integrity and optimizing the system's thermal performance. This research lays a solid foundation in the advancement of tissue engineering methodologies through cutting-edge bioprinting technology. The success of this project will help (i) explore novel bio-ink formulations with improved thermal stability and printability, (ii) develop multi-material printing strategies for complex tissue constructs, (iii) overcome compatibility issues with existing hardware and software, and (iv) resolve scalability concerns for large-scale production.

Author Contributions: A.H. conceived the idea. C.S. and S.C. created the preliminary design and fabrication. P.W. finalized the design and fabrication. All authors have read and agreed to the published version of the manuscript.

Funding: This research was supported by New Hampshire-EPSCoR through BioMade Award #1757371 from National Science Foundation, New Hampshire-INBRE through an Institutional Development Award (IDeA), P20GM103506, from the National Institute of General Medical and Sciences of the NIH. College of Engineering Technology, Rochester Institute of Technology.

Data Availability Statement: Data is contained within the article.

Acknowledgments: Department of Sustainable Product Design and Architecture, Keene State College, Keene, NH and Department of Manufacturing and Mechanical Engineering Technology of Rochester Institute of Technology.

Conflicts of Interest: The authors declare no conflicts of interest.

References

1. Available online: https://www.organdonor.gov/learn/organ-donation-statistics (accessed on 15 June 2024).
2. He, Y.; Yang, F.; Zhao, H.; Gao, Q.; Xia, B.; Fu, J. Research on the printability of hydrogels in 3D bioprinting. *Sci. Rep.* **2016**, *6*, 29977. [CrossRef] [PubMed]
3. Atala, A.; Kasper, F.K.; Mikos, A.G. Engineering complex tissues. *Sci. Transl. Med.* **2012**, *4*, 160rv112. [CrossRef]
4. Murphy, S.V.; Atala, A. 3D bioprinting of tissues and organs. *Nat. Biotechnol.* **2014**, *32*, 773–785. [CrossRef]
5. Pichler, R.; Rizzo, L.; Tröndle, K.; Bühler, M.; Brucker, H.; Müller, A.-L.; Grand, K.; Farè, S.; Viau, A.; Kaminski, M.M. Tuning the 3D microenvironment of reprogrammed tubule cells enhances biomimetic modeling of polycystic kidney disease. *Biomaterials* **2022**, *291*, 121910. [CrossRef]
6. Van Genderen, A.; Valverde, M.; Capendale, P.E.; Kersten, M.; Garví, E.S.; Schuurmans, C.C.; Ruelas, M.; Soeiro, J.T.; Tang, G.; Janssen, M.J. Co-axial printing of convoluted proximal tubule for kidney disease modeling. *Biofabrication* **2022**, *14*, 044102. [CrossRef] [PubMed]
7. Fransen, M.F.; Addario, G.; Bouten, C.V.; Halary, F.; Moroni, L.; Mota, C. Bioprinting of kidney in vitro models: Cells, biomaterials, and manufacturing techniques. *Essays Biochem.* **2021**, *65*, 587–602. [PubMed]
8. Rajalekshmi, R.; Shaji, A.K.; Joseph, R.; Bhatt, A. Scaffold for liver tissue engineering: Exploring the potential of fibrin incorporated alginate dialdehyde–gelatin hydrogel. *Int. J. Biol. Macromol.* **2021**, *166*, 999–1008. [CrossRef]
9. Bhise, N.S.; Manoharan, V.; Massa, S.; Tamayol, A.; Ghaderi, M.; Miscuglio, M.; Lang, Q.; Zhang, Y.S.; Shin, S.R.; Calzone, G. A liver-on-a-chip platform with bioprinted hepatic spheroids. *Biofabrication* **2016**, *8*, 014101. [CrossRef]
10. Maiullari, F.; Costantini, M.; Milan, M.; Pace, V.; Chirivì, M.; Maiullari, S.; Rainer, A.; Baci, D.; Marei, H.E.-S.; Seliktar, D.; et al. A multi-cellular 3D bioprinting approach for vascularized heart tissue engineering based on HUVECs and iPSC-derived cardiomyocytes. *Sci. Rep.* **2018**, *8*, 13532. [CrossRef]
11. Chung, J.H.; Naficy, S.; Yue, Z.; Kapsa, R.; Quigley, A.; Moulton, S.E.; Wallace, G.G. Bio-ink properties and printability for extrusion printing living cells. *Biomater. Sci.* **2013**, *1*, 763–773. [CrossRef]
12. Hsieh, C.-T.; Langrana, N.A. A system approach in extrusion-based multi-material CAD. In Proceedings of the Solid Freeform Fabrication Symposium, Austin, TX, USA, 6–8 August 2001; pp. 313–321.
13. Sakai, S.; Yoshii, A.; Sakurai, S.; Horii, K.; Nagasuna, O. Silk fibroin nanofibers: A promising ink additive for extrusion three-dimensional bioprinting. *Mater. Today Bio* **2020**, *8*, 100078. [CrossRef] [PubMed]

14. Quigley, C.; Limon, S.M.; Sarah, R.; Habib, A. Factorial Design of Experiment Method to Characterize Bioprinting Process Parameters to Obtain the Targeted Scaffold Porosity. *3D Print. Addit. Manuf.* **2023**. [CrossRef]
15. Tashman, J.W.; Shiwarski, D.J.; Feinberg, A.W. Development of a high-performance open-source 3D bioprinter. *Sci. Rep.* **2022**, *12*, 22652. [CrossRef] [PubMed]
16. Ovsianikov, A.; Yoo, J.; Mironov, V. *3D Printing and Biofabrication*; Springer International Publishing: Berlin/Heidelberg, Germany, 2018.
17. Chua, C.K.; Leong, K.F. *3D Printing and Additive Manufacturing: Principles and Applications (with Companion Media Pack)-of Rapid Prototyping*; World Scientific Publishing Company: Singapore, 2014.
18. Chua, C.K.; Yeong, W.Y. *Bioprinting: Principles and Applications*; World Scientific Publishing: Singapore, 2015.
19. Available online: https://www.cellink.com/bioprinting/bio-x-3d-bioprinter/ (accessed on 1 June 2024).
20. Gu, Z.; Fu, J.; Lin, H.; He, Y. Development of 3D bioprinting: From printing methods to biomedical applications. *Asian J. Pharm. Sci.* **2020**, *15*, 529–557. [CrossRef] [PubMed]
21. Deneault, J.R.; Chang, J.; Myung, J.; Hooper, D.; Armstrong, A.; Pitt, M.; Maruyama, B. Toward autonomous additive manufacturing: Bayesian optimization on a 3D printer. *MRS Bull.* **2021**, *46*, 566–575. [CrossRef]
22. Ioannidis, K.; Danalatos, R.I.; Champeris Tsaniras, S.; Kaplani, K.; Lokka, G.; Kanellou, A.; Papachristou, D.J.; Bokias, G.; Lygerou, Z.; Taraviras, S. A custom ultra-low-cost 3D bioprinter supports cell growth and differentiation. *Front. Bioeng. Biotechnol.* **2020**, *8*, 580889. [CrossRef]
23. Yilmaz, B.; Al Rashid, A.; Mou, Y.A.; Evis, Z.; Koç, M. Bioprinting: A review of processes, materials and applications. *Bioprinting* **2021**, *23*, e00148. [CrossRef]
24. Zhang, T.; Zhao, W.; Xiahou, Z.; Wang, X.; Zhang, K.; Yin, J. Bioink design for extrusion-based bioprinting. *Appl. Mater. Today* **2021**, *25*, 101227. [CrossRef]
25. Mironov, V.; Kasyanov, V.; Drake, C.; Markwald, R.R. Organ printing: Promises and challenges. *Regen. Med.* **2008**, *3*, 93–103. [CrossRef]
26. Gungor-Ozkerim, P.S.; Inci, I.; Zhang, Y.S.; Khademhosseini, A.; Dokmeci, M.R. Bioinks for 3D bioprinting: An overview. *Biomater. Sci.* **2018**, *6*, 915–946. [CrossRef]
27. Ozbolat, V.; Dey, M.; Ayan, B.; Povilianskas, A.; Demirel, M.C.; Ozbolat, I. 3D Printing of PDMS Improves its Mechanical and Cell Adhesion Properties. *ACS Biomater. Sci. Eng.* **2017**, *4*, 682–693. [CrossRef]
28. Dubbin, K.; Tabet, A.; Heilshorn, S.C. Quantitative criteria to benchmark new and existing bio-inks for cell compatibility. *Biofabrication* **2017**, *9*, 044102. [CrossRef]
29. Wang, F.; Tankus, E.B.; Santarella, F.; Rohr, N.; Sharma, N.; Märtin, S.; Michalscheck, M.; Maintz, M.; Cao, S.; Thieringer, F.M. Fabrication and Characterization of PCL/HA Filament as a 3D Printing Material Using Thermal Extrusion Technology for Bone Tissue Engineering. *Polymers* **2022**, *14*, 669. [CrossRef] [PubMed]
30. Jang, C.H.; Ahn, S.H.; Yang, G.-H.; Kim, G.H. A MSCs-laden polycaprolactone/collagen scaffold for bone tissue regeneration. *RSC Adv.* **2016**, *6*, 6259–6265. [CrossRef]
31. Nadernezhad, A.; Khani, N.; Skvortsov, G.A.; Toprakhisar, B.; Bakirci, E.; Menceloglu, Y.; Unal, S.; Koc, B. Multifunctional 3D printing of heterogeneous hydrogel structures. *Sci. Rep.* **2016**, *6*, 33178. [CrossRef] [PubMed]
32. Tariq, A.; Arif, Z.U.; Khalid, M.Y.; Hossain, M.; Rasool, P.I.; Umer, R.; Ramakrishna, S. Recent advances in the additive manufacturing of stimuli-responsive soft polymers. *Adv. Eng. Mater.* **2023**, *25*, 2301074. [CrossRef]
33. Sbirkov, Y.; Redzheb, M.; Forraz, N.; McGuckin, C.; Sarafian, V. High Hopes for the Biofabrication of Articular Cartilage—What Lies beyond the Horizon of Tissue Engineering and 3D Bioprinting? *Biomedicines* **2024**, *12*, 665. [CrossRef]
34. Chae, S.; Ha, D.-H.; Lee, H. 3D bioprinting strategy for engineering vascularized tissue models. *Int. J. Bioprint.* **2023**, *9*, 748. [CrossRef]
35. Choudhury, D.; Anand, S.; Naing, M.W. The arrival of commercial bioprinters–Towards 3D bioprinting revolution! *Int. J. Bioprint.* **2018**, *4*, 139. [CrossRef]
36. Abdulmaged, A.I.; Soon, C.F.; Talip, B.A.; Zamhuri, S.A.A.; Mostafa, S.A.; Zhou, W. Characterization of alginate–gelatin–cholesteryl ester liquid crystals bioinks for extrusion bioprinting of tissue engineering Scaffolds. *Polymers* **2022**, *14*, 1021. [CrossRef]
37. Quigley, C.; Sarah, R.; Hurd, W.; Clark, S.; Habib, M. Design and Fabrication of In-house Nozzle System to Extrude Multi-Hydrogels for 3D Bioprinting Process. *J. Manuf. Sci. Eng.* **2023**, *146*, 021003.
38. Derakhshanfar, S.; Mbeleck, R.; Xu, K.; Zhang, X.; Zhong, W.; Xing, M. 3D bioprinting for biomedical devices and tissue engineering: A review of recent trends and advances. *Bioact. Mater.* **2018**, *3*, 144–156. [CrossRef] [PubMed]
39. Falcone, G.; Mazzei, P.; Piccolo, A.; Esposito, T.; Mencherini, T.; Aquino, R.P.; Del Gaudio, P.; Russo, P. Advanced printable hydrogels from pre-crosslinked alginate as a new tool in semi solid extrusion 3D printing process. *Carbohydr. Polym.* **2021**, *276*, 118746. [CrossRef] [PubMed]
40. Diamantides, N.; Wang, L.; Pruiksma, T.; Siemiatkoski, J.; Dugopolski, C.; Shortkroff, S.; Kennedy, S.; Bonassar, L.J. Correlating rheological properties and printability of collagen bioinks: The effects of riboflavin photocrosslinking and pH. *Biofabrication* **2017**, *9*, 034102. [CrossRef]
41. Mouser, V.H.; Melchels, F.P.; Visser, J.; Dhert, W.J.; Gawlitta, D.; Malda, J. Yield stress determines bioprintability of hydrogels based on gelatin-methacryloyl and gellan gum for cartilage bioprinting. *Biofabrication* **2016**, *8*, 035003. [CrossRef]

42. Kolesky, D.B.; Homan, K.A.; Skylar-Scott, M.A.; Lewis, J.A. Three-dimensional bioprinting of thick vascularized tissues. *Proc. Natl. Acad. Sci. USA* **2016**, *113*, 3179–3184. [CrossRef]
43. Cidonio, G.; Glinka, M.; Dawson, J.; Oreffo, R. The cell in the ink: Improving biofabrication by printing stem cells for skeletal regenerative medicine. *Biomaterials* **2019**, *209*, 10–24. [CrossRef]
44. Habib, A.; Sathish, V.; Mallik, S.; Khoda, B. 3D printability of alginate-carboxymethyl cellulose hydrogel. *Materials* **2018**, *11*, 454. [CrossRef] [PubMed]
45. Habib, A.; Khoda, B. Development of clay based novel hybrid bio-ink for 3D bio-printing process. *J. Manuf. Process.* **2019**, *38*, 76–87. [CrossRef]
46. Habib, M.; Khoda, B. Fiber Filled Hybrid Hydrogel for Bio-Manufacturing. *J. Manuf. Sci. Eng.* **2020**, *143*, 041013. [CrossRef]
47. Nelson, C.; Tuladhar, S.; Launen, L.; Habib, M. 3D Bio-Printability of Hybrid Pre-Crosslinked Hydrogels. *Int. J. Mol. Sci.* **2021**, *22*, 13481. [CrossRef]
48. Ozbolat, I.T.; Hospodiuk, M. Current advances and future perspectives in extrusion-based bioprinting. *Biomaterials* **2016**, *76*, 321–343. [CrossRef]
49. Heo, D.N.; Alioglu, M.A.; Wu, Y.; Ozbolat, V.; Ayan, B.; Dey, M.; Kang, Y.; Ozbolat, I.T. 3D bioprinting of carbohydrazide-modified gelatin into microparticle-suspended oxidized alginate for the fabrication of complex-shaped tissue constructs. *ACS Appl. Mater. Interfaces* **2020**, *12*, 20295–20306. [CrossRef]
50. Masuma, R.; Kashima, S.; Kurasaki, M.; Okuno, T. Effects of UV wavelength on cell damages caused by UV irradiation in PC12 cells. *J. Photochem. Photobiol. B Biol.* **2013**, *125*, 202–208. [CrossRef]
51. Stanco, D.; Urbán, P.; Tirendi, S.; Ciardelli, G.; Barrero, J. 3D bioprinting for orthopaedic applications: Current advances, challenges and regulatory considerations. *Bioprinting* **2020**, *20*, e00103. [CrossRef]
52. Park, W.; Gao, G.; Cho, D.-W. Tissue-specific decellularized extracellular matrix bioinks for musculoskeletal tissue regeneration and modeling using 3D bioprinting technology. *Int. J. Mol. Sci.* **2021**, *22*, 7837. [CrossRef]
53. GhavamiNejad, A.; Ashammakhi, N.; Wu, X.Y.; Khademhosseini, A. Crosslinking strategies for 3D bioprinting of polymeric hydrogels. *Small* **2020**, *16*, 2002931. [CrossRef]
54. Zhang, M.; Vora, A.; Han, W.; Wojtecki, R.J.; Maune, H.; Le, A.B.; Thompson, L.E.; McClelland, G.M.; Ribet, F.; Engler, A.C. Dual-responsive hydrogels for direct-write 3D printing. *Macromolecules* **2015**, *48*, 6482–6488.

designs

MDPI

Article

Contribution of Artificial Intelligence (AI) to Code-Based 3D Modeling Tasks

Marianna Zichar * and Ildikó Papp

Department of Data Science and Visualization, Faculty of Informatics, University of Debrecen, H-4028 Debrecen, Hungary; papp.ildiko@inf.unideb.hu
* Correspondence: zichar.marianna@inf.unideb.hu

Abstract: The rapid advancement of technology and innovation is also impacting education across different levels. The rise of Artificial Intelligence (AI) is beginning to transform education in various areas, from course materials to assessment systems. This requires educators to reconsider how they evaluate students' knowledge. It is crucial to understand if and to what extent assignments can be completed using AI tools. This study explores two hypotheses about the risks of using code-based 3D modeling software in education and the potential for students to delegate their work to AI when completing assignments. We selected two tasks that students were able to successfully complete independently and provided the same amount of information (both textual and image) to AI in order to generate the necessary code. We tested the widely used ChatGPT and Gemini AI bots to assess their current performance in generating code based on text prompts or image-based information for the two models. Our findings indicate that students are not yet able to entirely delegate their work to these AI tools.

Keywords: 3D modeling; 3D printing; education; OpenSCAD; artificial intelligence; computational thinking; creativity; ChatGPT; Gemini

1. Introduction

The technological development of Industry 4.0 has dramatically impacted individual industries, resulting in a significant increase in productivity. Technological adaptation has become a critical factor because missing out on new technological achievements can cause a competitive disadvantage. The process is characterized by a high degree of digitization. It can be associated with many concepts known today, such as Additive Manufacturing, Artificial intelligence, Augmented reality, Autonomous robots, Big data and analytics, Blockchain, Cybersecurity, IoT (Internet of Things), and Simulation [1]. The shift towards Industry 5.0 can be considered a paradigm shift since instead of focusing on the role of technology, it places the cooperation between technology and people at the center of development. In addition, Industry 5.0 is closely linked to efforts for sustainable and inclusive economic growth [2].

Educational systems must reflect technical changes since they aim to bring up future employees (engineers, IT professionals) for the professional challenges ahead. For this reason, many institutions are motivated to teach 3D printing to students. However, creating a 3D model is necessary for 3D printing, so 3D modeling and printing often go hand in hand. Artificial intelligence, one of the latest emerging technologies, already provides significant assistance in various areas, such as writing texts, generating images, and even coding. If teachers want to assess the students' skills rather than those of the AI tools, then it is essential to be aware of the current capabilities of these systems. This study evaluates the abilities of two of the most popular AI bots in creating 3D models in OpenSCAD.

1.1. Related Work

Experience in 3D printing technologies helps to implement it in many aspects of their daily education process. The beginning can be a geometric modeling course, followed later by a CAD course to develop 3D models of the prototypes. This course can be suitable for introducing various technologies, applications, and practical experience related to additive manufacturing [3].

The best method is the learning-by-doing method when the motivation for acquiring technical knowledge is provided by processing a topic or solving a project task [4]. The goal may not only be that the students learn about and be able to apply modern technologies but also that an ancient and traditional subject, mathematics, be taught in a more modern form and be brought closer to the world of today's students [5]. Implementing STEAM (Science, Technology, Engineering, Arts, and Mathematics) lessons into regular university courses is difficult because of scheduled classes with time limits during the semester. To break these barriers, there are a lot of different suggested pedagogic instructions, methods, and case studies [6,7]. At the same time, it can already be seen that the targeted development of certain skills—for example, computational thinking—can be helpful even at a younger age [8]. It has been proven in many areas that project tasks carried out in groups would also have an effect on reducing social differences and help create an inclusive atmosphere [9].

Various excellent projects involve collaborations between scientists, engineers, and artists to provide innovative solutions that impact society. In accordance with the most essential guidelines of modern museology, the expectations connected to the museum have also changed. In the museum of the 21st century, an interactive exhibition is needed, in which digitization and some kind of creative activity for visitors play an increasingly important role [10].

Technological knowledge cannot be used for its own sake; it is important to use it to meet unique needs, so that it has a more significant impact on society [9,11].

Companies, educational institutions, and other associations can establish so-called Makerspaces, where people can use different tools to make their projects. The sessions help to reduce stress and increase creativity while introducing techniques that may not have been known before [12,13].

The critical aspects of 3D modeling for the authors are related to solid modeling, i.e., when a production process, typically 3D printing, can be started based on the model. Nowadays, it is typical for the model or a part of it to be made based on some 2D information. 3D solid modeling applications have some tools to transform 2D content into 3D objects. The 2D geometry content here comes from a drawing or a movement path. There are also online converters that are not connected to editing software, which can generate a 3D model in stl or obj format directly based on an image file [14]. In this case, the image's resolution greatly influences the output form's shape, accuracy, and resolution. Specialized model generation provided by artificial intelligence (AI) can be said to be novel; several services are available that can be used to create an interactively displayed product asset based on an image [15]. This way of getting the model is much cheaper and faster for the customer than using a professional 3D modeling service. We must mention that the model expectations of 3D production do not match the expectations necessary for visualization. BOOM Interactive's Bubbles application was developed for multiple operating systems to be widely used for the quick and accurate generation and furnishing of architectural interior design spaces [16]. This AI-assisted 3D space design tool is not only to create scenes from 2D floor plan images, scan their space, or build from scratch, but the real-time 3D editor allows them to make changes, such as moving walls, adding windows and doors, decorating, and making decisions. Meshy AI toolkit was developed to transform text or images into 3D models or texture to support artists, game developers, and creators with display-focused needs [17]. Some further online 3D model generators are overviewed in [18], but they aim to general 3D model that are not specific to 3D modeling and the file formats are also special ones which should be transform to stl with some other software.

1.2. Optional Course at Univesity of Debrecen

The Faculty of Informatics at the University of Debrecen offers students the opportunity to learn about 3D printing and modeling in the frame of an optional course. The fact that the Faculty of Informatics won a MakerBot Replicator 5th Generation 3D printer in a competition in 2016 gave an excellent motivation for working out the course syllabus [19]. Later, a Prusa MK2.5 printer and a Sense 3D scanner enriched the available tools during the classes. Both printers are of the Fused Deposition Modeling (FDM) type, which is the most popular printer type among hobby users. However, there are also many great examples of professional use of FDM devices in the industry. The main goal of the course is to introduce the 3D printing process, including designing a model, slicing it, and 3D printing. At the beginning of the course, we discuss the basics of FDM printing; the students get to know the printing process, the different materials that can be used for printing, and learn about the requirements for 3D printability. We can use various file repositories to obtain a model for 3D printing quickly and easily, but the course aims for students to create solid objects of varying complexity.

For our modeling purpose, we preferred applications that are available for free, so that everyone can participate in completing the course with equal chances.

1.2.1. TinkerCAD

In the first half of the course, we use the TinkerCAD application running in a web browser, which makes a large collection of primitives and improved primary bodies available after registration [20]. The complex, unique designs can be created by computing the union or difference of bodies, which means that limited CSG (Constructive Solid Geometry) modeling is applied (Figure 1). All private designs are stored in the personal account, and all the calculations run on servers of the owner Autodesk Inc. (San Francisco, CA, USA). The models can be downloaded for manufacturing, and projects can be shared and embedded in web pages. The TinkerCAD application has its shortcomings in terms of specifying all exact dimensions, the possibilities of redesigning models and engineering techniques, but it is nevertheless a perfect choice for learning entry-level modeling tools.

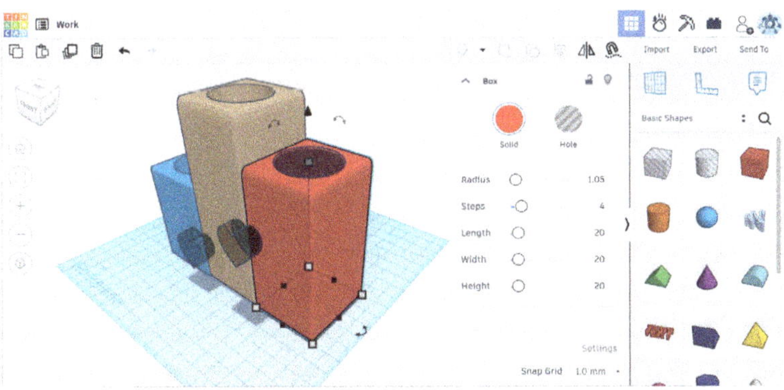

Figure 1. User interface of TinkerCAD application. Complex shapes can be designed from primitives of predefined libraries. All shapes can be used as solids or holes. Grouping is the tool to apply union or difference of the selected objects.

1.2.2. SolidWorks

The other modeling application chosen for the course is the SolidWorks engineering application, whose educational version is installed in our computer labs. At the same time, the Student Design Kit is available for students to download free of charge after completing a registration form [21]. SolidWorks is a CAD design system for mechanical engineering with many tools to help the user. Solids are created mainly from 2D drawings using various generation tools such as drawing, rotation, sweeping, or loft (Figure 2). Later, the basic shapes can be modified by cutting, chamfering, and filleting edges [22]. The modeling process is well documented, and each step can be reworked and modified afterward. As an engineering application, dimensioning is possible with a high degree of precision and parameterization, allowing model families to be created.

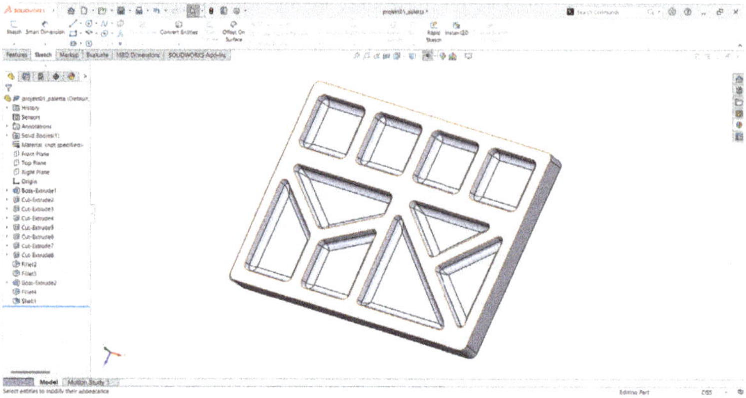

Figure 2. User interface of SolidWorks. The Command Manager Toolbar (horizontal bar at the top of the window) presents all available 3D Feature tools. The separated panel on the left shows tools building the model in the so-called design tree.

1.2.3. Print-in-Place—The Holy Grail of 3D Printing

Modeling is taught through targeted examples and exercises, and concepts specific to 3D printing are introduced in parallel. In preparation for 3D printing, students learn about the basic functionality of the slicing software and the different settings that affect print quality.

3D printing has unquestionable advantages over other manufacturing technologies. These include print-in-place object manufacturing, which is based on a model consisting of several precisely positioned parts connected with hinges and joints. The model is printed in a single printing process, still it will have moving parts without the need for any assembly (see Figure 3). The final model can contain even spinning parts.

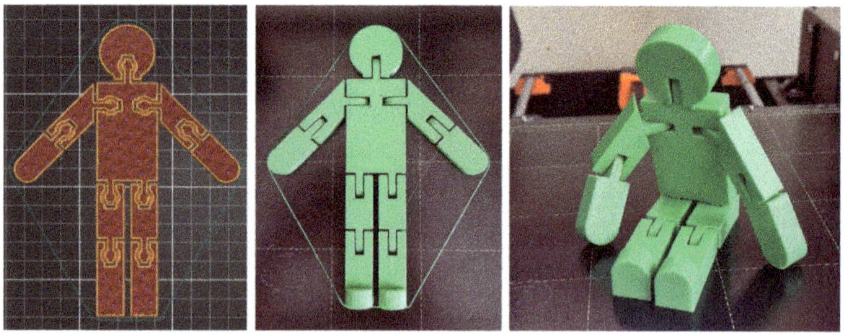

Figure 3. Sample print-in-place design. On the left is the model in the slicer software with removed top layers to illustrate separate parts with required gaps. The middle figure shows the printed object. The right figure shows the movement capability of this model immediately after the printing.

1.2.4. Evaluation of Students' Work

In the course, the evaluation of the students' work is based on three pillars: testing his theoretical knowledge, checking his modeling skills (one task per modeling software), and completing a project as a final assignment. The test on theoretical knowledge is realized in the E-learning (Moodle) system and focuses on basic facts, concepts, and terms and phrases of 3D printing and modeling. Basic skills gained using both 3D modelers are assessed in the modeling tests. Students must design models based on visual information (photos, images), instructions, and given dimensions within a specified time. They must upload the requested work files or shared links to the model, screenshots, and text supplements. The tests are organized in an exam-like environment, in a computer lab, with no supporting resources. By the end of the semester, students must complete their final assignment, which is the design of a 3D (FDM) printable object with detailed documentation. They can use all course materials and online content to demonstrate their proficiency and knowledge of the subject.

2. Research Hypotheses

Students who are offered the optional course mentioned above can program, so it is logical to consider whether code-based 3D modeling software should be included in their education or not. Fortunately, some free 3D modeling software, usually using their own scripting language, is available, thus enriching the curriculum is feasible.

2.1. OpenSCAD

3D modeling is most often done on a visual design interface, mainly with mouse manipulations and data entry, and the 3D model components can also be transformed and moved in the virtual space with mouse movements. OpenSCAD [23] takes a radically different approach to solid modeling. When we describe a component in some sentences, we need to answer some questions like what kind and what size of geometric elements it is made of, how they meet each other (whether they are in contact with each other or if there are gaps between them), are the components repeated or not, and if so, how many times and in what way, etc. The answers help to algorithmize the design process. OpenSCAD is a free, code-based, non-interactive modeler that does not focus on the artistic aspects of 3D modeling but instead on the CAD aspects. OpenSCAD uses its script language with control statements, named values (variables), modules, and functions (Figure 4). The software processes the code and generates a preview or rendered model version. It works with CSG modeling with primitives, 3D objects derived from 2D objects, and imported shapes.

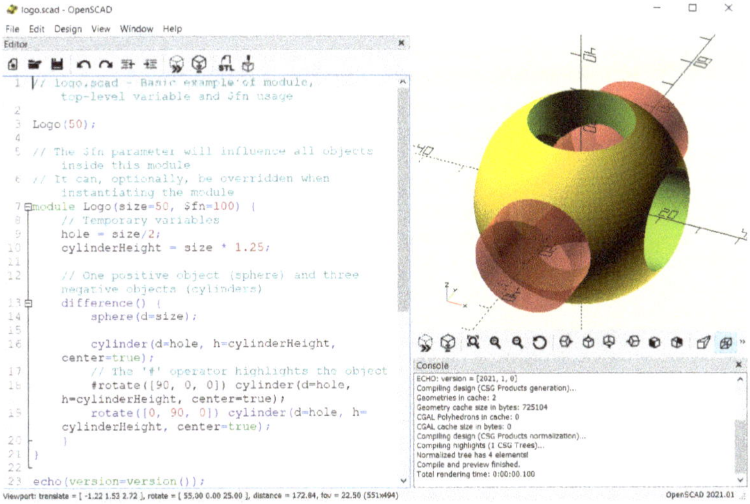

Figure 4. User interface of OpenSCAD application. Based on the sample code in the left window, the OpenSCAD logo is rendered.

2.2. Hypotheses

Lecturers must check students' knowledge, which can also be done as an assignment. In the third decade of the 21st century, when artificial intelligence chatbots are available with ever-expanding capabilities, we can wonder who will perform the tasks in the case of code-based 3D modeling software. Should we be afraid that the codes are not written by students but are generated by artificial intelligence? Can artificial intelligence help or hinder the work of students correcting partial solutions containing errors and misunderstood information?

In this study, we examine the following research hypotheses:

1. Introducing code-based 3D modeling software (such as OpenSCAD v.2021.01) for the students has the risk that they can generate the codes using only AI chatbots.
2. Currently available AI tools can provide relevant help in designing 3D models using a scripting language.

In the following, we will refer to these two hypotheses as H1 and H2, respectively.

3. Materials and Methods

In the frame of an ERASMUS+ project called INNOSID, we organized a workshop for students who were majored in IT-related programs at different universities in Croatia, France, Hungary, Portugal, and Spain [24]. The objective of our 5-h-long workshop was to introduce students with no previous knowledge of 3D printing and modeling to writing OpenSCAD scripts to form 3D printable models. Since the workshop took place in Valencia, Spain, we decided that the characteristics of the city would inspire the models. The idea promised multiple advantages:

- direct the students' attention to the architectural values of the city,
- develop their geometric skills,
- develop their algorithmic skills,
- motivate them to look for these shapes in the city after the workshop.

At the beginning of the workshop, the main principles and properties of 3D printing and the basics of OpenSCAD scripting (syntax rules, the most critical functions, flow control statements, and the role of modules) were discussed, followed by writing simple sample codes together. In the next stage, students received tasks to complete on their own. Each task consisted of a photo taken in the city of the object to be modeled and instructions about the primitives and functions to use. The 3D-printed models were also available in the classroom. At that time (May 2022), either ChatGPT [25], or another AI chatbot was not released; thus, the students worked alone.

To verify or drop the research hypotheses, we will check how two AI chatbots (ChatGPT from OpenAI LP (San Francisco, CA, USA) and Gemini [26] from Google AI (San Francisco, CA, USA)) can deal with two tasks of the students. There are other models, such as Grok-2, LLama-3.1, and Claude Sonnet, that we could test, but their popularity lags far behind the two leading ones. ChatGPT is considered the most well-known AI tool and is very popular; thus, its testing is unavoidable. Google, as a company with so many free services, is also very popular among students. The majority of the students have a Gmail account and use several Google tools. This was the reason to involve it in the testing. Since both companies make the temporary usage of their advanced AI models available, we used both the free versions and the advanced models. Prompts were given in English since the language of instruction in the INNOSID workshop was English. IT students are used to using English in their study and work, and we also run our courses at the university in English.

3.1. ChatGPT from OpenAI

ChatGPT is a chatbot based on the Generative Pre-trained Transformer (GPT) language model that was taught using texts from the Internet, Wikipedia, and digitized books. The system is based on deep learning technology and can communicate with humans in real time by allowing natural language processing and generating text responses to users' questions or requests. It mimics human-to-human communication at a high level, providing realistic and intelligent answers to user questions (prompts). The first version of ChatGPT was released in late 2022 based on the data collection until then. GPT-4 model has made ChatGPT more accurate, consistent, human-like, and detailed. The latest GPT-4o model, which can interpret images and browse the Internet, is available free of charge for a limited time only without a subscription.

3.2. Gemini from Google

Gemini is Google's AI-powered assistant, built right into Gmail, Docs, Sheets, and more. It offers enterprise-grade security and privacy. ChatGPT is the best solution for developer content and complex content requests. Gemini is a better solution for many images and creative content requests, as well as for built-in quality assurance tools.

4. Experiments

This section examines the workshop's two selected modeling tasks from the viewpoint of AI's applicability. Students solved these tasks, which proves that, based on the slides, the modeling tasks can be completed after a brief introduction to 3D modeling with OpenSCAD. We will also provide the slides containing the tasks and additional information that aided the students during the planning process. The two selected tasks have different characteristics. The first task involves a simple model geometry, but various modeling approaches can be applied. The second task has a much more complex model geometry, making it challenging to decipher the required solids and operators. To assist students with weaker geometry skills, the slides include a detailed explanation of the necessary steps to avoid failure.

Additionally, we will share our experience with the students' performance and our attempts to solve the tasks using artificial intelligence. We tested both the free and enhanced (image interpretation) versions. The ultimate goal was to generate the OpenSCAD code for both models.

4.1. Bridge Railing of Puente del Ángel Custodio

Although no river crosses the city of Valencia, it still has 18 bridges full of history over the Turia Garden, which was established in the former riverbed of Turia. Turia's course was diverted to prevent constant flooding in the city after a devastating flood in 1957. Turia Gardens is now a large urban park, running nine kilometers of green space through the city, with footpaths, leisure, and sports areas, making the area very popular with locals and tourists alike. We selected the bridge railing of Puente del Ángel Custodio as the first model to complete based on the instructions in Figures 5 and 6.

During the workshop, most students could complete the design of one column, and some could also apply the loop to generate a series of columns. Just a few students had enough time to connect the columns below and top.

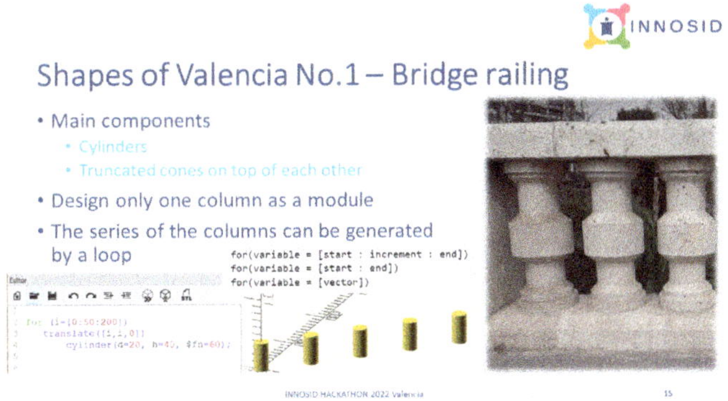

Figure 5. This slide includes a photo of the bridge railing and instructions on the main components of one item. Syntax forms of for loops and a sample code snippet about their usage are also available to help.

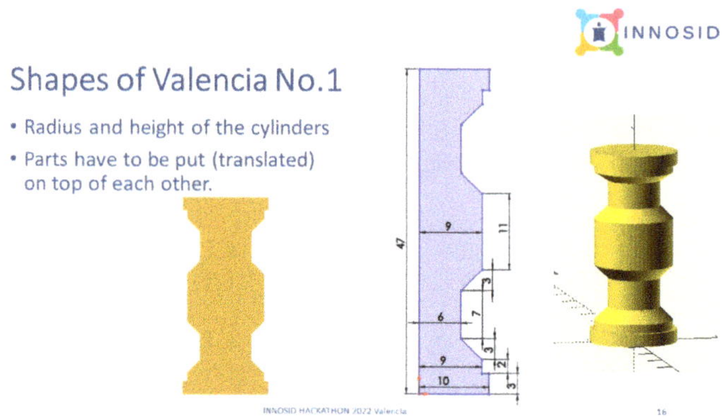

Figure 6. The second slide belonging to this task contains a CAD drawing with the dimensions and an image captured in OpenSCAD of the rendered item and its cross-section.

4.1.1. ChatGPT

First, the total free version was tested, where images cannot be used to explain the shape of the desired solid. To generate the script of one bridge rail column, we provided the list of the polygon vertices and asked ChatGPT to create a solid by rotating the polygon around the y-axis. The result was perfect, although the task was not complicated. Only the function *rotate_extrude()* should have been applied for the polygon created from the points.

In the next phase, we wanted to benefit from the visual information of the slides, too. Thus, we started to use the advanced GPT-4o model. First, the screenshot (right picture in Figure 6) was uploaded into the ChatGPT to generate the model. The result (left image in Figure 7) can be regarded as a good draft version, which was refined after a prompt: "Check the image again. The solid consists of more parts." (right image in Figure 7). Surprisingly, on the outmost level of the module, the function *difference()* was used without defining what to remove from the union of the set of cylinders. Note that in OpenSCAD (truncated), cones can be generated using the function *cylinder()* with different parametrization. We can claim that the result was similar to the required model, and after minor changes in the code, we got the appropriate model. ChatGPT also suggested: "You can fine-tune the dimensions and radii to better match the exact proportions of your model".

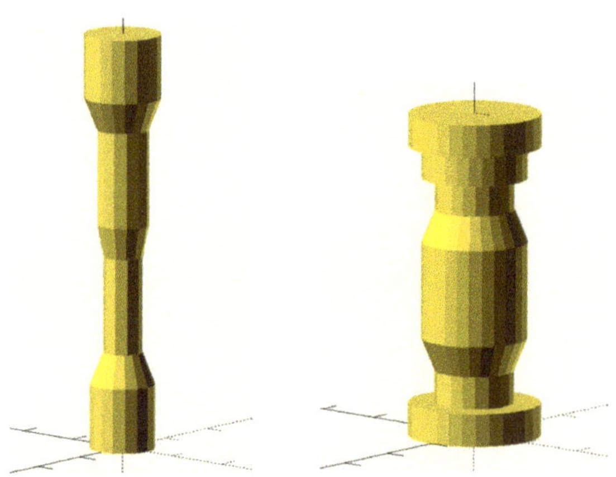

Figure 7. The models generated by ChatGPT based on the screenshot (located on the right-hand side of Figure 6). The left model is the first version, while code for the right-hand side column was proposed after a new prompt.

Since the second slide has a CAD drawing with the exact dimensions (middle picture in Figure 6), there were some students who decided to generate the column by rotating a polygon around the y-axis. This motivated us to ask AI to use the CAD drawing to generate a surface of rotation. As mentioned before, GPT-4o model can interpret not only text but images and videos. The left model in Figure 8 was created after the first prompt. ChatGPT defined a module that determined a polygon consisting of 16 points (whose positions were only partly correct). This polygon was the input shape for the *rotate_extrude()* function with an angle of 360 degrees. After the second prompt ("Please check the sketch again. The shape is not correct."), the model tended to be more similar to the desired one (see the middle solid in Figure 8). It is noticeable that there are no truncated cones at all, but the number of polygon points increased to 28. After a new prompt, the model became worse: it still only had cylinders and was missing the upper part (see the right solid in Figure 8).

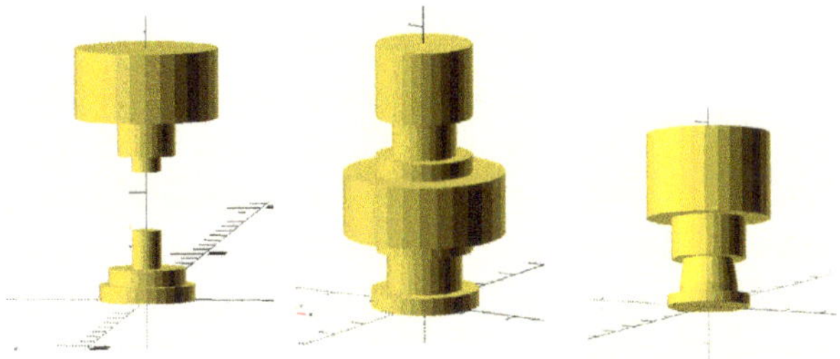

Figure 8. The models generated by ChatGPT based on the CAD drawing (located in the middle of Figure 6). The left model was proposed first, then the middle one after a second prompt. Since the middle model still consists of only cylinders, another prompt was asked, which resulted in the right model that is worse than the previous one.

If we are familiar with the basics of OpenSCAD syntax, we can alter the script to get the right model. All we have to do is insert some points into the list of polygon vertices and amend the coordinates if it is necessary.

This task cannot be completed without any OpenSCAD knowledge even if we use the current latest language model, GPT-4o.

4.1.2. Gemini

We followed the same scenario when testing Gemini; thus, no visual information was used at first. Gemini received the list of the polygon vertices and was asked to create a solid by rotating the polygon around the y-axis. Exactly the same prompt was used as in the case of ChatGPT. The proposed script and all the others, after providing further prompts to fix the syntax errors, could not be compiled in OpenSCAD. The script introduced variables to hold 2D and 3D shapes, which resulted in syntax errors. Our next prompt suggested using modules instead of variables. It is important to note that variables in OpenSCAD are set at compile-time, not run-time. Although the new script contained one module, it still had a variable preventing it from being compiled. At this stage, there was no more sense in enforcing the fixing of the code; it was faster and simpler to correct manually. Gemini does not insert comments into the script but gives an attempt to explain the recommended code. The final prompt was: "Do not use variables at all, but modules.", that resulted in the script in Figure 9, which still does not run.

In the next phase, the screenshot was first used as visual information to force Gemini Advanced to generate the OpenSCAD code of a railing column. As Figure 10 shows, the model is very simplified in contrast to the code, which is very complex. We can state that these 38 lines of code (containing seven lines of comments) are too much effort for these two concentric cylinders. The second ("Can you check the image again? The solid is more complicated.") and third prompts ("Try to create it from cylinders and truncated cones with different radii.") resulted in the same model visible right in Figure 10. At this point, there was no point in continuing prompting or trying to edit the code manually, which was rather complicated compared to the complexity of the solid.

```
module create_solid(points) {
  rotate_extrude(angle=360) polygon(points);
}

create_solid([[0, 0], [10, 0], [10, 3], [9, 3], [9, 5],
    [6, 8], [6, 15], [9, 18], [9, 29], [6, 32], [6, 39
  ], [9, 42], [9, 44], [10, 44], [10, 47], [0, 47], [
  0, 0]]);
```

Figure 9. The wrong script that Gemini generated after the prompt suggested using only modules instead of variables. The code directly passes the points to the function *polygon()* that is nested into *rotate_extrude()* function.

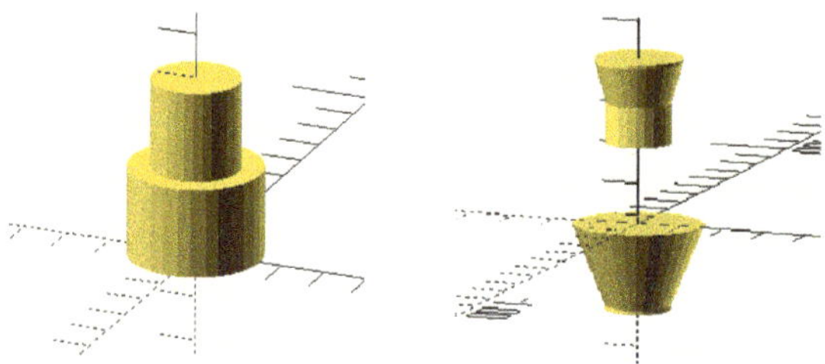

Figure 10. The solids designed by Gemini Advanced based on the OpenSCAD screenshot (right in Figure 6). Code for the left model was generated after the first prompt. The model on the right-hand side was rendered based on the code after the second and third prompts.

The experiment continued using the CAD drawing as a visual prompt for Gemini Advanced. The generated code was well-segmented: ten variables (note that ten numbers are visible in the drawing but did not match the variable values precisely) were introduced and used to create the list of vertices for the polygon to be revolved. Altogether, Gemini Advanced detected eleven vertices instead of 16, but their coordinates were not computed correctly, which resulted in an inappropriate solid (see Figure 11). The following prompts were about enforcing it to interpret the drawing more precisely (no change in the shape), taking the horizontal symmetry into account (new code with several syntax errors). Overall, we failed to use Gemini Advanced in this task. We have to mention at the same time that the first code version was capable of editing manually to have the desired OpenSCAD code.

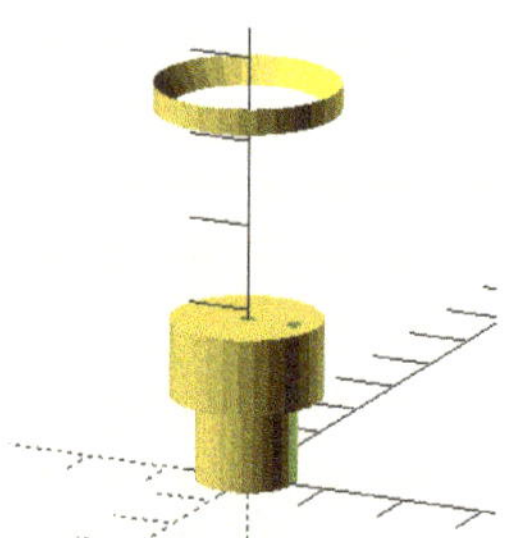

```
// Dimensions (based on the image)
total_height = 47;
width = 11;
step_1_height = 10;
step_1_width = 6;
step_2_height = 9;
step_2_width = 9;
step_3_height = 11;
notch_depth = 3;
notch_width = 2;
fillet_radius = 1;  // Optional: Add
```

Figure 11. The model generated by Gemini Advanced based on the CAD drawing (right in Figure 6), and the introduced variables from the script. Inappropriate calculations resulted in a self-intersecting polygon that caused the two disjoint parts after rotating.

4.2. Stone Ornament

While wandering in Valencia, you can see a particular solid used to decorate the top of barriers along the Turia Gardens' walls (right color photo in Figure 12). The form of this ornament is exciting, and its design requires some knowledge of geometry. That was why we selected this ornament's form as a task in the workshop. The students received the instructions on two slides (see Figures 12 and 13), where beyond the screenshot of the model, detailed help in text and numbers was available. The solid can be divided into three main parts, and only the middle one is complicated. To better understand how it can be designed, the second slide shows a screenshot taken in TinkerCAD (an online, free 3D modeling software). The orientation of the four ellipse-based cylinders is visible, and additional information on the necessary rotation and translation is given, too. This task was more challenging since the students (who majored in IT-related degree programs) had no advanced knowledge of geometry, but most of them could complete the design based on the slides and were satisfied with the result.

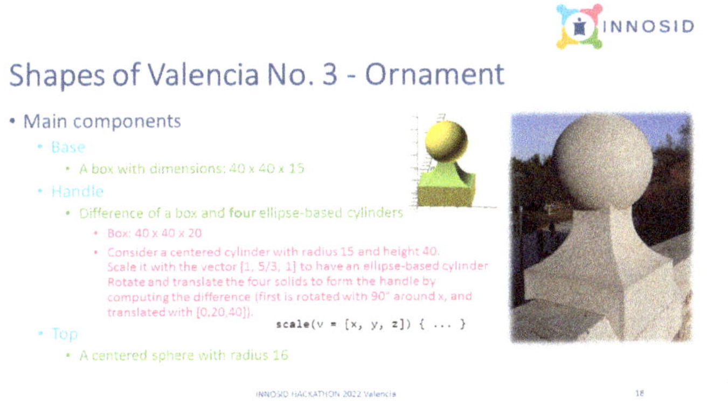

Figure 12. The three main components of the stone ornament, along with a photo and screenshot in OpenSCAD. The middle component has the most challenging shape, which justifies the need for more detailed instructions in the slide.

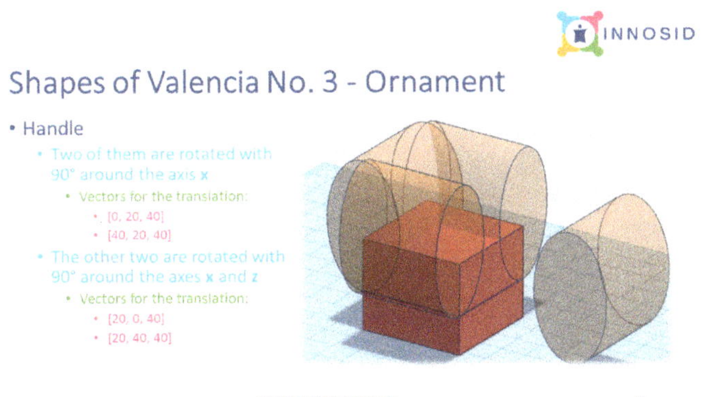

Figure 13. Detailed instruction on designing the middle part of the ornament. The right screenshot was taken in TinkerCAD to help understand the position and orientation of solids used to form the handle.

4.2.1. ChatGPT

Since the model to be designed is complex, and first, only textual information is used as prompts, we had to define the steps of modeling in detail. The following instructions were given: "Create a solid consisting of 3 parts. The first part is a box with dimensions 40 × 40 × 15. The second part is the difference of a box (40 × 40 × 20) located on the top of the first part and four ellipse-based cylinders deriving from a primitive: centered cylinder with a radius of 15 and height of 40 that is scaled with the vector [1, 5/3, 1]). The position of the cylinders: first: rotate the primitive with 90° around x, and translate it with [0, 20, 40], second: translate a copy of the first one with [40, 0, 0]. Third: rotate the primitive with 90° around x and z, and translate it with [20, 0, 40], Fourth: translate a copy of the third one with [0, 40, 0] The third part is a sphere with radius 16 placed on the second part top, but lower it a bit to intersect the handle, not only touch it". ChatGPT generated a script with modules and some comments but did not run. After specifying where OpenSCAD denoted the error, the enhanced version resulted in a solid. The issue was that ChatGPT nested two modules, which prevented the nested one from being accessed for free. Both version miscalculated the volume of translation along the z-axis for the cylinders. We also have to mention that the fragment number (parameter *$fn*) was not set for the cylinders or the sphere; the default values were used. The appropriate values were set after prompting the need for a smoother surface.

In the second turn, we uploaded the screenshot of the model taken in OpenSCAD into ChatGPT to give visual information for generating the OpenSCAD code. Based on the screenshot, it recognized three main parts whose mutual positions were acceptable (see left image in Figure 14). The middle part needed to be refined, so we used the prompt "The shape of the pedestal is more complicated. It fits exactly the base and has a curved surface from each side. Ellipsoids could be used to subtract". As the middle model (Figure 14) demonstrates, the textual information overwrote the image information since the model middle part (ChatGPT called it pedestal) shape became worse. As a last attempt, we defined the shape that could be used to form the middle part with this prompt: "Oh no, that is worse. Try to use elliptic cylinders to subtract". The result can be seen on the right in Figure 14, proving that we must provide detailed text information if we want to get the right shape. Since the model is vertically symmetric, an additional image could not provide valuable information on its shape.

Interestingly, ChatGPT assigned explicit colors to the different parts. Indeed, the screenshot taken in OpenSCAD uses not only one color to render the solid, but shading is the reason. This has no considerable significance in 3D modeling; it was just an interesting observation.

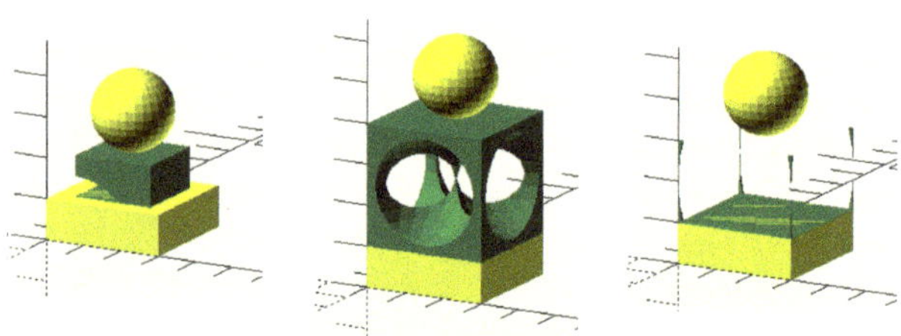

Figure 14. ChatGPT generated models based on the OpenSCAD screenshot of the stone ornament. The left image was created based only on the screenshot; in the case of the middle and right ones, textual information was also provided to refine the shape.

4.2.2. Gemini

The same detailed text instructions were given to Gemini to create the OpenSCAD script for the ornament. Gemini defined three modules for each part, just like ChatGPT, but it had issues interpreting the instructions precisely. The biggest mistake was that in the module responsible for the elliptic cylinder, the function *scale()* was misused, which resulted in no change in the dimensions. The positions of the four elliptic cylinders were calculated wrong, just like that of the sphere (see Figure 15). Gemini used the default smoothness for the cylinders and the spheres. Main errors in the code:

- wrong syntax for the function *scale()* which resulted no change in the dimensions
- the sphere was not aligned with the other parts (the default position for the cube is not centralized but for the sphere)
- wrong position for the four elliptic cylinders.

After the prompt "The cylinders are not scaled" there was no change in calling the function, and as you can see in Figure 15, the generated solid is less similar to the original ornament. There was no sense in giving further prompts at this point since it took less effort to make the necessary changes manually than to define several prompts to correct all the wrong calculations.

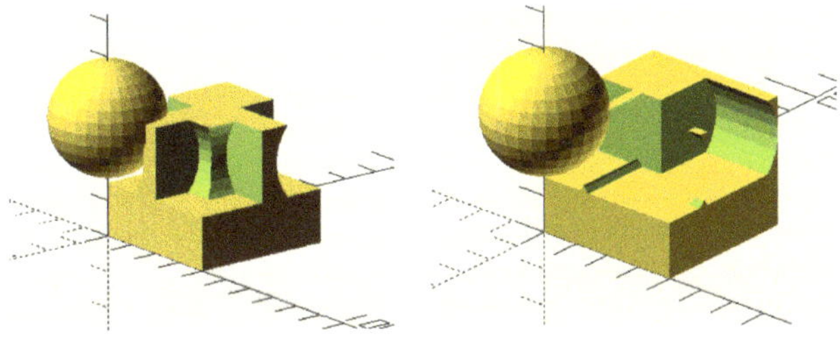

Figure 15. The solids generated by Gemini using only text prompts for the ornament. The same instructions were given as ChatGPT received in the first round. The left solid is the first model that is far from the desired one. The right-hand side model was created after a second prompt but resulted in no relevant changes.

Since we have a screenshot of the model, we tried to use it in Gemini Advanced as in the ChatGPT case. The left model in Figure 16 represents what Gemini Advanced designed based on the screenshot. Gemini Advanced detected four parts of the solid: base, cone, sphere, and cutouts. The role of the first three components is straightforward (see the right-hand side model in Figure 16) and could be matched to the parts of the ornaments. Without the cutouts, it could be considered a fair draft version, although the parts are not aligned and are not placed on top of each other. We have to notice that the most complicated middle part has become significantly simplified.

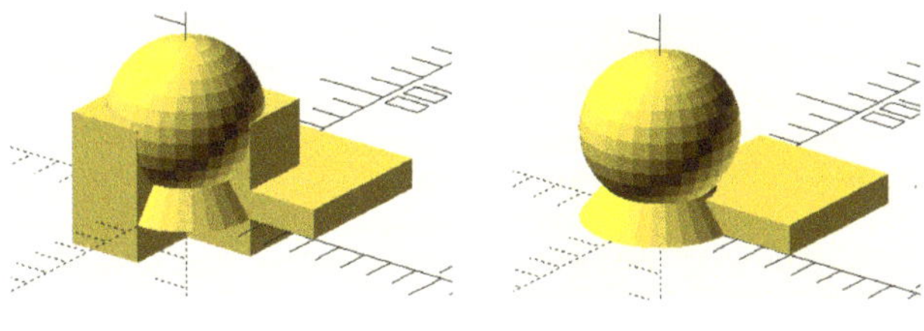

Figure 16. The model generated by Gemini Advanced based on the screenshot of the ornament model. The left model represents all four parts, while the right one does not show the cutouts (whose role is unclear).

Figure 17 shows two further versions generated after new prompts. First, we wanted a smoother surface for the sphere and a more realistic shape for the middle part (Gemini Advanced called it 'cone' in the code). The prompt was: "The sphere is not smooth enough, and the part between the sphere and base is different". The generated model (see left model in Figure 17) was improved since the sphere meets the requirement, but the shape of the middle part did not change significantly. The following prompt "The different parts are not aligned. The cone should be on the base, and the sphere should be on the cone." resulted in better alignment of the three components, but it still suffers from some inaccuracy: the sphere is disjoint, the cutouts removed some parts from the cone, and the unique shape of the middle part is only approximated. If we define the shape of the middle part as text, we can improve it, but we already tested that Gemini could not interpret our detailed description correctly.

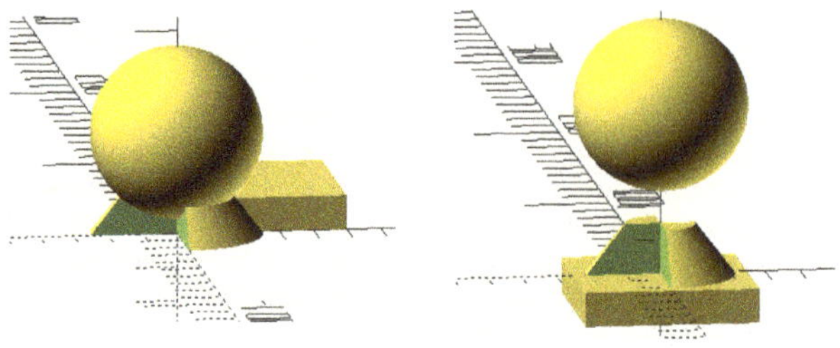

Figure 17. Models generated by Gemini Advanced after two further prompts. The first prompt resulted in the left model with a smooth circle but no improvement in the middle part. The right-hand side model has better-aligned parts but is still far from the desired shape.

To explore every possibility, we decided to check how Gemini Advanced can deal with the ornament's very detailed text prompt already used in Gemini. Unfortunately, the first offered code had the same syntax error (inappropriate usage of variables) as Gemini produced earlier. We used the prompt "Define the cylinder primitive as a module since now it has a syntax error." to overcome this issue, which already resulted in a valid code although the solid is not correct (see left image in Figure 18):

- the base cube is centered (along each axis), but its position is misused later
- positions of the four elliptic cylinders are miscalculated.

The right image in Figure 18 demonstrates the wrong positions of the four elliptic cylinders. Their actual locations, not only along the x and y but also along the z-axis, are incorrect, resulting in a wrong-shaped solid. However, the miscalculations can be fixed manually, although this requires knowledge thus we could not spare much effort with Gemini Advanced.

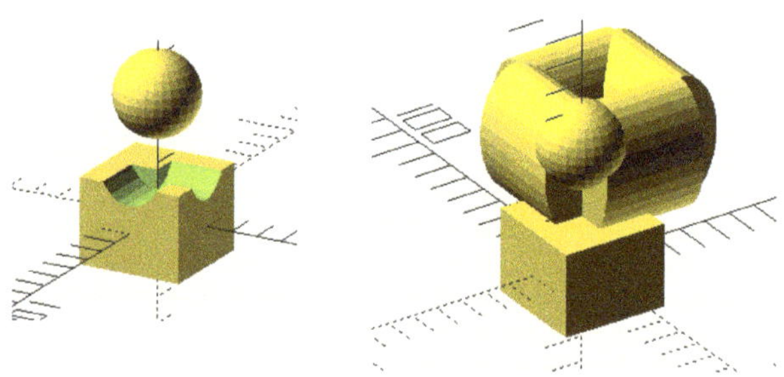

Figure 18. Gemini Advanced generated the right-hand side model after resolving the syntax errors. The left image shows the wrong positions of the cylinders caused by using the centered cube.

5. Results

We can return to our two hypotheses and assess them based on the experiments. Before doing so a table was created to provide instant visual information on the performances of the two AI chatbots (see Table 1). The assessment of the artificial intelligence's ability was stringent because we wanted to compare the AI's skills with those of the students who could successfully complete the tasks. Therefore, we established three categories:

- The model generated by the code met all the requirements.

- The proposed code generated a model that was far from the desired one and could not be used even as a sketch.
- The proposed code could be used as a draft version of the desired model. After some manual editing, the code can result in the model. The amount of needed editing was divided into two additional subcategories.

In Table 1, we used different symbols to refer to the three categories:

- (+) denotes total success
- (%) denotes relevant help where additional manual coding is necessary
- (-) denotes failure where AI could not provide help.

Results about the performance of the two tested AI tools (free and commercial) are listed in the Table 1, considering the two models separately. If the model's shape is complex, then code-based reconstruction using a picture is not feasible for either of the two tools. Without complex geometry skills, this task is challenging for humans as well. As we can see, AI has the potential to develop in the future if we want to use it in OpenSCAD to assist us in coding.

Table 1. Overview of how the ChatGPT and Gemini performed [1] while trying to use them to generate OpenSCAD scripts using textual and different visual prompts.

	Railing Column			Stone Ornament	
	Only Text	Only Screenshot	CAD Drawing	Only Text	Only Screenshot
ChatGPT	+	-	-	%	-
ChatGPT (GPT-4o)	+	%	%	%	-
Gemini	%	-	-	%%	-
Gemini Advanced	%	-	%	%	-

[1] Legend: +: success, -: failed, %: could be improved by editing the code manually, which requires knowledge of OpenSCAD syntax rules. The number of symbols reflects the volume of the further editing.

5.1. Testing the First Hypothesis (H1)

Hypothesis H1 is about whether the students can totally delegate their 3D modeling assignments to an AI chatbot or not. According to our experiments, the risk exists, but only if the student can precisely describe the model (which requires some modeling knowledge). It could also be achieved if images provide enough input to generate the code. Currently, the two tested AI chatbots' free versions do not support interpreting image-based information. The advanced versions are available for free as short-term trials, but they cannot even produce the desired model based only on images. The risk exists, but currently, its volume is meager.

5.2. Testing the Second Hypothesis (H2)

Hypothesis H2 concerns whether AI can provide relevant help if we design a 3D model using code-based modeling software. In this hypothesis, we do not distinguish between the prompt types. According to Table 1, in the case of both models, ChatGPT and also Gemini could provide some help in generating the code, saving some effort. Nevertheless, the chance is meager that we will not have to edit the code, which requires knowledge again. Hypothesis H2 is verified according to the results of the experiments.

6. Discussion

Emerging technologies reshape our lives in many areas. Artificial Intelligence, one of the latest achievements of the human mind, does the same and shows its impact on education, too. Many lecturers and educators at every level try to follow what is happening in the field of AI since it may have several roles in education, too. There are also some concerns, like the possibilities of delegating homework and assignments to AI tools, preventing the students from doing their own work. In this study, our objective was to examine the possible effects of introducing a code-based 3D modeling software into the curricula of our course *3D printing and modeling* at the University of Debrecen.

The key findings from the study are as follows. Teaching a code-based 3D modeling software does not increase the possibility of delegating the students' work to AI. In our experiment, we considered only single solids since the assignments to be completed by the end of the semester have to be more complicated. In some cases, AI can save some coding but cannot complete the whole code without further editing. Image-based prompts could be very convenient, but AI tools must be developed.

We conducted a test on two chatbots using their current models. It is uncertain how much they will develop in the future or when the results of this study will become outdated. These uncertainties can be seen as limitations of the study and the technology being tested. We believe that AI can be introduced in the classroom; it can save some effort but cannot totally replace the skills (either geometry or coding). Plagiarism is still a more significant threat when assessing student performance.

There are additional concerns related to applying AI in education. Institutions must introduce (some of them may have already done) new regulations about allowing or blocking its usage. Prohibition is often counterproductive, and we have no appropriate safe tool to check whether AI was used. With a more permissive regulation, a big question is how to ensure equal access to AI tools. The best, most innovative models are not free of charge. The students' financial background should not affect the range of available tools. However, someone has to pay for the development of the models. Further challenges of using AI such as critical exploration whether ChatGPT and related AI technologies serve as assets or disadvantages in education, shedding light on the unavoidable challenges encountered during their incorporation are discussed in [27].

In the future, our research will focus on testing other AI models (such as Grok-2, LLaMa-3.2, Claude Sonnet), and AI's applicability in other fields of 3D modeling and testing, including its straightforward application possibilities. With deploying image and voice capabilities of the models, they should be involved into the testing to cover all the modality of the tools. It could be interesting to test different prompting technologies with involvement students too.

7. Conclusions

Artificial intelligence tools cannot be omitted in any field of life. What we could not imagine yesterday can be real tomorrow. We must periodically reevaluate and redesign our teaching material and testing methods for students' skills in education. This study defined two hypotheses about the possible applicability of AI in code-based 3D modeling.

The first hypothesis (*Introducing code-based 3D modeling software (such as OpenSCAD) for the students has the risk that they can generate the codes using only AI chatbots.*) could not be verified thoroughly. As the experiment showed, the tested AI models in their current status can provide some help but cannot take the burden of work from the students. Thus, this hypothesis is only partially verified. Further research could give a more nuanced picture, while we also have to monitor the development of the AI tools.

The second hypothesis (*Currently available AI tools can provide relevant help in designing 3D models using a scripting language.*) could be verified with the experiment. Changes related to this hypothesis's validity can be expected in the future since the development of large language models does not stop.

Author Contributions: Conceptualization, M.Z. and I.P.; methodology, M.Z.; software, M.Z. and I.P.; validation, M.Z. and I.P.; investigation, M.Z.; resources, I.P.; writing—original draft preparation, M.Z. and I.P.; writing—review and editing, M.Z. and I.P.; visualization, M.Z. and I.P.; supervision, M.Z. and I.P. All authors have read and agreed to the published version of the manuscript.

Funding: This research received no external funding.

Data Availability Statement: All relevant data is included within the article.

Conflicts of Interest: The authors declare no conflicts of interest.

References

1. Bai, C.; Dallasega, P.; Orzes, G.; Sarkis, J. Industry 4.0 technologies assessment: A sustainability perspective. *Int. J. Prod. Econ.* **2020**, *229*, 107776. [CrossRef]
2. Barata, J.; Kayser, I. Industry 5.0—Past, Present, and Near Future. *Procedia Comput. Sci.* **2023**, *219*, 778–788. [CrossRef]
3. Alhamad, I.M.; Ahmed, W.K.; Ali, H.Z.; AlJassmi, H. Chapter 6: 3D Printing Applications in Mechanical Engineering Education. In *Integrating 3D Printing into Teaching and Learning*; Ali, N., Khine, M., Eds.; Brill: Leiden, The Netherlands, 2020; pp. 90–131.
4. Özeren, Ö.; Özeren, E.B.; Top, S.M.; Qurraie, B.S. Learning-by-Doing using 3D printers: Digital fabrication studio experience in architectural education. *J. Eng. Res.* **2023**, *11*, 1–6. [CrossRef]
5. Li X.; Gould, T.; Zaki, R. The Impact of Artificial Intelligence on Learners and Teachers: A Mathematics Education Perspective. In *Education, Development and Intervention: Toward Participatory and Integrated Solutions*; Papadakis, S., Kalogiannakis, M., Eds.; Springer: Cham, Switzerland, 2024; pp. 133–148.
6. Laksmiwati, P.A.; Lavicza, Z.; Cahyono, A.N.; Alagic, M.; Mumcu, F. When engineering design meets STEAM education in hybrid learning environment: Teachers' innovation key through design heuristics. *Asia Pac. J. Educ.* **2024**, 1–19. [CrossRef]
7. Yunianto, W.; Cahyono, A.N.; Prodromou, T.; El-Bedewy, S.; Lavicza, Z. CT integration in STEAM learning: Fostering students' creativity by making Batik stamp pattern. *Sci. Act. Proj. Curric. Ideas STEM Classr.* **2024**, 1–27. [CrossRef]
8. Wang, F.; Huang, J.; Zheng, X.L.; Wu, J.Q.; Zhao, A.P. STEM activities for boosting pupils' computational thinking and reducing their cognitive load: Roles of argumentation scaffolding and mental rotation. *J. Res. Technol. Educ.* **2024**, 1–20. [CrossRef]
9. Kiliánová, K.; Kočková, P.; Biernatová, K.; Hurný, T.; Kostolányová, K. STEAM Labs: 3D Printing and Inclusive Education in Primary Schools. In Proceedings of 16th International Conference on Education and New Learning Technologies (EDULEARN24), Palma, Spain, 1–3 July 2024; pp. 7670–7675.
10. Colorado, H.A.; Mendoza, D.E.; Valencia, F.L.; Perdomo, J.M. Production of Teaching Materials Through 3D Printing as Support for Educational Processes Related to the Sciences, Heritage, and Culture. In *TMS 2020 149th Annual Meeting & Exhibition Supplemental Proceedings*; The Minerals, Metals & Materials Series; Springer, Cham, Switzerland, 2020; pp. 257–263. [CrossRef]
11. Babic, J.; Car, Z.; Gace, I.; Lovrek, I.; Podobnik, V.; Topolovac, I.; Vdovic, H.; Zilak, M.; Gourves, A.; Sable, C.; et al. *Analysis of Emerging Technologies for Improving Social Inclusion of People with Disabilities*, 1st ed.; University of Zagreb Faculty of Electrical Engineering and Computing: Zagreb, Croatia, 2020; 168p, ISBN 9789531842624. Available online: http://sociallab.fer.hr/wordpress/wp-content/uploads/2021/02/INNOSID-Report-O1.pdf (accessed on 31 July 2024).
12. Eldebeky, S.M. High School Students' Experience of a 3D Printing Station at a Bilingual School Makerspace in Kuwait. *Stud. Technol. Enhanc. Learn.* **2021**, *1*, 371–386. [CrossRef]
13. PrusaLab—Hi-Tech Prototyping Workshop from Prusa Research. Available online: https://prusalab.cz/en/ (accessed on 31 July 2024).
14. Free File Converter Online—AnyConv. Available online: https://anyconv.com/ (accessed on 30 September 2024).
15. Transform Text and 2D Images into 3D Assets with Generative AI for Free—Alpha3D. Available online: https://www.alpha3d.io// (accessed on 30 September 2024).
16. Design with Bubbles 2D to 3D in Seconds. Available online: https://www.designwithbubbles.com/ (accessed on 30 September 2024).
17. Meshy—Free 3D AI Model Generator. Available online: https://www.meshy.ai/ (accessed on 30 September 2024).
18. Jiayuan, L.; Xinyi, S.; Jeanhun, C. A Study on Process of Creating 3D Models Using the Application of Artificial Intelligence Technology. *Int. J. Adv. Cult. Technol.* **2023**, *11*, 346–351. [CrossRef]
19. Papp, I.; Tornai, R.; Zichar, M. What 3D technologies can bring to education: The impacts of acquiring a 3D printer. In Proceedings of 7th IEEE International Conference on Cognitive Infocommunications (CogInfoCom), Wroclaw, Poland, 16–18 October 2016; pp. 257–261.
20. TinkerCAD—From Mind to Design in Minutes. Available online: https://www.tinkercad.com (accessed on 31 July 2024).
21. SOLIDWORKS Community Download Instructions. Available online: https://www.solidworks.com/support/community-download#no-back (accessed on 31 July 2024).
22. Onwubolu, G.C. *Introduction to SolidWorks: A Comprehensive Guide with Applications in 3D Printing*, 1st ed.; Taylor & Francis Group, CRC Press: Boca Raton, FL, USA, 2017; p. 1193.
23. OpenSCAD—The Programmers Solid 3D CAD Modeller. Available online: https://openscad.org/ (accessed on 31 July 2024).
24. Solutions Based on Emerging Technologies for Improving Social Inclusion of People with Disabilities. Available online: https://sociallab.fer.hr/innosid/ (accessed on 31 July 2024).
25. ChatGPT | OpenAI. Available online: https://openai.com/chatgpt/ (accessed on 31 July 2024).
26. Gemini—Chat to Supercharge Your Ideas. Available online: https://gemini.google.com (accessed on 31 July 2024).
27. Vaughan, G.; Kovács, Á.; Szűts, Z. The educational challenges of ChatGPT. *Ann. Math. Informat.* **2023**, *58* 170–181. [CrossRef]

designs

MDPI

Article

4D Printing: Research Focuses and Prospects

Yuran Jin [1,*] and Jiahui Liu [1,2]

1 School of Business Administration, University of Science and Technology Liaoning, Anshan 114051, China; ljh981817@163.com
2 School of Architectural Engineering Management, Hebei College of Science and Technology, Tangshan 063200, China
* Correspondence: jinyuran@ustl.edu.cn; Tel.: +86-13941289982

Abstract: As an emerging technology in the field of additive manufacturing, 4D printing is highly disruptive to traditional manufacturing processes. Therefore, it is necessary to systematically summarize the research on 4D printing to promote the development of related industries and academic research. However, there is still an obvious gap in the visual connection between 4D printing theory and application research. We collected 2070 studies from 2013 on 4D printing from the core collection of Web of Science. We used VOSviewer 1.6.20 and CiteSpace software 6.3.3 to visualize the references and keywords to explore focuses and trends in 4D printing using scientometrics. In addition, real-world applications of 4D printing were analyzed based on the literature. The results showed that "tissue engineering applications" is the most prominent focus. In addition, "shape recovery", "liquid crystal elastomer", "future trends", "bone tissue engineering", "laser powder bed fusion", "cervical spine", "4D food printing", "aesthetic planning" are also major focuses. From 2013 to 2015, focuses such as "shape memory polymers" and "composites" evolved into "fabrication". From 2015 to 2018, the focus was on "technology" and "tissue engineering". After 2018, "polylactic acid", "cellulose", and "regenerative medicine" became emerging focuses. Second, emerging focuses, such as polylactic acid and smart polymers, have begun to erupt and have become key research trends since 2022. "5D printing", "stability" and "implants" may become emerging trends in the future. "4D + Food", "4D + Cultural and Creative", "4D + Life" and "4D + Clothing" may become future research trends. Third, 4D printing has been widely used in engineering manufacturing, biomedicine, food printing, cultural and creative life, and other fields. Strengthening basic research will greatly expand its applications in these fields and continuously increase the number of applicable fields.

Keywords: 4D printing; CiteSpace; VOSviewer; scientometrics; visualization

1. Introduction

Additive manufacturing technology is based on designed three-dimensional CAD model data and transforms various forms of materials, such as powders, wires, liquids, and sheets, into three-dimensional entities through digitally-driven layer-by-layer accumulation [1–3]. Since the 1980s, additive manufacturing technology, particularly 3D printing technology, has developed rapidly [4,5]. With Professor Tibbits of the Massachusetts Institute of Technology proposing the concept of 4D printing in 2013, the "3D + time" manufacturing method entered people's vision, and then there was a craze for research on 4D printing [6–8]. Currently, the main 4D printing process technologies used by researchers are fused deposition molding (FDM), stereolithography (SLA), selective laser sintering (SLS)\laser melting (SLM), directed energy deposition (DED), and other process technologies, such as photosensitive polymer jetting (PolyJet) and direct ink writing (DIW) [7,9–12]. 4D printing is an emerging, intelligent additive manufacturing technology developed based on 3D printing technology. Its intelligent characteristics are reflected in the fact that printed products have automatic stress response characteristics that can achieve a corresponding evolution in form and nature under different incentive conditions. This is a "dynamic" product [13,14]. There are at least two stable states in a 4D printed structure, and under the stimulation of the corresponding medium, the structure can transition from one state to another [9]. By combining it with ordinary materials, the structure can undergo dynamic changes under the excitation of certain media [15]. Many scholars have explored dynamic changes in 4D printing under different conditions from multiple perspectives, such as shape memory [16], residual stresses [17], phase transformations [18], and springback [19]. Dynamic changes in structures include changes in shape and material properties or functions [20]. The self-responsive stress characteristics of 4D-printed products have great development potential and have attracted the interest of many scholars worldwide [7,20–22]. Cerbe et al. [17] investigate the relationship between stress and strain during 4D printing, proposing that printing parameters affect structures' heating and cooling process. Fu et al. [23] reviewed important 4D printing technologies in conjunction with basic polymer science and engineering and discussed the challenges and future opportunities faced by 4D printing. Chen et al. [24] proposed that 4D bioprinting will become the next generation of bone repair technology, providing more inspiration for clinical medicine of bone regeneration. At present, scholars have carried out extensive research on 4D printing technology and initially formed a theoretical system of "smart materials" + "smart structures" + "3D printing" [25]. With the continuous deepening of scientific research, 4D-printed smart products have broad application prospects in many fields, such as future robot development, intelligent equipment manufacturing, aerospace, and biomedicine [26].

Although academic research on 4D printing technology has been greatly enriched, in the process of reviewing existing research results, we found three obvious gaps in 4D printing technology. First, owing to the numerous research results on 4D printing, there has been an obvious lack of systematic research since the advent of 4D printing technology and a more comprehensive review is required to reveal the research focuses and application prospects of 4D printing technology. Second, most existing studies have explored the development of 4D printing technology from a technical perspective. In general, overall research on 4D printing technology is insufficient. Third, single-disciplinary research on 4D printing has been dominant. Comprehensive research must be conducted from a multidisciplinary perspective to help understand the focuses and trends of 4D printing and further popularize the application of 4D printing technology. Therefore, in terms of content, we consider 4D printing as a research object and discuss its research focuses and application prospects so that readers can better understand the key points and development directions of 4D printing technology. In terms of methods, we are based on scientific metrology and use visualization technology to better display the research focuses and trends of 4D printing technology to readers quantitatively and intuitively. Finally, by analyzing the visual information of 4D printing technology, we discover the future application prospects of 4D printing technology, help readers quickly understand the development direction and application prospects of 4D printing technology, and find key entry points for future research. Specifically, our research focuses on the research and application prospects of 4D printing technology to solve the following problems:

- RQ1: What research hotspots have existed in 4D printing since its inception?
- RQ2: What have been the research trends in 4D printing in the last decade?
- RQ3: What is the current status of 4D printing applications?
- RQ4: How should 4D printing be developed further in the future?

The remainder of this paper is organized as follows. In Section 2, described the sources of the literature and briefly analyzed the collected literature to explicitly visualize and analyze it using scientometrics. In Section 3, based on the collected literature, VOSviewer and CiteSpace are used to visualize and analyze references and keywords and explore their research focuses and emerging trends. In Section 4, combined with the previous research focus on 4D printing, the application prospects of 4D printing technology in four fields—engineering manufacturing, biomedicine, food printing, and cultural and creative life—are analyzed. Section 5 summarizes the paper, presents new findings of the research, highlights the theoretical and practical significance, and explains the limitations of the future outlook.

2. Data Collection and Methods

2.1. Data Sources

We selected SCI, SSCI, AHCI, CPCI-S, CPCI-SSH, ESCI, CCR-EXPANDED, and IC from the core collection of Web of Science and screened the literature related to 4D printing. To obtain comprehensive data and guarantee the validity of the literature screening, we chose to combine "all fields", "topics" and "title" searches. First, considering that the concept of 4D printing was proposed by Prof. Tibbits of MIT in 2013, we delineated the time range from 1 January 2013 to 30 June 2024 as the time to collect literature, which is consistent with the formal proposal of the concept of 4D printing and in line with the need to collect comprehensive and up-to-date literature. Secondly, we used "all fields" to collect references from the time range from January 2013 to June 2024 and then used the combination of "subject" and "title" search methods to remove duplicate references. Considering that 4D printing remains within the scope of additive manufacturing and is a rapid prototyping technology, a variety of expressive keywords for 4D printing are listed in the search. The advanced search format created was Ts1 = (4D (all fields) and (additive manufacturing (subject) or 4D printing (subject))). Ts2 = (4D (all fields) and (additive manufacturing (title) or 4D printing (title))). Ts1 obtained a total of 2054 references, Ts2 obtained a total of 1064 references, and after the deduplication process, a total of 2070 references were obtained (search time was 10 July 2024); the publication types and search results are shown in Table 1.

Table 1. Publication types and search results.

Search Field	Sum.	Research Papers	Review Paper	Conference Papers	Editorial Material	Book Reviews	Other
TS_1	2054	1393	463	146	28	3	21
TS_2	1064	771	180	69	19	3	22

2.2. Literature Analysis

To better study the research hotspots and application prospects of 4D printing, we classified the publication years and citations of the collected references. The results are shown in Figure 1. Judging from the publication year, since the concept of 4D printing was first proposed in 2013, relevant research results have increased annually until 2023 (from 2024 to 30 June 2024, it is impossible to infer the trend in 2024), reaching 467 articles by 2023. The year-on-year growth of 4D printing technology is due to the gradual establishment of a complete theoretical framework by 4D printing technology and an interdisciplinary cooperation model, which promotes the rapid iteration and innovation of new technologies [25]. Simultaneously, 4D printing technology has attracted great attention from the public, and 4D printing has great potential [27]. From 2013 to 2023, the number of citations of 4D printed references has increased rapidly, indicating that 4D printing has received increasing attention from academics. From the perspective of the Web of Science, materials science is the main research category of 4D printing, with a total of 873 references. This is mainly because the implementation of 4D printing technology is highly dependent on the special properties and the responsiveness of materials. The core of 4D printing technology is to enable a printed object to change its shape or structure under external stimulation [28]. The autonomous deformation ability of 4D printing is mainly owing to the special properties of the materials, such as temperature, humidity, and electromagnetic responses [6]. Therefore, the research and development of materials are key to realizing 4D printing technology [14,29,30]. At the same time, it can also be seen from the top ten categories of Web of Science that current research on 4D printing technology mainly focuses on materials, technologies, and applications. From the perspective of research direction, materials science is also the main research direction of 4D printing, with the total number of references reaching 1104, accounting for more than half of the total references. Research on smart materials plays a central role in the development of 4D printing [28]. The smart materials required for 4D printing not only provide the basic physical form but also give printed products the ability to change dynamically, giving them the "fourth dimension"—the changing characteristics of time. Currently, these are the main themes in 4D printing research direction.

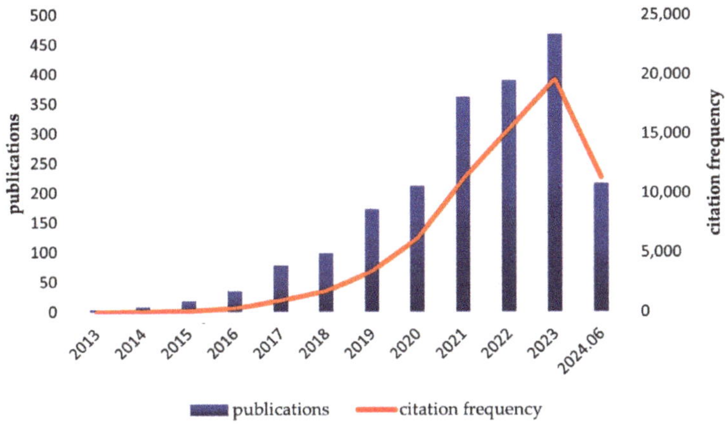

Figure 1. The number of times publications related to 4D printing are cited.

2.3. Visualization Analysis Using Scientific Metrology

Scientometrics is a discipline in which quantitative research on scientific activities is conducted. It focuses on the relationship between the investment of scientific researchers and research funds and the number of papers, citations, and other outputs, as well as the dissemination and formation of scientific information exchange networks process [31]. Scientometrics was initiated by D. Price and others in the early 1960s and adopted quantitative research methods to study the nature of science itself. Garfield laid the foundation for the development of scientometrics by creating a large SCI database [32]. After years of development, scientific metrology has expanded from simple literature measurements to analyzing scientists' research activities, scientific and technological innovation models, and scientific communication networks [33]. As the latest research direction of additive manufacturing, 4D printing technology has attracted widespread attention from scholars worldwide since its inception [6,22,34]. Since it was proposed in 2013, 4D printing technology has produced rich research results in the past ten years, providing us with a solid foundation for conducting visual analysis using scientific metrology [35]. As effective software for scientific metrology, VOSviewer can visualize and analyze large-scale, high-dimensional datasets [36]. In addition, VOSviewer provides various technologies, such as rotation, scaling, and filtering, for research on browsing and manipulating datasets in real-time, and the appearance and behavior of VOSviewer can be customized to suit specific needs and preferences [37]. CiteSpace is a citation visualization analysis software that focuses on the analysis of underlying knowledge in the scientific literature [38]. CiteSpace helps researchers sort past research trajectories and predict the prospects of future research by visualizing the relationships between references in a scientific knowledge map [39]. CiteSpace can reveal the structural changes and evolution laws of scientific knowledge and uses a network diagram composed of nodes and connections to show the co-occurrence relationships between different references, authors, or keywords [40]. CiteSpace can also identify hot themes and development trends in the research field and help users track cutting-edge issues [41]. CiteSpace can generate keyword co-occurrence maps, which visualize the frequency and co-occurrence of keywords in the literature, thus intuitively showing past and present hotspots in the field [40]. Keywords with a high frequency of occurrence typically represent core themes in the field [42]. The change in the keyword co-occurrence graph in the timeline is a concentrated reflection of the development trend. Newly appearing keywords or keyword combinations indicate emerging research directions [41]. Simultaneously, the keyword co-occurrence diagram can also determine the overall pulse of this field [38]. By analyzing the connection between different keywords, the relationship and evolution process between various themes can be understood [39]. CiteSpace provides a variety of parameter setting options, such as time slicing, selection criteria, and clipping methods, which allow users to conduct flexible analyses based on their research requirements [42]. CiteSpace is widely used for writing reviews. We can use the knowledge maps generated to comprehensively outline theoretical views and evolutionary paths in specific fields [41]. Therefore, we chose VOSviewer 1.6.20 and CiteSpace 6.3.3.

3. Results

3.1. Focuses and Trends Based on References

To explore the knowledge network structure of the 4D printing technical literature, we used VOSviewer 1.6.20 and CiteSpace 6.3.3 to conduct cocitation literature analysis on the collected references.

We used VOSviewer software to date the literature from January 2013 to June 2024, selected a map based on bibliographic data, then read the data from bibliographic database files and imported the data from the Web of Science core collection. References in citations were selected for the type of analysis, and 10, 50, 100, and 200 were selected for the minimum number of citations of a document. Finally, we found that it is more realistic to select 50 to produce the VOSviewer-created density, network, and overlay visualizations of the cited literature map. The final landscape map is shown in Figure 2, and the 19 most-cited references are listed in Table 2. In the cited literature map created by VOSviewer, 389 items were included, generating 24 clusters and 4316 links. In Figure 2, the difference in the number of reference citations is represented by the size of the node, and the different clustering is distinguished by color. We observed the visual landscape map in Figure 2 and found that in 2016, 4D printing technology produced many references with high influence and a high citation rate. In particular, the bionic 4D printing coding method put forward by [43] has obvious advantages compared to other cited literature. As 4D printing technology raw material research continues to deepen, coding and manufacturing methods have gradually become the focus of scholarly research. Among the 19 most-cited references created by VOSviewer, all were related to the manufacturing methods and applications of 4D printing. Therefore, external stimulation, modeling simulation design, and application areas of 4D printing technology have greater potential for development in the future and may become a research hotspot for 4D printing technology.

We utilized CiteSpace software with a time slice set from January 2013 to June 2024 and selected cited literature on node type and g-index coefficients at K = 25, K = 5, K = 50, and K = 100. The CiteSpace results were the most realistic when K = 25, and the final landscape is shown in Figure 3, where the 25 most cited references are shown in Figure 3. Figure 3 shows a 4D printout of the overall network landscape of co-cited literature, which contains 1006 nodes and 4869 connectors. The color bar at the top of the figure indicates when the literature was co-cited, with each color representing one year. The node with the large outline indicates that it is a landmark paper, and the black text indicates the name of the generated cluster, which is preceded by a smaller sequence number, indicating the greater importance of the cluster. Red indicates the outbreak degree of the cluster node. The brighter red the color and the larger the node outline, the higher the cited intensity of the literature. According to the size of the serial numbers of the clusters in the figure, "emerging applications", "4D-printed shape memory polymers" and "4D-printed active composite structures" are the three largest clusters, reflecting that they are the greatest research focuses of 4D printing technology. Moreover, such as "3D printing", "liquid crystal elastomers", "geometry-driven finite elements", "biomedical applications", etc., are also hot research areas for 4D printing.

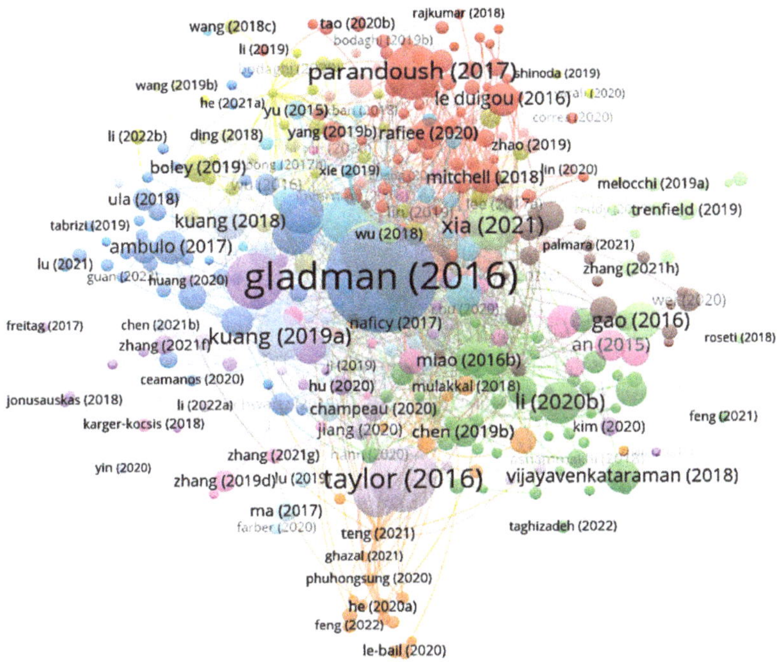

Figure 2. The visual landscape map of co-cited references of 4D printing.

Table 2. The 19 most cited studies.

Document	Theme	Citations	Links
Gladman et al. [43]	Biomimetic printing	2088	176
Taylor [44]	Self-Healing hydrogels	968	10
Tibbits [45]	Programmable materials	814	141
Ge et al. [46]	Tailorable features	742	113
Parandoush et al. [47]	Review: methods	723	14
Momeni et al. [48]	Review: concepts and tools	663	122
Xia et al. [49]	Review: emerging directions	611	27
Kuang et al. [50]	Shape-shifting materials	588	109
Luo et al. [51]	Polyion Complex	552	7
Ge et al. [52]	Programmed action	550	116
Ge et al. [53]	Origami	524	99
Li et al. [54]	Review: biomedical	479	29
Ding et al. [55]	Active composite materials	477	94
Gao et al. [56]	Bioprinting	421	50
Bakarich et al. [57]	Thermally Actuating Hydrogels	367	82
Pan et al. [58]	Artificial intelligence	360	1
Kuang et al. [59]	novel ink	353	50
Vijayavenkataraman et al. [60]	Organ transplantation	350	3
Ambulo et al. [61]	Liquid Crystal Elastomers	340	45

To further explore the focuses and trends in the above three largest clusters, we used CiteSpace to specifically cluster the following three themes. "Emerging applications", "4D-printed shape memory polymers" and "4D-printed active composite structures". The results are shown in Table 3. It can be seen that "comparative review", "critical review" and "typical application" are the main research directions of "emerging applications". Other themes, such as "Biological interface" and "Emerging direction" also occupy an important position in the future research of "emerging applications". "Polymeric material", "smart polymeric composite" and "tissue engineering" are the main research themes of "4D-printed shape memory polymers" and account for a high proportion in the overall research results. Together with "controlled sequential shape" "advanced properties", "comprehensive review", "biomedical application" and "democratic 4D printing" they form the main body of the "4D-printed shape memory polymers" research network. The number of cluster themes formed by "4D-printed active composite structures" is smaller than that of "emerging applications" and "4D-printed shape memory polymers". Cluster "robotics application" is the main research direction of "4D-printed active composite structures", which has an absolute advantage over other clusters. With the maturity of 4D printing technology, "Active composite material", "On-demand local actuation," and "Permanent shape reconfigurability" will also become important research directions. To further demonstrate the focuses and trends in the three largest clusters mentioned above, we have chosen to visually analyze the cluster "4D-printed shape memory polymers" with the closest internal cluster relationship, and the results are shown in Figure 4. The arrow represents the situation where one cluster is referenced by another cluster, the dependencies between clusters. It can be seen that "Polymeric material" is at the center of "4D-printed shape memory polymers" and has a close dependence on other clusters. Other clusters are also interdependent and together form a research network of "4D-printed shape memory polymers".

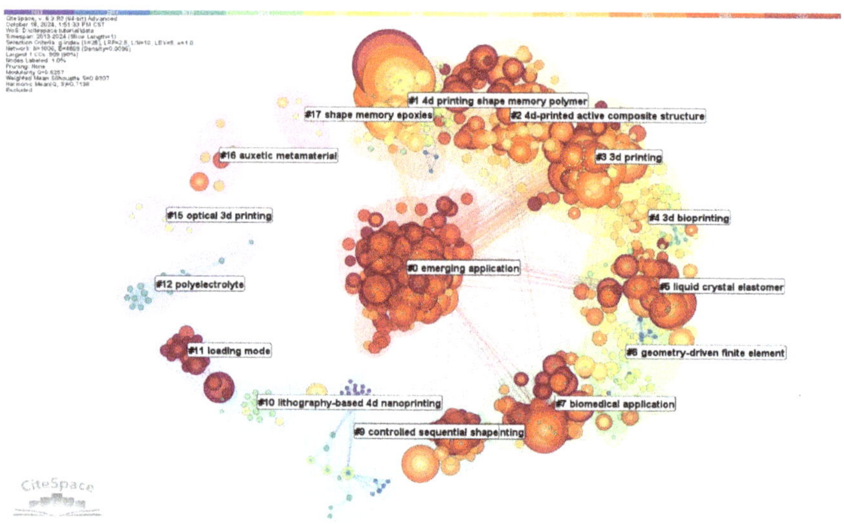

Figure 3. The intellectual landscapes of the 4D printing based on co-citation of the literature.

The network contains 2070 files from January 2013 to June 2024 and produces 17 clusters, the results of which are shown in Table 4. The silhouette in Table 4 reflects the quality of the cluster configuration. The higher the silhouette value, the better the quality of the clustering. The mean year is the average publication year of all literature in a cluster. The closer the average year, the more it represents the research frontier of 4D printing. The red line indicates the burst time interval of the reference. The dark blue line represents the publication time of the reference. In terms of the size of the clusters, the largest cluster, "#0 emerging applications" has 241 members, which indicates that "Emerging Applications" has a central position in the research on 4D printing technology. Moreover, "#1 4D-printed shape memory polymers" and "#2 4D-printed active composite structures" are important. In addition, there are 4 clusters with a scale of more than 100, which indicates a greater concentration of research and applications in 4D printing in terms of materials, structures, applications, and their relationship to 3D printing. In terms of the average year, "#0 emerging application, 2020", "#11 loading modo, 2022" and "#20 robotics application, 2020" undoubtedly represent the latest R&D directions of 4D printing technology. It is worth noting that "#6 geometry-driven fine element, 2013", "#9 controlled sequential shape, 2011", "#10 lithography-based 4D nanoprinting, 2013" and "#12 polyelectrolyte, 2012" were once popular research areas for 4D printing technology. However, few scholars have paid attention to this in recent years, and relevant research has shown a shrinking state, which reflects that these themes may not have much development potential in 4D printing technology.

Table 3. Themes of the largest three clusters.

Cluster Name	Emerging Applications	4D-Printed Shape Memory Polymers	4D-Printed Active Composite Structures
Theme	Comparative review Critical review Typical application Biological interface Emerging direction Shape memory material Fabrication material Advance Additive manufacturing Tissue engineering development Dawn	Polymeric material Smart polymeric composite Tissue engineering Controlled sequential shape Advanced properties Comprehensive review Biomedical application Democratizing 4D printing Loading mode Shape memory epoxies Reentrant honeycomb	Robotics application Active composite material On-demand local actuation Permanent shape reconfigurability Graphene-based polymer bilayer Ultrafast inverse design Controllable deformation design Bioinspired 4D printing 3D printing

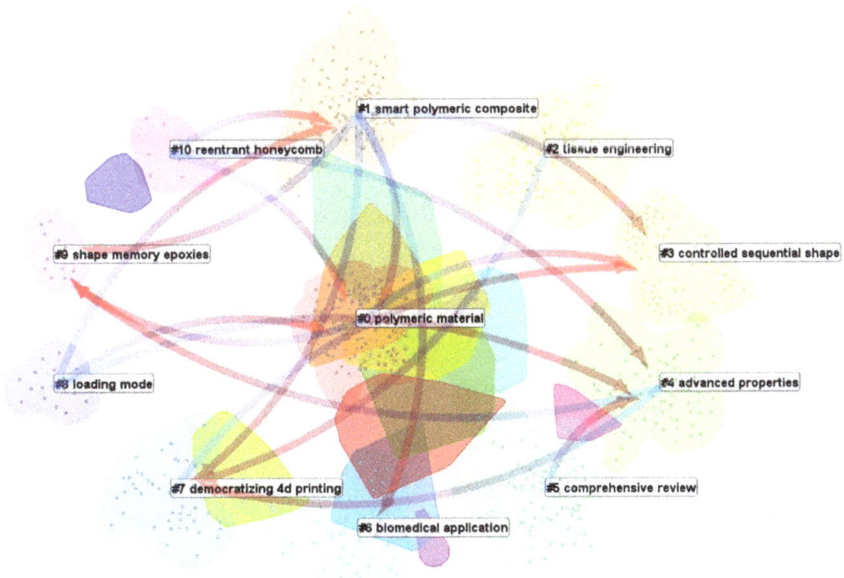

Figure 4. The co-cited reference network of the cluster "shape memory polymers".

Table 4. Summary of the 17 largest clusters.

Cluster ID	Size	Silhouette	Label (LLR)	Theme	Average Year
0	241	0.767	Emerging application	Application areas	2020
1	107	0.861	4D printing shape memory polymer	Shape memory polymer	2015
2	104	0.699	4D-printed active composite structure	Composite properties and structures	2018
3	102	0.726	3D printing	Medical device	2017
4	68	0.876	3D bioprinting	Hydrogel	2015
5	58	0.934	Liquid crystal elastomer	Bionic material	2019
6	48	0.920	Geometry-driven finite element	4D programming	2013
7	46	0.885	Biomedical application	Biomedical	2019
8	35	0.976	6D printing	Future trend	2019
9	29	0.982	Controlled sequential shape	Programming method	2011
10	18	0.995	Lithography-based 4D nanoprinting	Printing technology	2013
11	16	0.997	Loading mode	Design	2022
12	16	0.963	Polyelectrolyte	Smart materials	2012
15	6	0.996	Optical 3D printing	Optical applications	2014
16	6	0.998	Auxetic metamaterial	Artificial composite structural material	2017
17	6	0.985	Shape memory epoxies	Advanced materials	2016
20	3	0.998	Robotics application	Robotics	2020

3.2. Focuses and Trends Based on Keywords

Keywords can highlight the key points and core content of a paper, allowing readers to quickly grasp its theme [31]. By analyzing keywords, we can quickly grasp the dynamics of academic research and guide academic research. Therefore, visual analysis of keywords in reference documents can be used as an indicator to identify hotspots and emerging trends.

In VOSview, we determined the literature time from January 2013 to June 2024, selected "Create a map based on bibliographic data" then selected "Read data from bibliographic database files" and imported the data from the core collection of Web of Science. All keywords in co-occurrence according to the type of analysis were selected, and 10, 50, 100, and 200 were selected for the minimum number of citations of a document. Finally, we found that selecting 50 was more in line with the actual situation. We created a density visualization keyword map using VOSviewer. The final landscape map is shown in Figure 5, and the most highlighted keywords are listed in Table 5. The keyword mutation map created by VOSviewer contains a total of 15 projects, resulting in 4 clusters and 104 links, with a total link length of 4209. Figure 5 shows that 4D printing has obvious advantages for all keywords. Keywords such as 3D printing, 3D, and AM have higher citation rates than the other keywords. This is mainly because research on 4D printing has just gone through the initial stages of concept definition and scope determination. Technological development and applied research are still booming. In the early days of research on 4D printing technology, scholars mainly conducted research on the similarities and differences between 4D printing, 3D printing technology, and additive manufacturing technology. Most research has focused on the basic content of 4D printing technology, characteristics, and application methods. In the middle of the research on 4D printing technology, that is, the current stage, technological development and applied research are gradually emerging. The top keywords, such as "fabrication", "design", "composition" and "scaffolds" all represent scholars' research on promoting 4D printing technology. Future research on the application and technological development of 4D printing technology is an area with great potential.

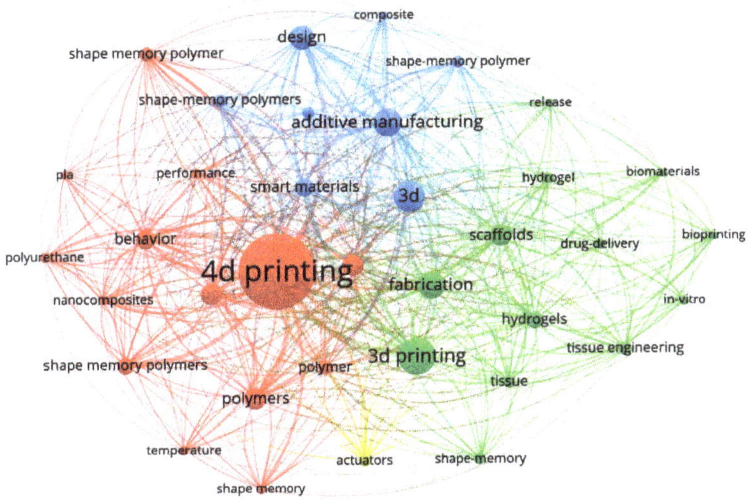

Figure 5. Co-occurrence keyword networks of the 4D printing.

Table 5. VOSviewer includes the 15 most prominent keywords.

Keyword	Occurrences	Total Link Strength
4D printing	1025	1680
3D printing	382	753
3D	296	752
additive manufacturing	268	657
fabrication	248	627
design	214	503
composites	183	476
scaffolds	182	472
polymers	182	468
mechanical-properties	178	404
behavior	154	361
smart materials	150	434
hydro-gels	133	298
shape memory polymers	129	292
shape memory polymer	107	241

We used CiteSpace to generate a keyword hotspot landscape map, as shown in Figure 6. The network contains 581 nodes and 3072 connectors. The color bar at the top of the figure indicates the time the literature was co-cited, with each color representing one year. The black text indicates the name of the generated theme, and the smaller the ordinal number in front of it, the greater the importance of the cluster. As shown in Figure 6, the keywords of 4D printing technology cover many fields, such as engineering and manufacturing, biomedicine, household products, food, culture, and creativity. Among the generated clusters, "#0 tissue engineering application" is still the most researched field, which indicates that tissue engineering applications are the most basic and important content in 4D printing research. The "#1 shape recovery" and "#2 liquid crystal elastomer" as basic research on the properties and materials of 4D printing, also have an extremely important position in the overall research field. This indicates that 4D printing technology has not yet been perfected in terms of technology, materials, and properties; that the technology itself still has great potential for development; and that research on 4D printing technology itself will still be the focus of future development. Notably, the "#4 future trend" has become an important area in the keyword research hotspot of 4D printing technology. This indicates that with the continuous improvement of the definition, characteristics, and technology of 4D printing, the prospect of its application has been gradually emphasized by scholars. In particular, "#3 bone tissue engineering", "#5 laser powder bed fusion" and "#7 cervical spine" were directly related to biomedicine and were among the most common keyword clusters. This is consistent with the results that, at the beginning of the advent of 3D printing technology, attracted attention in the medical field and resulted in numerous research results [62]. Biomedicine has become a key application in 3D printing. 4D printing is an emerging industry of "3D + time". Compared with 3D printing, it has the characteristics of autonomous response and intelligent evolution and has broader development prospects in the biomedical field. In the future, with the continuous improvement and promotion of 4D printing technology, there are revolutionary possibilities for the development and application of biomedical engineering medical devices, engineering instruments, and other fields. At the same time, "#6 4D food printing" and "#9 aesthetic planning" have gradually attracted increasing attention, expanding the prospects for the application of 4D printing technology.

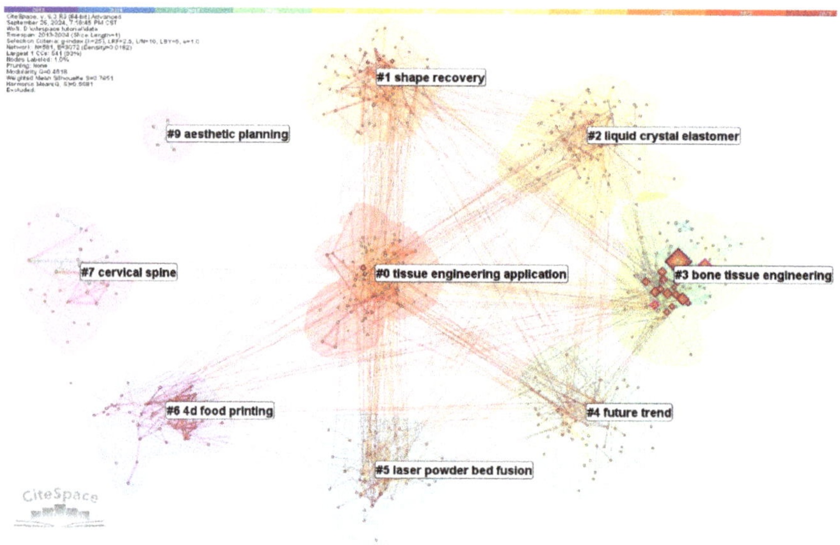

Figure 6. Keyword hotspot network.

Figure 7 lists the 25 main keywords with the strongest citation bursts generated by CiteSpace. The red line represents the outbreak period. The dark blue line represents the time of keywords from publication to now in the field of 4D printing. The results show that "design" (2015–2016) and "deposits" (2015–2017) broke out earliest and attracted much attention from scholars in 2015. Basic research is a hotspot for 4D printing in the early stages of research. At the same time, we also see that "shape", "stereo-lithography", "motion", "recovery force", "metadata" and so on, have relatively long outbreak periods. There is no doubt that 4D printing technology is still in the process of continuous improvement and in-depth research, and has attracted great attention from all walks of life. However, in recent years, these hotspots have gradually subsided. In contrast, emerging hot spots such as "polylactic acid" and "smart polymers" have begun to erupt and become key research trends since 2022. From the perspective of burst intensity, "shape memory polymers" have significantly greater intensity than other keywords, indicating that smart materials are still the main problem plaguing the development of 4D printing technology. However, the outbreak period of "shape memory polymer" research was relatively short, indicating that shape memory polymers attracted the focuses of research by scholars during 2016–2017 and made breakthrough progress.

Keywords	Year	Strength	Begin	End	2013 - 2024
design	2013	3.49	**2015**	2016	
shape memory polymers	2013	5.69	**2016**	2017	
shape	2016	3.77	**2016**	2019	
geometry	2016	2.49	**2016**	2017	
model	2016	2.46	**2016**	2018	
reconstruction	2016	2.45	**2016**	2018	
active structures	2016	2.41	**2016**	2018	
complex	2017	3.76	**2017**	2019	
growth	2017	3.23	**2017**	2019	
origami	2017	3.14	**2017**	2018	
light	2017	2.99	**2017**	2019	
films	2017	2.35	**2017**	2018	
soft	2018	4.71	**2018**	2020	
recovery force	2018	3.09	**2018**	2021	
shape memory polymer	2016	2.74	**2018**	2019	
constructs	2019	4.6	**2019**	2021	
robust	2017	2.4	**2019**	2020	
photopolymerization	2019	2.38	**2019**	2022	
strain sensors	2020	3.51	**2020**	2021	
motion	2018	3	**2020**	2021	
3d and 4d printing	2020	2.63	**2020**	2021	
surface modification	2021	3.26	**2021**	2022	
shape memory alloys	2019	2.69	**2021**	2022	
polylactic acid	2022	3.9	**2022**	2024	
smart polymers	2022	3.08	**2022**	2024	

Figure 7. Top 25 burst keywords on 4D printing in 2013–2024.

To further delve into the research directions of different keyword clusters, Figure 8 was generated with CiteSpace to depict a visual timeline of 4D printing keywords from 2013 to 2024, which contains 581 nodes and 3072 links. The "diamond" represents the emergent keyword nodes of each cluster on the Timeline, and the outline size represents the degree of keyword explosion. In this section, we found nine keywords, such as "tissue engineering application" and "shape recovery" which are the most prominent hot themes. In addition, we show the evolution process of some keywords' hot spots.

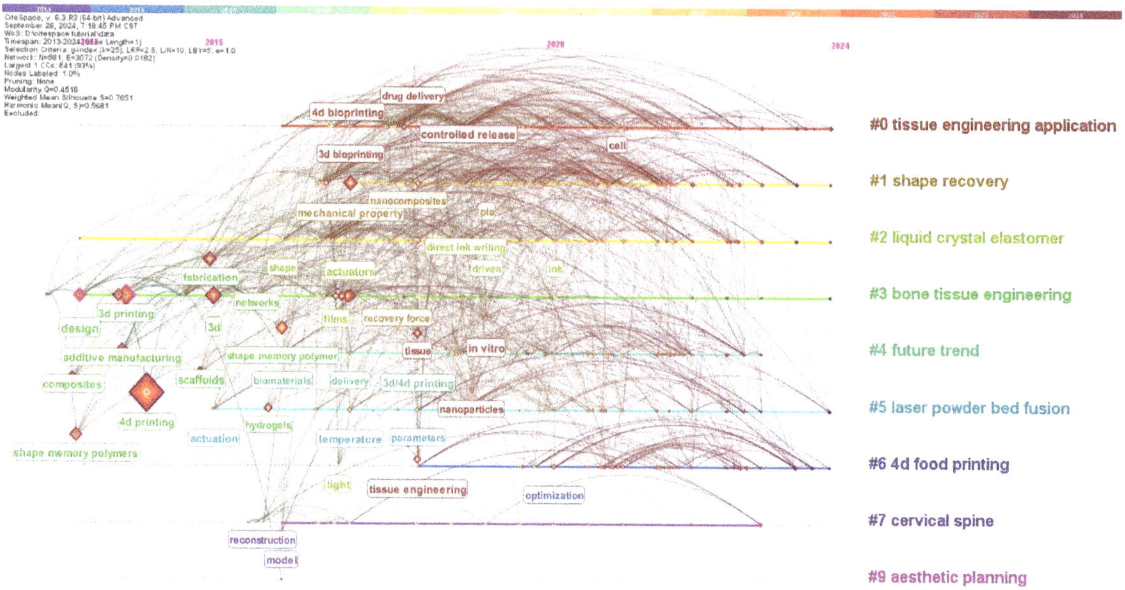

Figure 8. Timeline view of 4D printing keywords.

From 2013 to 2015, hot spots such as "shape memory polymers" and "composites" evolved into "fabrication". From 2015 to 2018, hot spots became "technology", "tissue engineering", etc. After 2018, "poly-lactic acid", "cellulose", "regenerative medicine", etc., became emerging hot spots. In addition, "elastomer", "4D food printing", "time", "biomedical application", "orientation", "energy absorption", "smart polymers" and so on are also significant hot spots. "5D printing", "stability", "implants" etc., may become emerging trends in the future. The specific visualization results are shown in Figure 8. The elements in a map include nodes, timelines, links, and clusters. Nodes represent keywords; the color of connections represents the time when the connected nodes first co-occurred; and to the right of each timeline are clusters containing nodes. The smaller the sequence number is, the larger the cluster and the more nodes it contains. We can see that "tissue engineering application", "shape recovery", "liquid crystal elastomer", "bone tissue engineering", "future trend", "laser powder bed fusion", "4D food printing", "cervical spine" and "aesthetic planning" are the top nine keyword focuses. "Tissue engineering application" is the greatest research hotspot. From a timeline perspective, "4D food printing" is an emerging trend. We will focus our research on the evolution of the keywords "#0 tissue engineering application" and "#1 shape recovery". In the cluster "#0 tissue engineering application", keywords evolved from "3D printing" and "matrix" before 2016 to "membrane" and "shape memory composites" and then to "cellulose" by 2024. By 2016, 4D printing had completed the first stage of evolution. This confirms that 4D printing is an emerging and constantly improving innovative process. The emergence of the keyword "3D printing" shows that in the development of 4D printing, it is necessary to learn from the existing research results of 3D printing and combine the needs of development in various fields to continuously innovate and develop new technologies. From 2015 to 2020, the vigorous development of 4D printing technology occurred. At this stage, 4D printing technology was not limited to the initial biomedical and engineering manufacturing fields, but "4D + food", "4D + cultural creation", "4D + life" and "4D + clothing" have gradually become active in academic research [24,63,64]. The original 4D printing technology was influenced by the demands of different fields, and its technical development gradually began to diversify under the guidance of demand [26]. Biomedical care, as the initial and most fruitful field of 4D printing technology, began to develop explosively after 2015. After 2015, hot keywords such as "hyaluronic acid", "poly-lactic acid" and "regenerative medicine" appeared frequently in Cluster "#0 tissue engineering application" which also reflects that interdisciplinary research has become more frequent and in-depth since 2015. This is strong evidence of the application prospects of 4D printing technology. Finally, in recent years, the popular keywords of 4D printing technology have increasingly diversified and been used in multiple domains. Especially in the cluster "#4 Future Trends" some scholars have put forward the idea of "5D printing". After a comprehensive analysis, we believe that research on 4D printing is developing according to the route of "basis-technology-application". Its application prospects gradually present the development trend of "3D application field-multifield field-cross field". In the future, with the maturity of 4D printing technology, its research results will gradually affect many fields, such as the military industry, art, culture, and clothing, and will ultimately lead to a situation in which multiple disciplines and multiple fields cross and develop and promote each other. In addition, we used CiteSpace to generate Figure 9 to better present the evolutionary relationship of keywords. Each circle in the graph represents a keyword that is the year that first appeared in the analyzed dataset. The lines represent connections between keywords. This connection makes the time zone chart have a time factor. As indicated by Figure 9a, the stage of formation for 4D printing was prior to 2020. At this time, research on 4D printing focused mainly on its essence and application. In order to better study the relationship between 4D printing, 3D printing and additive manufacturing, and to explore the research hotspots of 4D printing at different times, selecting terms such as "3D printing", "4D printing" and "additive manufacturing" respectively (Highlighted red text in Figure 9), Figure 9b–d are obtained. Figure 9b–d respectively reflect the hot keywords of "additive manufacturing", "4D printing" and "3D printing" in different periods. It can be seen that the early research on 4D printing focused mostly on the comparison between 3D printing and additive manufacturing. However, since 2020, research on the correlation between 4D printing "3D printing" and "additive manufacturing" has gradually decreased, and more emerging hotspots, such as "6D printing" and "lacor", have been formed.

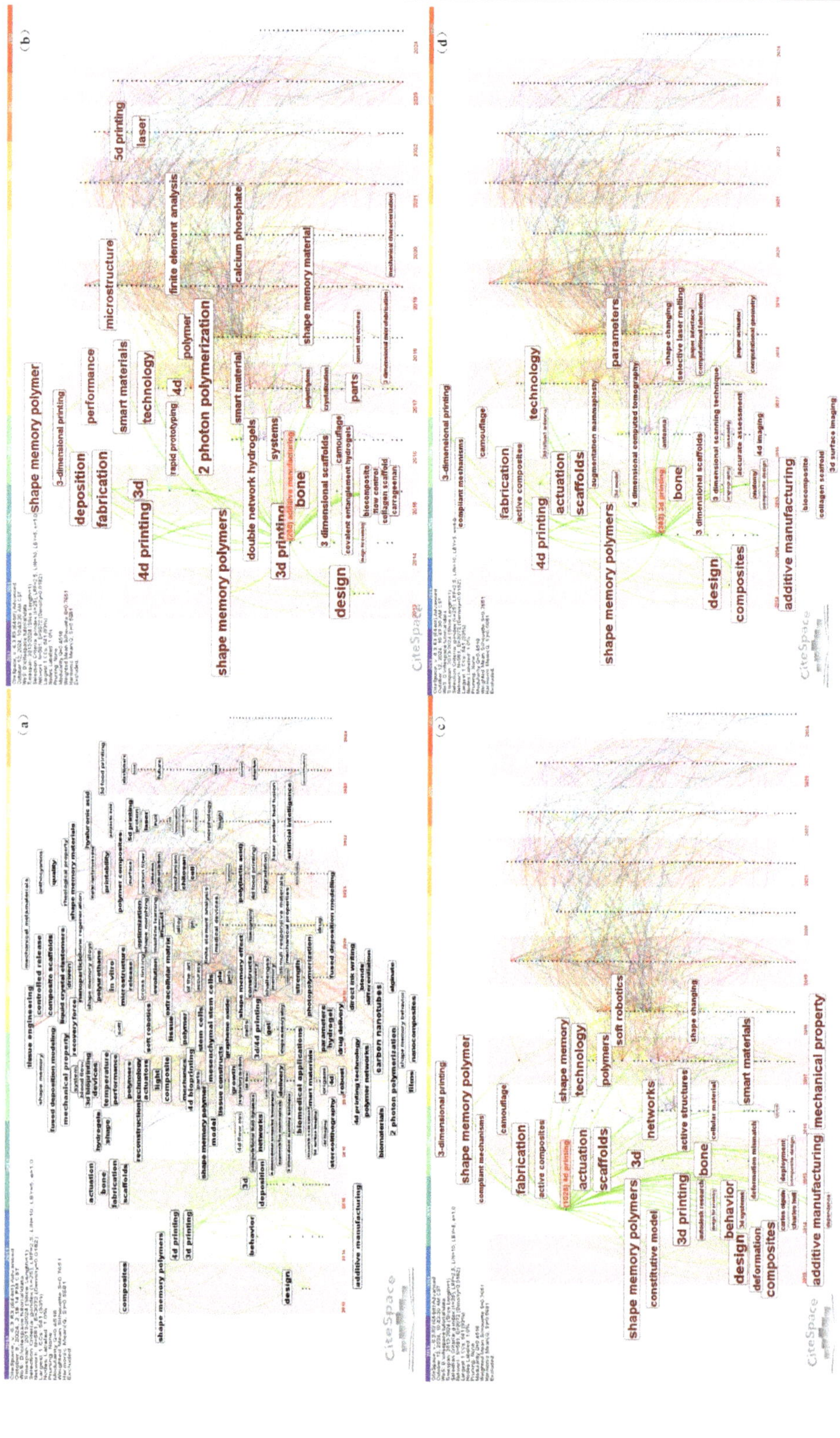

Figure 9. Time zone view of 4D printing keywords.

4. Prospects

Through the previous analysis of the research focuses of 4D printing, we found that the current research on 4D printing has gone through the basic research stage and developed into the technology application stage. Therefore, studying the application prospects of 4D printing has research significance and theoretical value. Analysis using VOSviewer and Citespace software revealed that engineering manufacturing, biomedicine, food and cultural and creative life accounted for a relatively high proportion of the overall research cluster. We will analyze the application prospects of 4D printing from four aspects: engineering manufacturing, biomedicine, food and cultural and creative life.

4.1. Engineering Manufacturing

As an emerging manufacturing technology, 4D printing is opening up the possibility of intelligent manufacturing in the future, heralding major changes in production methods. In the field of engineering manufacturing, 4D printing focuses on basic research. At the beginning of 4D printing research, the main research focused on 4D printing itself, such as technology [65], characteristics [64], materials [55], etc. As 4D printing technology has matured, scholars have also begun to pay attention to its application prospects in engineering manufacturing. For example, Haleem et al. [64] studied the characteristics and workflow of 4D printing technology and proposed that 4D printing technology is a more practical and better production method in engineering manufacturing and can minimize manufacturing labor costs. In the future, these materials can be used in the manufacturing of smart goods, comfort car seats and airbags, smart plumbing devices, transformations, design freedom, enhanced production possibilities, smart water valves, complex assemblies, enhanced market growth of manufacturing, and wide use in many fields, such as improving reliability and performance [64]. 4D printing can self-deform objects under specific environmental conditions, allowing large-scale engineering products to automatically adapt to environmental changes, which is considered by some scholars to be a revival of technology and manufacturing [7]. At present, 4D printing is playing an increasingly important role in the most cutting-edge fields of engineering and manufacturing, such as robots [12], aerospace [66], national defense and military [67], sensors [21], and the automobile industry [14]. In particular, 4D printing can process smart materials into dynamic structures with various stimulus-responsive behaviors, making it possible for scientists to create single-material soft robots that do not require any other processing procedures and can move unfettered materials, thereby ensuring stable and powerful robot rolling [68]. However, some scholars have also suggested that 4D printing applications still face technical, material and design challenges [69]. For example, 4D printing materials are limited to several types of materials, such as hydrogels, thermally responsive, photoresponsive, electrically responsive, magnetically responsive materials, piezoelectric materials, and pH-responsive materials [69]. This undoubtedly limits the application areas of 4D printing technology. In the future, if we can further strengthen basic research on 4D printing, especially AM processes, materials, stimuli, interaction mechanisms and modeling, it will greatly expand the application and applicable fields of 4D printing in the field of engineering and manufacturing.

4.2. Biomedical

After reviewing the research results of 4D printing through VOSviewer and CiteSpace software, we found that biomedicine plays a very important role in its application. Among the largest clusters generated by CiteSpace, "#3 bone tissue engineering", "#5 laser powder bed fusion", and "#7 cervical spine clusters" are all directly related to biomedicine and rank among the top clusters formed. In addition, other top-ranked clusters also indirectly intersect with biomedicine, and their results have had an important impact on biomedicine. Notably, among the top three clusters in biomedicine, "#3 bone tissue engineering" has great research results. For example, Chen et al. [24] regarded 4D bioprinting as a representative next-generation bone repair technology, and in their work, they used 4D-printed biological scaffolds to achieve advanced orthopedic surgery, proposing that 4D bioprinting combined with programmable biomaterials, living cells and bioactive factors provide greater potential for building dynamic, personalized and precise bone tissue engineering scaffolds through the complex structural formation and functional maturity. You et al. [34] used 4D printing technology to create a multi-response bilayer deformable membrane composed of a shape memory polymer (SMP) layer and a hydrogel layer by remotely adjusting the fate of stem cells, which can be accurately switched between proliferation and differentiation to promote bone formation and match specific macro bone shapes in clinical scenarios, thereby improving the formation of deformable membranes in new bone and accelerating bone healing in vivo. In other fields of biomedicine, 4D printing technology also plays an important role [30]. In the field of stomatology, 4D printing technology has expanded the application of dental materials, such as shape memory alloys, smart ceramics, smart composites, glass ion-based cement, etc., improving dental restoration technology, which can rely on the biocompatibility of 4D printing, the elastic modulus and the thermal layer expansion coefficient to simulate the body structure to prevent adverse physical reactions such as swelling, inflammation or ischemic reactions [70]. In the field of implantable organs, some scholars have carried out theoretical research in the areas of cardiovascular and cerebrovascular [6] systems, regenerative organ tissues [71], etc., but this technology is still in its infancy, and the application of 4D printing in manufacturing human organs requires further research to achieve clinical application in the neural field. Currently, Cui et al. [28] have used 4D printing technology to develop a neurogenic nerve conduit with self-driving capabilities by immobilizing neurogenic factors in situ on a printed structure with aligned microgrooves. Neurogenic guidance can be used for nerve regeneration. In addition, 4D printing technology has also played an increasingly important role in other fields, such as minimally invasive surgery [8,72], drug delivery [73], tissue engineering [22], and skin reconstruction [74]. In the future, with the continuous improvement of 4D printing technology, it will also have broad application prospects in other fields, such as smart textiles (such as orthopedic plaster), manufacturing actuators and sensors [62].

4.3. Food Printing

Food printing is an emerging field in 4D printing applications, but it quickly attracted the attention of scholars, and numerous research results have erupted. At present, research on 4D printing technology in the field of food printing has focused mainly on food interactivity, food characteristics (including flavor, nutrition or color), food production, and food transportation and storage [35]. In terms of food interactivity, 4D printing can be used to construct suspended structures, such as blooming flowers. The transformation of flowers from closed to blooming makes eating more interesting, thus attracting more consumer attention; for example, for picky eaters, 4D-printed food provides new ideas for the design of interactive food and enhances the interaction between diners and food materials [75]. In terms of food characteristics, the color, taste, texture and shape of 4D-printed food samples can lead to food characteristics such as color, taste, aroma, texture and shape in response to various stimuli (such as temperature, pH, light, and ionic strength). Change, giving consumers a better eating experience [13]. In terms of food production, 4D printing can effectively control structural changes during the drying process, thereby making some healthy snacks [35]. In addition, while realizing the automated production of food, 4D printing technology is combined with digital cooking to customize it according to individual needs for food shape, structure and flavor, realizing the combination of automated production and customized production [63]. In terms of food transportation and storage, companies can prepare deformed food through 4D printing to achieve flat packaging, reducing transportation costs and storage space [29].

4.4. Cultural and Creative Life

With its astonishing development speed, 4D printing technology has spread from pure engineering and medical applications to the textile and fashion industries [10]. In particular, for fashion goods such as textiles and jewelry, researchers have prepared products that make the wearer feel comfortable based on the adaptability of the environment or requirements [11]. Compared with a traditional cultural and creative life, the main advantages of 4D printing technology include: first, geometric flexibility, which allows for the improvement and optimization of product functions and structural characteristics; second, modification of microstructure characteristics through metamaterial methods; third, use less raw materials; fourth, cost and resource efficiency of small-scale production; fifth, cloud manufacturing shortens the supply chain due to more localized production, which has important sustainability benefits [76]. Despite advances in 4D printing technology, applying 4D printing to textiles remains challenging. This is mainly due to the technical gap between specialty prototypes and manufacturing scalability. Unfortunately, research on textile applications of 4D printing is still in the infancy of research and technology development (R&TD) [77]. Although 4D printing has full potential in different applications in the apparel field, such as the development of functional filament fibers/wires, 4D printing of textiles, and 4D printing of finished garments and 4D textiles, further development is still needed [78]. If 4D printing technology can be applied to the clothing field, the cost and resource efficiency of small-scale production can be optimized through localized production, and the supply chain and demand-driven manufacturing can be shortened, which can be customized and expanded, taking into account cost and environmental sustainability [79]. 4D printing also plays an important role in other fields of cultural and creative life [27]. For example, when 4D printing is combined with culture and education, there will be different kinds of chain reactions occur. In addition to the function of the product itself, cultural and creative products printed in 4D add the story of the product because of its shape memory effect, which plays an auxiliary role in learning historical and cultural knowledge [69]. When 4D printing is combined with VR, AR and other augmented reality technologies, it will provide more immersive experience services to users [80]. The application and development of new 4D printing technology in the literature and entertainment has a broad market.

5. Conclusions

In this research, we studied the research focuses and emerging trends of 4D printing. We used VOSviewer and CiteSpace software to visually analyze 4D printed documents collected from SCI, SSCI, AHCI, CPCI-S, CPCI-ssh, ESCI, CCR-EXPANDED and IC collections on the Web of Science. Through the analysis of co-cited document networks, sudden keywords, and other indicators, we generated a knowledge landscape of 4D printing technology and explored hot spots and trends in related fields. We answered the following questions surrounding 4D printing technology: hot spots, trends, applications, and future development; we made some new findings.

5.1. New Findings

In terms of the hotspots of 4D printing, we found that as the research on the raw materials of 4D printing technology continues to deepen, the encoding method and the manufacturing method have gradually become the focus of scholars' research direction. This is consistent with Cerbe et al. [17] research on the technical parameters of fused deposition modeling (FDM) for 4D printing. Of the 19 most cited references we created using VOSview, all of them are related to the manufacturing methods and applications of 4D printing, and these findings further support Ahmed et al. [69]. Therefore, the external stimulation, modeling, and simulation design of 4D printing technology has great development potential and may become a research hotspot in 4D printing technology in the future [17,34]. Meanwhile, we intuitively found through the co-cited reference network and hotspot clusters created by CiteSpace that "emerging applications", "4D-printed shape memory polymers", and "4D-printed active composite structures" are the three largest clusters reflecting that they are the largest research hotspots for 4D printing technology. Others, such as liquid crystal elastomers [81], geometry-driven finite elements [82], and biomedical applications [30], are also hot research areas for 4D printing. In addition, we have noticed that the research on materials, structures, applications of 4D printing, and its relationship with 3D printing has become more concentrated. Moreover, by visualizing the hotspots and emerging trends in keywords, we found that keywords such as "3D printing", "3D" and "additive manufacturing" have higher citation rates compared to other keywords through the web visualization keyword mutation map and highlighted keywords created by VOSview. At the current stage of 4D printing research, technology development and application research are gradually coming to the forefront, and the keywords of "fabrication", "design", "composites" and "scaffolds" are at the top of the list, all show that scholars are researching on advancing 4D printing technology. Finally, by using the keyword hotspot Timeline view created by CiteSpace, we analyzed that "tissue engineering application" and "shape recovery" are the most prominent hotspots in 4D printing.

In terms of research trends in 4D printing: By visualizing the hot and emerging trends in the references, we find that "#0 Emerging Applications, 2020", "#11 Loading Patterns, 2022" and "#20 Robotic Applications, 2020" in the clusters represent the latest research and development directions in 4D printing technology. And more importantly, "#6 geometry-driven finite elements, 2013", "#9 controlled sequential shapes, 2011", "#10 4D nano-printing based on laser technology, 2013" and "#12 polyelectrolytes, 2012" used to be the hot areas of 4D printing technology. However, few scholars have paid attention to them in recent years, and the related research has shown a contraction, which reflects that the development potential of these themes in 4D printing research may not be large. At the current stage of 4D printing research, technology development and application research are gradually emerging, and the keywords of "fabrication", "design", "composites" and "scaffolds" in the top rankings all show that scholars are interested in advancing 4D printing technology, which is in line with Bajpai et al. [62] and Kabirian et al. [6]. Therefore, we infer that the application prospects and technological development around 4D printing technology will be a very promising area of research. Meanwhile, we analyzed the keyword hotspot network and highlighted keywords created by CiteSpace, and the keywords of 4D printing technology cover many fields such as engineering and manufacturing, biomedical, lifestyle products, food, and cultural and creative industries. Cluster "#0 tissue engineering application" is still the most researched field, which is the most basic and important content in 4D printing research. Cluster "#1 shape recovery" and "#2 liquid crystal elastomer", as the basic research on the characteristics and materials of 4D printing, are also extremely important in the overall research field. Cluster "#4 future trend" has become an important area in the keyword research hotspot of 4D printing technology, indicating that the technical application prospect of 4D printing has also been gradually emphasized by scholars. In addition, the medical-related clusters in the formed keyword clusters account for a relatively large proportion and rank high, such as "#3 bone tissue engineering", "#5 laser powder bed fusion", "#7 cervical spine" and so on. Some emerging fields such as "#6 4D food printing" and "#9 aesthetic planning" are also gaining attention.

In terms of the current status of 4D printing applications: One hand, by analyzing the application prospects of 4D printing technology in the four fields of engineering manufacturing, biomedicine, food printing, and cultural and creative life, we found that 4D printing is a very disruptive emerging manufacturing technology because of its adaptability and proactive response characteristics, showing great application potential in these fields. On the other hand, there is still a large space for research on the technology, materials, and design of 4D printing, which limits the prospect of its application in existing fields and its expansion to more fields [12,27,30].

In terms of the future development of 4D printing, over time, the initial hotspots of 4D printing, such as "shape memory polymers" and "composites" have gradually evolved into "poly-lactic acid" and "composites". "Poly-lactic acid", "cellulose", "regenerative medicine" and "5D printing" have become emerging hotspots. "5D printing", "stability" and "implants" may become emerging trends in the future. Moreover, combining the time evolution of 4D printing keywords hotspots, we infer that: first, the original 4D printing technology is affected by the needs of different fields, and its technological development has gradually begun to diversify under the guidance of demand. Secondly, 4D printing has become more frequent and in-depth in cross-disciplinary and cross-field research after 2015. Thirdly, the hot keywords of 4D printing technology in recent years are more diversified and multidisciplinary.

5.2. Theoretical and Practical Significance

By collecting research results on 4D printing from databases SCI, SSCI, AHCI, CPCI-S, CPCI-SSH, ESCI, CCR-EXPANDED and IC in the core collection of Web of Science and visualizing and analyzing them by using VOSviewer and CiteSpace software, we can provide readers with a more intuitive and comprehensive understanding of the research hotspots and emerging trends in 4D printing rather than simply analyzing the research results of 4D printing technology in a particular field [8,34,75]. This facilitates scholars who wish to conduct research in this field. For example, by analyzing the prospects of 4D printing applications, we can enable scholars in the field of biomedicine to focus on bone tissue engineering [34] and further minimally invasive surgery [8,65], drug delivery [66], tissue engineering [15], skin reconstruction [67], and other emerging fields. Moreover, by using scientometrics to visualize and analyze 4D printing, the development direction of 4D printing can be clarified, which leads to the conclusion that the application prospects and technological development will be the future promising research focuses of 4D printing, which is in line with other scholars' research on 4D printing [7,17,35] and provides references for other scholars' subsequent research. In addition, we obtained cluster maps and landscape maps by visualizing and analyzing the hotspots and emerging trends in references and keywords related to 4D printing technology. For the first time, many hotspots and emerging trends were identified through co-cited reference networks and exploded keywords, which fills the gap in the academic community's systematic research on existing research results related to 4D printing through visualization. The relevant conclusions can not only further enrich the related theories of 4D printing but also provide new insights for the development of 4D printing.

5.3. Limitations and Future Research

Although we used multiple perspectives and multiple indicators to obtain more comprehensive and accurate conclusions, there is no guarantee that all the results will be obtained. Therefore, there may be omissions or deviations in the conclusions regarding the research focuses and emerging trends of 4D printing. In addition, although the database we have chosen well covers the current high-quality literature around the world, there may still be some important studies that have been missed because they are not in these databases, which will also have a certain impact on our research. In summary, we will further optimize the relevant parameters in subsequent research, expand the data range, and continuously improve the quality of the research.

Author Contributions: Conceptualization, J.L.; Data curation, J.L.; Formal analysis, J.L.; Funding acquisition, Y.J.; Investigation, J.L.; Methodology, Y.J. and J.L.; Project administration, Y.J. and J.L.; Resources, Y.J.; Software, Y.J. and J.L.; Supervision, Y.J.; Validation, J.L.; Visualization, J.L.; Writing—original draft, J.L.; Writing—review and editing, Y.J. and J.L. All authors have read and agreed to the published version of the manuscript.

Funding: This work was supported by the 2024 Special Funding for Basic Research Operating Expenses of Universities in Liaoning Province (No: LJ122410146071) and the Social Science Foundation of Liaoning Province (No: L22BJY040).

Institutional Review Board Statement: Not applicable.

Informed Consent Statement: Not applicable.

Data Availability Statement: The raw data supporting the conclusions of this article will be made available by the authors upon request.

Acknowledgments: We sincerely thank all participants for their support of this article.

Conflicts of Interest: The authors declare no conflicts of interest.

References

1. Wong, K.V.; Hernandez, A. A review of additive manufacturing. *Int. Sch. Res. Not.* **2012**, *2012*, 208760. [CrossRef]
2. Abdulhameed, O.; Al-Ahmari, A.; Ameen, W.; Mian, S.H. Additive manufacturing: Challenges, trends, and applications. *Adv. Mech. Eng.* **2019**, *11*, 1687814018822880. [CrossRef]
3. Gardan, J. Additive manufacturing technologies: State of the art and trends. In *Additive Manufacturing Handbook*; Springer: Cham, Switzerland, 2017; pp. 149–168.
4. Herzog, D.; Seyda, V.; Wycisk, E.; Emmelmann, C. Additive manufacturing of metals. *Acta Mater.* **2016**, *117*, 371–392. [CrossRef]
5. Boonhaijaroen, N.; Sitthi-amorn, P.; Srituravanich, W.; Suanpong, K.; Ekgasit, S.; Pengprecha, S. Alignment Control of Ferrite-Decorated Nanocarbon Material for 3D Printing. *Micromachines* **2024**, *15*, 763. [CrossRef] [PubMed]
6. Kabirian, F.; Mela, P.; Heying, R. 4D printing applications in the development of smart cardiovascular implants. *Front. Bioeng. Biotechnol.* **2022**, *10*, 873453. [CrossRef]
7. Khalid, M.Y.; Arif, Z.U.; Ahmed, W. 4D printing: Technological and manufacturing renaissance. *Macromol. Mater. Eng.* **2022**, *307*, 2200003. [CrossRef]
8. Langford, T.; Mohammed, A.; Essa, K.; Elshaer, A.; Hassanin, H. 4D printing of origami structures for minimally invasive surgeries using functional scaffold. *Appl. Sci.* **2020**, *11*, 332. [CrossRef]
9. Aldawood, F.K. A comprehensive review of 4D printing: State of the arts, opportunities, and challenges. *Actuators* **2023**, *12*, 101. [CrossRef]
10. Ali, M.H.; Abilgaziyev, A.; Adair, D. 4D printing: A critical review of current developments, and future prospects. *Int. J. Adv. Manuf. Technol.* **2019**, *105*, 701–717. [CrossRef]
11. Biswas, M.C.; Chakraborty, S.; Bhattacharjee, A.; Mohammed, Z. 4D printing of shape memory materials for textiles: Mechanism, mathematical modeling, and challenges. *Adv. Funct. Mater.* **2021**, *31*, 2100257. [CrossRef]
12. Khalid, M.Y.; Arif, Z.U.; Ahmed, W.; Umer, R.; Zolfagharian, A.; Bodaghi, M. 4D printing: Technological developments in robotics applications. *Sens. Actuators A Phys.* **2022**, *343*, 113670. [CrossRef]
13. Navaf, M.; Sunooj, K.V.; Aaliya, B.; Akhila, P.P.; Sudheesh, C.; Mir, S.A.; George, J. 4D printing: A new approach for food printing; effect of various stimuli on 4D printed food properties. A comprehensive review. *Appl. Food Res.* **2022**, *2*, 100150. [CrossRef]
14. Raina, A.; Haq, M.I.U.; Javaid, M.; Rab, S.; Haleem, A. 4D printing for automotive industry applications. *J. Inst. Eng. Ser. D* **2021**, *102*, 521–529. [CrossRef]
15. Jeong, H.Y.; Lee, E.; An, S.-C.; Lim, Y.; Jun, Y.C. 3D and 4D printing for optics and metaphotonics. *Nanophotonics* **2020**, *9*, 1139–1160. [CrossRef]
16. Khalid, M.Y.; Arif, Z.U.; Noroozi, R.; Zolfagharian, A.; Bodaghi, M. 4D printing of shape memory polymer composites: A review on fabrication techniques, applications, and future perspectives. *J. Manuf. Process.* **2022**, *81*, 759–797. [CrossRef]
17. Cerbe, F.; Mahlstedt, D.; Sinapius, M.; Hühne, C.; Böl, M. Relationship between programming stress and residual strain in FDM 4D printing. *Prog. Addit. Manufac.* **2024**, *9*, 123–132. [CrossRef]
18. Yang, Y.; Zhang, R.; Zhao, S.; Deng, Y.; Yu, Q.; Zeltmann, S.; Yin, S.; Ciston, J.; Ophus, C.; Asta, M.; et al. In situ observations and measurements of plastic deformation, phase transformations and fracture with 4D-STEM. *Microsc. Microanal.* **2021**, *27* (Suppl. S1), 1494–1495. [CrossRef]
19. Wu, J.; Zhang, Z.; Shang, Q.; Li, F.; Hui, Y.; Fan, H. A method for investigating the springback behavior of 3D tubes. *Int. J. Mech. Sci.* **2017**, *131*, 191–204. [CrossRef]
20. Bodaghi, M.; Damanpack, A.; Liao, W. Self-expanding/shrinking structures by 4D printing. *Smart Mater. Struct.* **2016**, *25*, 105034. [CrossRef]
21. Tang, Y.; Dai, B.; Su, B.; Shi, Y. Recent advances of 4D printing technologies toward soft tactile sensors. *Front. Mater.* **2021**, *8*, 658046. [CrossRef]
22. Zeenat, L.; Zolfagharian, A.; Sriya, Y.; Sasikumar, S.; Bodaghi, M.; Pati, F. 4D printing for vascular tissue engineering: Progress and challenges. *Adv. Mater. Technol.* **2023**, *8*, 2300200. [CrossRef]
23. Fu, P.; Li, H.; Gong, J.; Fan, Z.; Smith, A.T.; Shen, K.; Khalfalla, T.O.; Huang, H.; Qian, X.; McCutcheon, J.R.; et al. 4D printing of polymers: Techniques, materials, and prospects. *Prog. Polym. Sci.* **2022**, *126*, 101506. [CrossRef]
24. Chen, X.; Han, S.; Wu, W.; Wu, Z.; Yuan, Y.; Wu, J.; Liu, C. Harnessing 4D printing bioscaffolds for advanced orthopedics. *Small* **2022**, *18*, 2106824. [CrossRef] [PubMed]
25. Zhu, P.; Yang, W.; Wang, R.; Gao, S.; Li, B.; Li, Q. 4D printing of complex structures with a fast response time to magnetic stimulus. *ACS Appl. Mater. Interfaces* **2018**, *10*, 36435–36442. [CrossRef]
26. Mohol, S.S.; Sharma, V. Functional applications of 4D printing: A review. *Rapid Prototyp. J.* **2021**, *27*, 1501–1522. [CrossRef]
27. Manshor, M.R.; Alli, Y.A.; Anuar, H.; Ejeromedoghene, O.; Omotola, E.O.; Suhr, J. 4D printing: Historical evolution, computational insights and emerging applications. *Mater. Sci. Eng. B* **2023**, *295*, 116567. [CrossRef]

28. Cui, H.; Zhu, W.; Miao, S.; Sarkar, K.; Zhang, L.G. 4D Printed Nerve Conduit with In Situ Neurogenic Guidance for Nerve Regeneration. *Tissue Eng. Part A* **2023**, *30*, 293–303. [CrossRef]
29. Maraveas, C.; Bayer, I.S.; Bartzanas, T. 4D printing: Perspectives for the production of sustainable plastics for agriculture. *Biotechnol. Adv.* **2022**, *54*, 107785. [CrossRef]
30. Sahafnejad-Mohammadi, I.; Karamimoghadam, M.; Zolfagharian, A.; Akrami, M.; Bodaghi, M. 4D printing technology in medical engineering: A narrative review. *J. Braz. Soc. Mech. Sci. Eng.* **2022**, *44*, 233. [CrossRef]
31. Brown, R.J. Measuring measurement–What is metrology and why does it matter? *Measurement* **2021**, *168*, 108408. [CrossRef]
32. Plant, A.; Hanisch, R. Reproducibility in science: A metrology perspective. *Harv. Data Sci. Rev.* **2020**, *2*. [CrossRef]
33. Tsimidou, M.Z.; Ordoudi, S.A.; Mantzouridou, F.T.; Nenadis, N.; Stelzl, T.; Rychlik, M.; Belc, N.; Zoani, C. Strategic priorities of the scientific plan of the European Research Infrastructure METROFOOD-RI for promoting metrology in food and nutrition. *Foods* **2022**, *11*, 599. [CrossRef] [PubMed]
34. You, D.; Chen, G.; Liu, C.; Ye, X.; Wang, S.; Dong, M.; Sun, M.; He, J.; Yu, X.; Ye, G.; et al. 4D printing of multi-responsive membrane for accelerated in vivo bone healing via remote regulation of stem cell fate. *Adv. Funct. Mater.* **2021**, *31*, 2103920. [CrossRef]
35. Teng, X.; Zhang, M.; Mujumdar, A.S. 4D printing: Recent advances and proposals in the food sector. *Trends Food Sci. Technol.* **2021**, *110*, 349–363. [CrossRef]
36. Xie, L.; Chen, Z.; Wang, H.; Zheng, C.; Jiang, J. Bibliometric and visualized analysis of scientific publications on atlantoaxial spine surgery based on Web of Science and VOSviewer. *World Neurosurg.* **2020**, *137*, 435–442.e4. [CrossRef]
37. Xu, S.; Zhang, X.; Feng, L.; Yang, W. Disruption risks in supply chain management: A literature review based on bibliometric analysis. *Int. J. Prod. Res.* **2020**, *58*, 3508–3526. [CrossRef]
38. Chen, C.; Hu, Z.; Liu, S.; Tseng, H. Emerging trends in regenerative medicine: A scientometric analysis in CiteSpace. *Expert Opin. Biol. Ther.* **2012**, *12*, 593–608. [CrossRef]
39. Chen, X.; Liu, Y. Visualization analysis of high-speed railway research based on CiteSpace. *Transp. Policy* **2020**, *85*, 1–17. [CrossRef]
40. Wei, F.; Grubesic, T.H.; Bishop, B.W. Exploring the GIS knowledge domain using CiteSpace. *Prof. Geograph.* **2015**, *67*, 374–384. [CrossRef]
41. Zhang, D.; Xu, J.; Zhang, Y.; Wang, J.; He, S.; Zhou, X. Study on sustainable urbanization literature based on Web of Science, scopus, and China national knowledge infrastructure: A scientometric analysis in CiteSpace. *J. Clean. Prod.* **2020**, *264*, 121537. [CrossRef]
42. Sood, S.K.; Rawat, K.S.; Kumar, D. Analytical mapping of information and communication technology in emerging infectious diseases using CiteSpace. *Telemat. Inform.* **2022**, *69*, 101796. [CrossRef] [PubMed]
43. Sydney Gladman, A.; Matsumoto, E.; Nuzzo, R.; Mahadevan, L.; Lewis, J.A. Biomimetic 4D printing. *Nat. Mater.* **2016**, *15*, 413–418. [CrossRef]
44. Taylor, D.L.; in het Panhuis, M. Self-healing hydrogels. *Adv. Mater.* **2016**, *28*, 9060–9093. [CrossRef]
45. Tibbits, S. 4D printing: Multi-material shape change. *Archit. Des.* **2014**, *84*, 116–121. [CrossRef]
46. Ge, Q.; Sakhaei, A.H.; Lee, H.; Dunn, C.K.; Fang, N.X.; Dunn, M.L. Multimaterial 4D printing with tailorable shape memory polymers. *Sci. Rep.* **2016**, *6*, 31110. [CrossRef]
47. Parandoush, P.; Lin, D. A review on additive manufacturing of polymer-fiber composites. *Compos. Struct.* **2017**, *182*, 36–53. [CrossRef]
48. Momeni, F.; Liu, X.; Ni, J. A review of 4D printing. *Mater. Des.* **2017**, *122*, 42–79. [CrossRef]
49. Huang, J.; Xia, S.; Li, Z.; Wu, X.; Ren, J. Applications of four-dimensional printing in emerging directions: Review and prospects. *J. Mater. Sci. Technol.* **2021**, *91*, 105–120. [CrossRef]
50. Kuang, X.; Roach, D.J.; Wu, J.; Hamel, C.M.; Ding, Z.; Wang, T. Advances in 4D printing: Materials and applications. *Adv. Funct. Mater* **2019**, *29*, 1805290. [CrossRef]
51. Luo, F.; Sun, T.L.; Nakajima, T.; Kurokawa, T.; Ihsan, A.B.; Li, X. Free reprocessability of tough and self-healing hydrogels based on polyion complex. *ACS Macro Lett.* **2015**, *4*, 961–964. [CrossRef]
52. Ge, Q.; Qi, H.J.; Dunn, M.L. Active materials by four-dimension printing. *Appl. Phys. Lett.* **2013**, *103*, 131901. [CrossRef]
53. Ge, Q.; Dunn, C.K.; Qi, H.J.; Dunn, M.L. Active origami by 4D printing. *Smart Mater. Struct.* **2014**, *23*, 094007. [CrossRef]
54. Li, Y.; Zhang, F.; Liu, Y.; Leng, J. 4D printed shape memory polymers and their structures for biomedical applications. *Sci. China Technol. Sci.* **2020**, *63*, 545–560. [CrossRef]
55. Ding, Z.; Yuan, C.; Peng, X.; Wang, T.; Qi, H.J.; Dunn, M.L. Direct 4D printing via active composite materials. *Sci. Adv.* **2017**, *3*, e1602890. [CrossRef]
56. Gao, B.; Yang, Q.; Zhao, X.; Jin, G.; Ma, Y.; Xu, F. 4D bioprinting for biomedical applications. *Trends Biotechnol.* **2016**, *34*, 746–756. [CrossRef]
57. Bakarich, S.E.; Gorkin, R.; Spinks, G.M. 4D Printing with Mechanically Robust, Thermally Actuating Hydrogels. *Macromol. Rapid Commun.* **2015**, *36*, 1211–1217. [CrossRef]
58. Pan, Y.; Zhang, L. Roles of artificial intelligence in construction engineering and management: A critical review and future trends. *Autom. Constr.* **2021**, *122*, 103517. [CrossRef]
59. Kuang, X.; Chen, K.; Dunn, C.K.; Wu, J.; Li, V.C.; Qi, H.J. 3D printing of highly stretchable, shape-memory, and self-healing elastomer toward novel 4D printing. *ACS Appl. Mater. Interfaces* **2018**, *10*, 7381–7388. [CrossRef]

60. Vijayavenkataraman, S.; Yan, W.-C.; Lu, W.F.; Wang, C.-H.; Fuh, J.Y.H. 3D bioprinting of tissues and organs for regenerative medicine. *Adv. Drug Deliv. Rev.* **2018**, *132*, 296–332. [CrossRef]
61. Ambulo, C.P.; Burroughs, J.J.; Boothby, J.M.; Kim, H.; Shankar, M.R.; Ware, T.H. Four-dimensional printing of liquid crystal elastomers. *ACS Appl. Mater. Interfaces* **2017**, *9*, 37332–37339. [CrossRef]
62. Bajpai, A.; Baigent, A.; Raghav, S.; Brádaigh, C.Ó.; Koutsos, V.; Radacsi, N. 4D printing: Materials, technologies, and future applications in the biomedical field. *Sustainability* **2020**, *12*, 10628. [CrossRef]
63. Chen, Y.; Zhang, M.; Sun, Y.; Phuhongsung, P. Improving 3D/4D printing characteristics of natural food gels by novel additives: A review. *Food Hydrocoll.* **2022**, *123*, 107160. [CrossRef]
64. Haleem, A.; Javaid, M.; Singh, R.P.; Suman, R. Significant roles of 4D printing using smart materials in the field of manufacturing. *Adv. Ind. Eng. Polym. Res.* **2021**, *4*, 301–311. [CrossRef]
65. Subeshan, B.; Baddam, Y.; Asmatulu, E. Current progress of 4D-printing technology. *Prog. Addit. Manufac.* **2021**, *6*, 495–516. [CrossRef]
66. Ntouanoglou, K.; Stavropoulos, P.; Mourtzis, D. 4D printing prospects for the aerospace industry: A critical review. *Procedia Manuf.* **2018**, *18*, 120–129. [CrossRef]
67. Shao, L.-H.; Zhao, B.; Zhang, Q.; Xing, Y.; Zhang, K. 4D printing composite with electrically controlled local deformation. *Extrem. Mech. Lett.* **2020**, *39*, 100793. [CrossRef]
68. Zhai, F.; Feng, Y.; Li, Z.; Xie, Y.; Ge, J.; Wang, H.; Qiu, W.; Feng, W. 4D-printed untethered self-propelling soft robot with tactile perception: Rolling, racing, and exploring. *Matter* **2021**, *4*, 3313–3326. [CrossRef]
69. Ahmed, A.; Arya, S.; Gupta, V.; Furukawa, H.; Khosla, A. 4D printing: Fundamentals, materials, applications and challenges. *Polymer* **2021**, *228*, 123926. [CrossRef]
70. Zarek, M.; Mansour, N.; Shapira, S.; Cohn, D. 4D printing of shape memory-based personalized endoluminal medical devices. *Macromol. Rapid Commun.* **2017**, *38*, 1600628. [CrossRef]
71. Wang, Y.; Cui, H.; Esworthy, T.; Mei, D.; Wang, Y.; Zhang, L.G. Emerging 4D printing strategies for next-generation tissue regeneration and medical devices. *Adv. Mater.* **2022**, *34*, 2109198. [CrossRef]
72. Vinnakota, M.; Bellur, K.; Starnes, S.L.; Schulz, M.J. Scaling a Hydraulic Motor for Minimally Invasive Medical Devices. *Micromachines* **2024**, *15*, 131. [CrossRef] [PubMed]
73. Osouli-Bostanabad, K.; Masalehdan, T.; Kapsa, R.M.; Quigley, A.; Lalatsa, A.; Bruggeman, K.F.; Franks, S.J.; Williams, R.J.; Nisbet, D.R. Traction of 3D and 4D printing in the healthcare industry: From drug delivery and analysis to regenerative medicine. *ACS Biomater. Sci. Eng.* **2022**, *8*, 2764–2797. [CrossRef] [PubMed]
74. Kamolz, L.-P.; Kotzbeck, P.; Schintler, M.; Spendel, S. Skin regeneration, repair, and reconstruction: Present and future. *Eur. Surg.* **2022**, *54*, 163–169. [CrossRef]
75. Ghazal, A.F.; Zhang, M.; Liu, Z. Spontaneous color change of 3D printed healthy food product over time after printing as a novel application for 4D food printing. *Food Bioprocess Technol.* **2019**, *12*, 1627–1645. [CrossRef]
76. Farid, M.I.; Wu, W.; Liu, X.; Wang, P. Additive manufacturing landscape and materials perspective in 4D printing. *Int. J. Adv. Manuf. Technol.* **2021**, *115*, 2973–2988. [CrossRef]
77. Manaia, J.P.; Cerejo, F.; Duarte, J. Revolutionising textile manufacturing: A comprehensive review on 3D and 4D printing technologies. *Fash. Text.* **2023**, *10*, 20. [CrossRef]
78. Leist, S.K.; Gao, D.; Chiou, R.; Zhou, J. Investigating the shape memory properties of 4D printed polylactic acid (PLA) and the concept of 4D printing onto nylon fabrics for the creation of smart textiles. *Virtual Phys. Prototyp.* **2017**, *12*, 290–300. [CrossRef]
79. Franco Urquiza, E.A. Advances in Additive Manufacturing of Polymer-Fused Deposition Modeling on Textiles: From 3D Printing to Innovative 4D Printing—A Review. *Polymers* **2024**, *16*, 700. [CrossRef]
80. Mallakpour, S.; Tabesh, F.; Hussain, C.M. 3D and 4D printing: From innovation to evolution. *Adv. Colloid Interface Sci.* **2021**, *294*, 102482. [CrossRef]
81. Ambulo, C.P.; Ford, M.J.; Searles, K.; Majidi, C.; Ware, T.H. 4D-Printable liquid metal–liquid crystal elastomer composites. *ACS Appl. Mater. Interfaces* **2020**, *13*, 12805–12813. [CrossRef]
82. Zolfagharian, A.; Durran, L.; Gharaie, S.; Rolfe, B.; Kaynak, A.; Bodaghi, M. 4D printing soft robots guided by machine learning and finite element models. *Sens. Actuators A Phys.* **2021**, *328*, 112774. [CrossRef]

designs

MDPI

Article

Hybrid Multi-Criteria Decision Making for Additive or Conventional Process Selection in the Preliminary Design Phase

Alessandro Salmi, Giuseppe Vecchi *, Eleonora Atzeni and Luca Iuliano

Department of Management and Production Engineering (DIGEP), Politecnico di Torino, Corso Duca degli Abruzzi 24, 10129 Torino, Italy; alessandro.salmi@polito.it (A.S.); eleonora.atzeni@polito.it (E.A.); luca.iuliano@polito.it (L.I.)

* Correspondence: giuseppe_vecchi@polito.it; Tel.: +39-011-090-7263

Abstract: Additive manufacturing (AM) has become a key topic in the manufacturing industry, challenging conventional techniques. However, AM has its limitations, and understanding its convenience despite established processes remains sometimes difficult, especially in preliminary design phases. This investigation provides a hybrid multi-criteria decision-making method (MCDM) for comparing AM and conventional processes. The MCDM method consists of the Best Worst Method (BWM) for the definition of criteria weights and the Proximity Index Value (PIV) method for the generation of the final ranking. The BWM reduces the number of pairwise comparisons required for the definition of criteria weights, whereas the PIV method minimizes the probability of rank reversal, thereby enhancing the robustness of the results. The methodology was validated through a case study, an aerospace bracket. The candidate processes for the bracket production were CNC machining, high-pressure die casting, and PBF-LB/M. The production of the bracket by AM was found to be the optimal choice for small to medium production batches. Additionally, the study emphasized the significance of material selection, process design guidelines, and production batch in the context of informed process selection, thereby enabling technical professionals without a strong AM background in pursuing conscious decisions.

Keywords: additive manufacturing; DfAM; PBF-LB; CNC machining; HPDC; hybrid MCDM; BWM; PIV

1. Introduction

The activity of producing a component is a crucial step in the manufacturing workflow, beginning with the conception of an idea and culminating with its realization [1]. As concerns manufacturing, it has traditionally been divided into *mass conserving* and *mass reducing* processes, depending on whether they retain the initial provided mass or not [1]. Nowadays, these categories should be expanded to include *mass increasing* processes, typical of the Additive Manufacturing (AM) industry. AM fabricates parts by adding material layer by layer until the final desired shape is met [2]. AM originated in the late 1980s as Rapid Prototyping (RP), primarily concerning the fast production of polymeric prototypes. Over the decades, it has evolved into an actual manufacturing process able to produce market-ready metallic parts [3]. The AM family of manufacturing processes can overcome many constraints of conventional manufacturing (CM) processes that have long limited designers' concepts [4]. The most critical of these limits are the need for specific tools for each manufacturing step, the cost of a part being strictly dependent on its geometrical complexity [5], and the need for many sequential processes to achieve the net shape of a component [6]. However, designers should be aware that new possibilities also bring new constraints and limitations. AM systems are strongly limited by the scarcity of dedicated materials, modest working volumes, and prolonged fabrication times [7]. Additionally, AM processes cannot provide the same quality ensured by machining operations in terms of dimensional tolerances, geometrical tolerances, and surface roughness [8]. Although AM processes have been previously proposed as holding several competitive advantages over conventional ones, it is not straightforward to decide if a component should or should not be realized by AM, and which AM process to consider [9].

Each manufacturing process requires tailored design considerations. Therefore, it is straightforward that the manufacturing process should be uniquely defined during the design phase to be fed with an appropriately shaped component. Understanding which is the most suitable manufacturing process for the production of a component is still a demanding activity, requiring high level knowledge by the operator in charge. A powerful tool supporting the process selection is represented by Multiple-Criteria Decision Making (MCDM) methods [10], enabling the comparison of different conflicting criteria coming from different fields [11]. Currently, several methods have been already profitably used in MCDM field, such as the Analytic Hierarchy Process (AHP), Technique for Order of Preference by Similarity to Ideal Solution (TOPSIS), and VIKOR methods [12], whereas new MCDM methods such as the Best Worst Method (BWM) and the Proximity Index Value (PIV) are receiving increased attention [13,14].

MCDM methods have been successfully implemented in the AM industry for various objectives, including material selection [15], part design selection [16,17], and part orientation [18]. Moreover, in the AM field, MCDM methods have been extensively utilized for selecting the most suitable AM process. Mançanares et al. [19] proposed a two-step procedure to identify the most suitable AM process based on the requirements of the part. The manufacturability of the component was evaluated based on its size and material, followed by an AHP process selection step which provided the final ranking of AM processes under investigation. Similarly, Liu et al. [20] assessed the manufacturability of the selected component using AM processes, only considering the functional specifications of the part. Subsequently, the remaining AM processes were ranked from the most suitable to the least using the AHP method. Zaman et al. [21] applied the AHP method to define the best solution for producing an aerospace component, considering AM materials, AM processes, and AM machine systems. Ghaleb et al. [22] conducted a comparative analysis on the behavior of the AHP, TOPSIS, and VIKOR methods to assess the best manufacturing process for the production of a hydraulic pump casing. The study directly compared casting and AM processes, representing the first study in which these two manufacturing paradigms were directly compared.

Furthermore, the proposal of hybrid MCDM methods has significantly increased the reliability of the results obtained. Different MCDM methods can successfully cover various phases of the process selection framework, leveraging their strengths and minimizing their weaknesses at the same time. Wang et al. [23] developed a hybrid process selection method to compare different polymeric AM processes. The AHP method was used to weight the considered criteria, and the TOPSIS method was used to compile the final ranking. Wang et al. [24] used a nonlinear fuzzy geometric mean (FGM) approach followed by a fuzzy VIKOR to evaluate the best AM system for the production of an aircraft component, choosing between fused deposition modeling (FDM), PBF-LB, and MultiJet Fusion. Grachev et al. [25] assembled a hybrid AHP-TOPSIS method for material selection in AM dental applications. Finally, Raigar et al. [26] employed a hybrid BWM-PIV method to identify the most appropriate AM machine for a given component. The authors compared various polymeric AM processes, such as vat photopolymerization, material extrusion, and material jetting, with metal AM processes, specifically powder bed fusion. The methodology proposed was evaluated on the case study of a conceptual model of spur gear.

Although a clear interest of AM shareholders is demonstrated by the reported studies, no hybrid MCDM methods have been applied to compare AM processes to conventional ones, limiting the investigation to the only AM environment. Most of published investigations have yielded helpful results by means of largely established MCDM methods, AHP, TOPSIS, and VIKOR above all. Most recent MCDM, such as the BWM and the PIV method, have been underutilized and never applied to compare AM processes to conventional ones. BWM is claimed to reduce the number of pairwise comparisons between considered elements, increasing the reliability of the results. PIV might be of great interest in the field of process selection as it claims to minimise the vulnerability of the proposed ranking to the rank reversal phenomenon.

This paper confidently answers a common question every company faces when first considering AM, namely "Can this component be produced by AM, and is it advantageous to do so?". We suggest that a hybrid MCMD method could be used to compare AM with CM processes, expanding its application to a broader range of technologies. Section 2 presents the adopted methodology. The chosen hybrid MCDM method consists of a first linearized BWM method to define attribute weights and PIV method to rank the processes. The BWM guarantees the minimal number of pairwise comparisons during the definition of criteria weights, thereby simplifying the procedure. Furthermore, the PIV method is designed to mitigate the rank reversal problem, thereby ensuring a more robust outcome at the conclusion of the procedure. The resilience of the PIV method to rank reversal is of paramount importance in the proposed methodology, as it accounts for the potential introduction or removal of manufacturing processes during the evaluation, which could occur in a real industrial setting. Finally, an inspiring topology optimization (TO) phase is also proposed for improving the design of the component, able to improve its suitability in the AM scenario. Section 3 presents a case study coming from the aerospace sector to demonstrate the applicability of the proposed methodology in a real scenario. Finally, Section 4 draws the conclusions of the study, emphasizing the most relevant findings.

2. Materials and Methods

The proposed framework is intended to empower industrial figures, without a strong AM background, in evaluating the suitability and convenience of AM processes for the production of a given part out of additive and conventional manufacturing processes. The proposed hybrid MCDM method can easily identify the issues associated with the component at an early stage of the design, prior to its finalization. This allows for the incorporation of modifications that could enhance its manufacturability, and therefore, allowing engineers and designers to be completely aware of process requirements even at early design stages.

An overview of the whole methodology is presented in Figure 1. At first, candidate processes are identified based on the functional specifications of the part concept. Both conventional manufacturing processes and AM processes are considered. Subsequently, in the process exploration phase, a first screening is performed to discard unsuitable processes, then the most appropriate process is identified in the process selection phase, through the application of a MCDM method. As results, the manufacturability by AM and its convenience is established, or the AM process is rejected. Details of each phase of the methodology are presented in the following subsections.

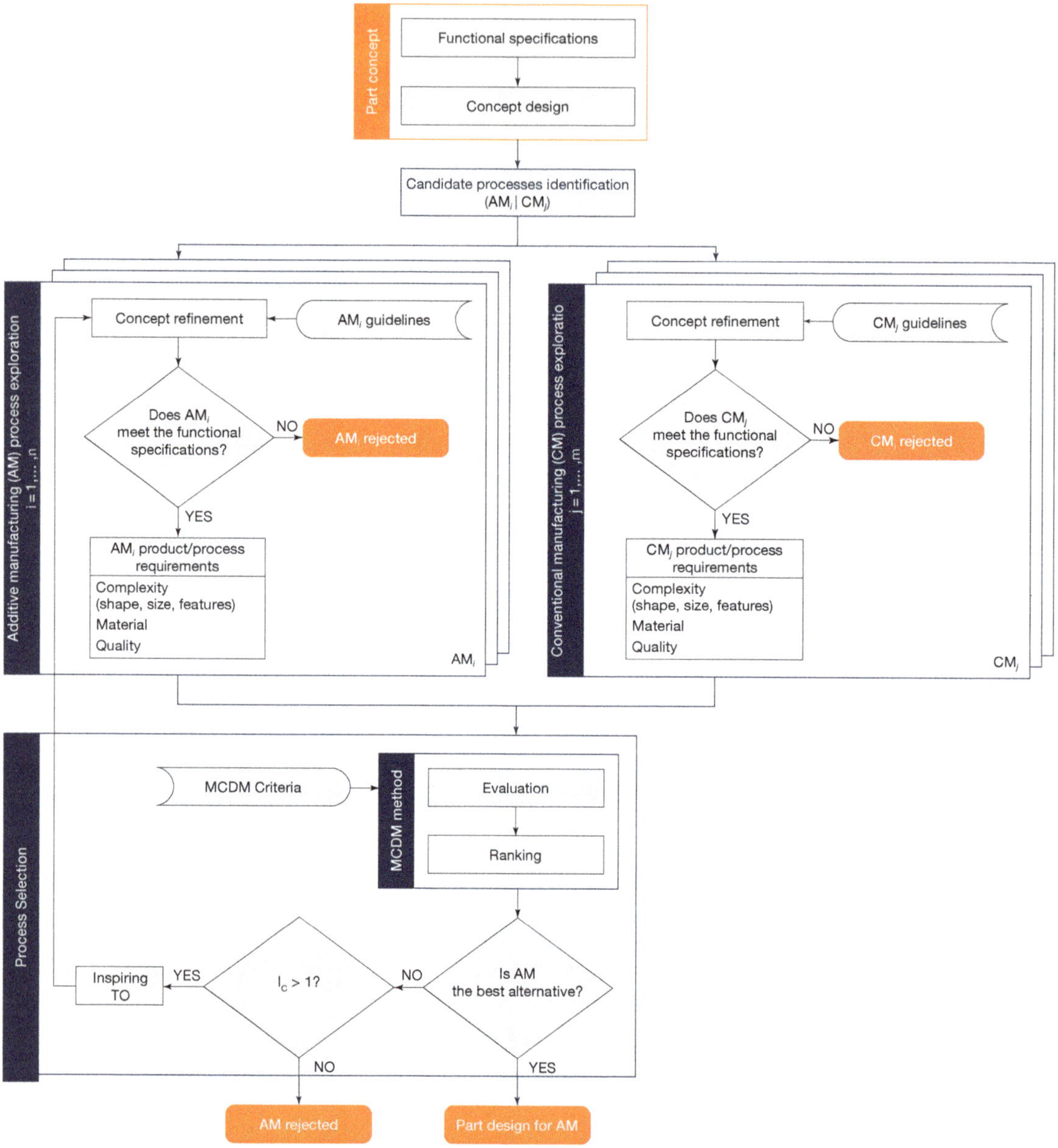

Figure 1. Methodology flowchart.

2.1. Process Exploration Phase

Once the process candidates have been identified, the initial task is to refine the concept design of the part by applying the process guidelines in order to improve its manufacturability. This is followed by verification of the consistency of the design with the functional specifications. At this stage, the use of software packages may be necessary to perform the numerical simulations required to assess if functional specifications are met. If the compliance with part functional specifications is verified, this phase leads to product/process requirements. Conversely, the process is rejected. These tasks are carried out in parallel for each candidate process. For instance, in the case of an AM process, the basic considerations in the design refinement are:

- A commercially available material can be used;
- Overall dimensions of the part fit the building volume (to avoid assembly operations);
- The minimum wall thickness can be achieved;
- The process tolerances meet the required tolerances, or tolerances can be achieved with post-processing operations.

It is possible that some modifications may be required at this stage. Minor details may be altered or a nonprocessable material may be replaced with a similar one, thereby enhancing the manufacturability of the part. The refined part concept is now capable of being produced by AM. However, in order for the part to be considered for AM, it must also meet the functional specifications in order to properly undergo the requisite working loads during its intended operational lifetime.

2.2. Process Selection Phase

Once the manufacturability of a component has been established for a given set of processes that have successfully completed the exploration phase, the most suitable manufacturing process must be identified. A hybrid MCDM method is employed during the process selection phase. This involves selecting criteria and then assessing the convenience of each manufacturing process based on these criteria. Specifically, when defining criteria, geometry metrics, sustainability, production time, and costs are considered. The necessity of exploiting different software packages arises also during the process selection phase. For instance, the definition of the waste material and of the energetic demand, which contribute to the aforementioned sustainability criterion, may require the utilization of specific software packages with the objective of achieving higher estimate accuracy.

The complexity of the part plays a major role in the process selection framework, especially when dealing with AM processes. Geometrical complexity is often regarded as "for free" in AM applications [27], meaning that the same machine system can be used to manufacture parts of varying geometrical complexity without, or with minimal, additional costs. In this paper, part complexity is computed based on three main parameters, as shown below.

- *Volumetric index*, which is a measure of the amount of the volume occupied by the part within a regular bounding box in which it is contained:

$$I_V = \frac{V}{V_{box}} \quad (1)$$

  where V is the volume of the part and V_{box} is the volume of the bounding box.
- *Detail index*, which measures the complexity of the part by taking into account the connected features by looking at the number of vertices and edges:

$$I_D = \frac{0.07}{\sqrt{N_v^2 + N_e^2}} \quad (2)$$

  where N_v is the number of vertices, N_e is the number of edges, and the coefficient 0.07 is the value obtained for a conical part that has one vertex and one edge. I_D is assumed equal to 1 in the limit case of a spherical part.
- *Freeform index*, which represents the complexity of the surfaces, measured in terms of the ratio of the number of freeform surfaces to the total number of surfaces (including regular surfaces):

$$I_F = 1 - \frac{N_{ff}}{N_{tot}} \quad (3)$$

  where N_{ff} is the number of freeform surfaces and N_{tot} is the total number of surfaces.

All three parameters are bounded between 0 and 1, where values close to 0 suggest a complex geometry, and values close to 1 a simple one. The complexity index (I_C) is defined as the sum of the three parameters, i.e., $I_C = I_V + I_D + I_F$. If I_C approaches 3, the geometry of the part becomes extremely simple. However, as the I_C approaches zero, the geometry becomes increasingly complex. From the perspective of sustainability, material waste is a key factor. Material waste considers all the accessory material that must be processed alongside the part, such as machining allowances, sprues, and supports. A significant increase in material waste can lead to higher operational costs and broaden production times. In addition, surface finishing, usually expressed in terms of average roughness, is relevant in ensuring high-quality parts. Low surface quality is detrimental not only for aesthetic reasons but also because it could reduce the corrosion resistance and the fatigue life of the part [28]. Finally, it is important to consider the energy required by the manufacturing process, particularly in the light of the current European GHG reduction plan [29]. The overall cost of the part should always be considered in process selection frameworks. A process that guarantees high technical performances at an enormous cost might not be convenient for all industrial sectors. Evaluating the time-to-market of a part can provide significant competitive advantages over competitors. Based on the above considerations, the criteria identified for this methodology are:

- Complexity index;
- Surface finishing;
- Material waste;
- Energy consumption;
- Time to market;
- Overall cost.

The relative weights of the aforementioned criteria are attributed by BWM, in the relatively new MCDM method proposed by Rezaei [13] in 2014. As opposed to previous MCDM methods such as the AHP method, BWM only compares alternatives with the best and worst ones, not in between them. In this manner, results reliability is improved, and number of comparisons to perform is minimized. The linear version of the BWM model Rezaei [13], easier to use and providing a unique solution, is implemented in the current study.

The final ranking of the alternative is provided by the PIV method. The PIV method is built on the pillar that the chosen option should be the one with the shortest distance from a fictitious best alternative [14]. The closeness to the best ideal solution is given by the overall proximity value computed during the process. Although this method seems close to the TOPSIS one, which is well known and established, it minimizes the problem of rank reversal, strongly undesired in engineering applications [14]. PIV method allows to remove and/or add alternative to the ranking without meaningfully altering preference order yet defined.

The final ranking allows to identify the most suitable process for fabricating the component. In the event that AM is the best solution, the designer can apply the principles of DfAM and send the component design for engineerization. Otherwise, if AM did not result in the most promising manufacturing option, and if the complexity of the part is considered relatively low (complexity index greater than 1), an additional TO step might be considered. TO could suggest meaningful design changes to enhance the suitability of the component for AM, helping the user understanding if it is worth to invest time in more complex redesign activities. The implied hypothesis, already presented, is that a complex geometry holds a higher added value, making TO an appealing alternative. AM profitability could be increased by entry-level TO tools at this stage. After TO is performed, its result is again ranked by means of the MCDM method.

3. Case Study—Bracket for Aerospace Applications

The methodology described above was applied to a case study, a bracket for aerospace applications, the geometry of which was taken from the GrabCAD open library [30], and considered as a part concept (Figure 2). The bracket is a structural component, typically produced in the AA2024 aluminum alloy by machining operations [31]. The AA2024 aluminum alloy is widely used in aircraft structures due to its high strength to weight ratio, good stiffness, and corrosion resistance [32,33]. Moreover, the same AA2024 alloy has also been largely investigated in the scientific literature, providing a comprehensive knowledge on its processability [34,35]. The four holes on the base of the bracket allow its fastening to the underlying structure using bolted connections, whereas the through hole in the upper part of the bracket accommodates a rotating shaft, as schematically depicted in Figure 2. The tolerances and functional requirements of the part were determined using the Geometric Dimensioning and Tolerancing (GD&T) system, as outlined in UNI EN-ISO 22768 [36] (Figure 3). Tolerances of the order of a hundredth of a millimeter should be reached on mating surfaces to ensure correct assembly. A production batch of 50 pieces was assumed. All bracket functional specifications are reported in Table 1.

Table 1. Functional specifications.

Specification	Value
Maximum overall dimensions	$10 \times 10 \times 10\,\text{mm}^3$
Minimum wall thickness	5 mm
Maximum surface roughness, Ra	10 µm
Tolerances on mating surfaces	0.01 mm
General tolerances	ISO 2768-mK
Maximum weight	0.5 kg
Working load	4000 N
Minimum Safety Factor	1.5
Maximum deformation (magnitude)	0.5 mm

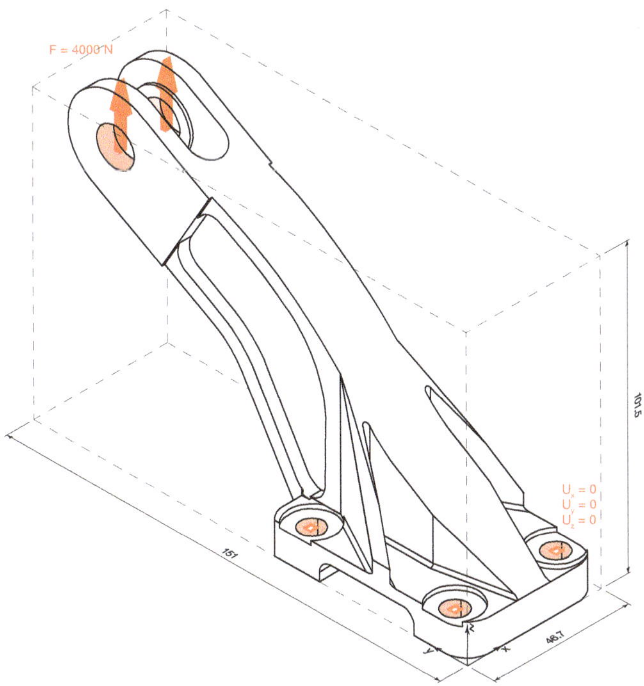

Figure 2. Isometric view of the aerospace bracket initial concept, mechanical loads and constraints highlighted. Bounding box represented as a dashed line.

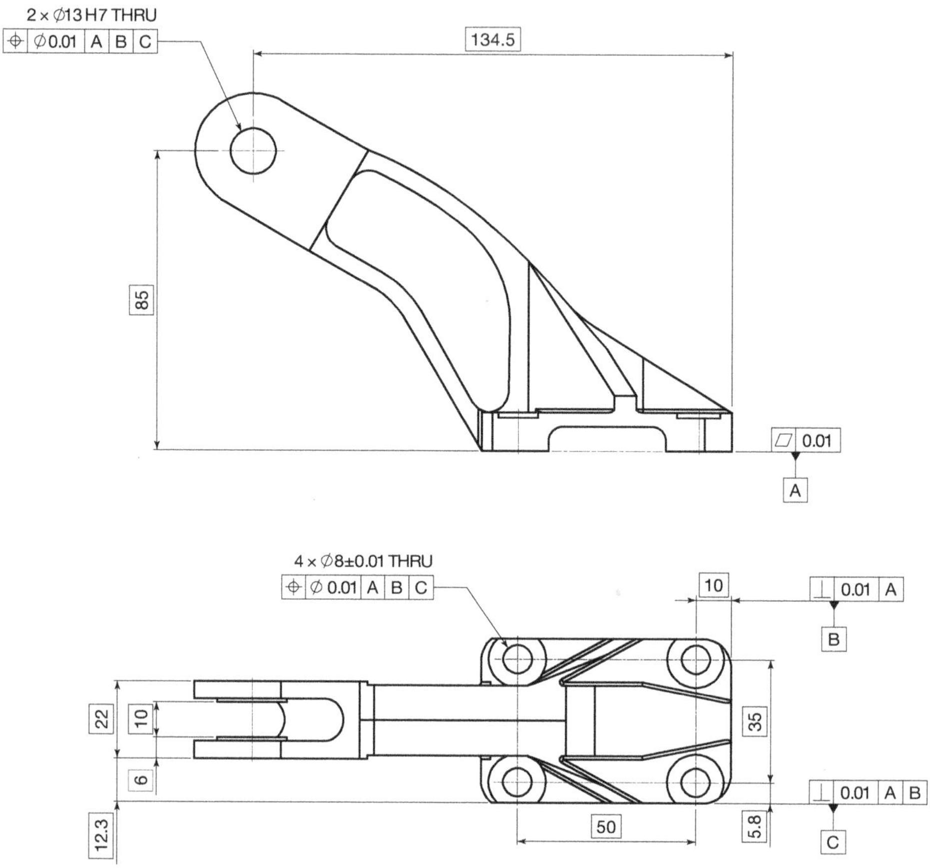

Figure 3. Dimensioned technical drawing of the aerospace bracket.

3.1. Process Exploration

In alternative to CNC machining, the traditional high pressure die casting (HPDC) process and the powder bed fusion with laser beam and metallic powder (PBF-LB/M) were selected as candidate processes. The three alternatives were compared in an MCDM framework to define the best fitting solution. It is worth noting that both PBF-LB/M and HPDC processes will require additional machining operations, to reach the desired net shape and tolerances.

3.1.1. CNC Machining Process Exploration

Nowadays, machining processes such as drilling, turning, and milling represent a common route for processing complex-shaped aluminum components [37–39]. Limiting to the current case study, the selected AA2024 aluminum alloy, the dimensions of the bracket, its minimum wall thickness, and the required surface quality do not represent an issue for part machinability, being well beneath the capability of commercial CNC machining centers. Only one enhancement was proposed to improve the machinability of the part concept, by increasing the minimum internal radius to 5 mm to avoid unnecessary finishing operations with custom tools. The refined design concept in AA2024 results in a mass of

0.260 kg, which is consistent with the specified limit. In order to evaluate the static response of the machined bracket and ascertain whether the maximum deformation is respected under the working load, Fusion 360, produced by Autodesk (San Francisco, CA, USA), was utilized. Fusion 360 was selected over other similar software packages primarily due to its relatively straightforward learning curve, which enables users to readily set up and launch structural analyses in an intuitive environment. In light of the fact that the intended user of the methodology is a technically minded individual with limited experience of computer-aided engineering (CAE), the simplicity of the software package was identified as the primary factor to be taken into account. The resulting maximum deformation of 0.49 mm was below the set threshold of 0.5 mm (Figure 4). As result of this exploration, the CNC machining was considered eligible for the process selection phase. Updates to the product and process requirements for CNC machining are limited to increasing the minimum radius, as the part concept has been verified without any material changes.

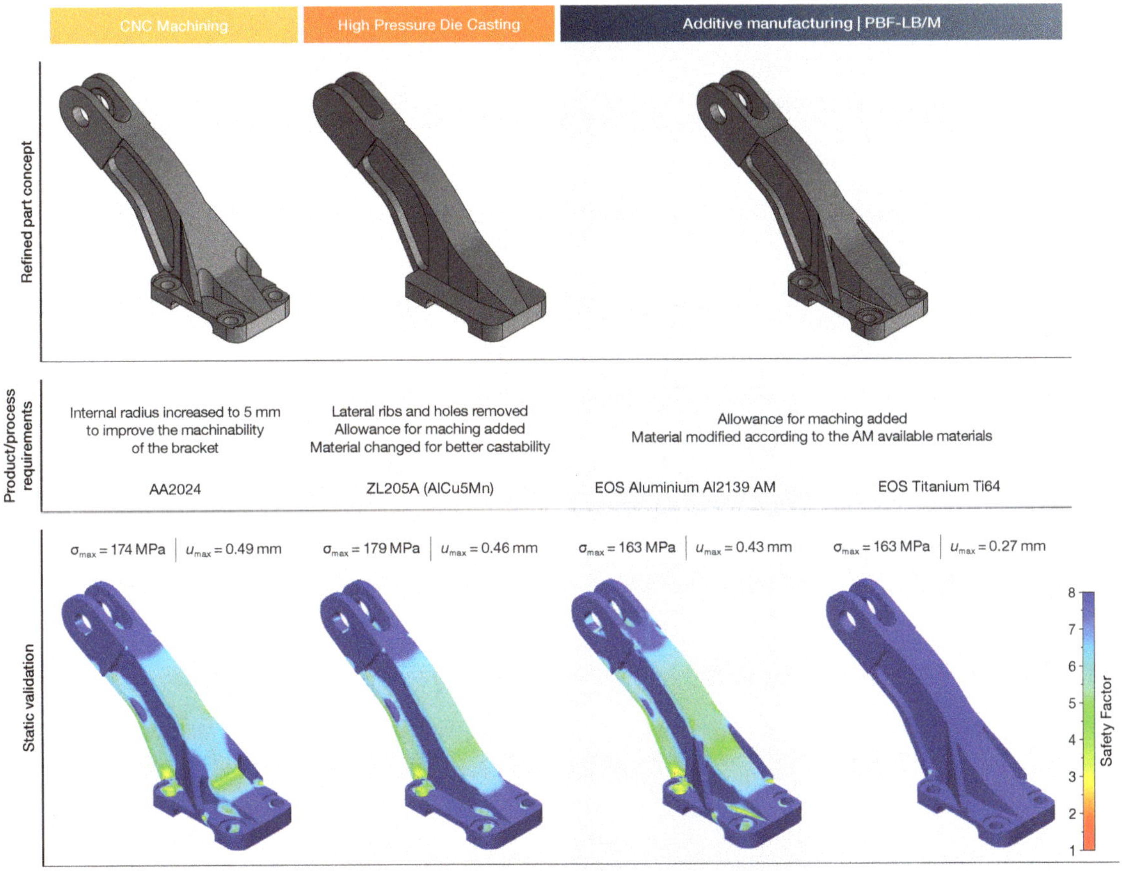

Figure 4. Concept refinements of the aerospace bracket, product requirements, and subsequent FE static validation. Colored maps refer to the Safety Factor computed during static validation. Maximum stress and maximum deformation were reported for each refined concept.

3.1.2. High Pressure Die Casting Process Exploration

High pressure die casting is a widespread manufacturing process allowing the fabrication of complex shaped components at high production rate [40]. Aluminum, zinc, and magnesium alloys are the most used materials, as excellent alloy castability is a mandatory prerequisite for a successful HPDC [40]. Although AA2024 exhibits excellent mechanical and corrosion resistance properties, it is not commonly casted, especially if complex shapes are required. Therefore, the ZL205A (AlCu5Mn) aluminum alloy was proposed as alternative material for the HPDC process. The ZL205A is an Al–Cu–Mn–Ti alloy already profitably used in casting operations for aircraft frame components [41–43]. The dimensions of the brackets were considered well inside the capabilities of HPDC systems, as well as its minimum wall thickness and surface quality. The minimum wall thickness producible by HPDC goes from 2 mm in the case of large castings to 1 mm for smaller ones [44,45]. Wall thicknesses below this threshold may hinder the material flow resulting in unfilled voids in the mold cavity. Similarly, the presence of holes in the components should be carefully considered as they could induce vorticity in molten material, preventing a correct cavity filling. In light of the HPDC guidelines here synthetically exposed, the manufacturability of the concept of the bracket was asserted. However, some elements of the bracket might be easily modified to improve its manufacturability. In particular, the lateral ribs naturally create undercuts, requiring complex mold solutions with sensible higher costs. Therefore, they were removed from the part concept to allow for an easier processing. Moreover, the holes were also removed from the design considering that they can be easily produced in the subsequent CNC finishing operations. These refinements served to reduce the complexity of the geometry, allowing the part to be realized by orienting the larger dimension normal to the die closing, with only one undercut in correspondence with the fork of the bracket. The concept refinement is shown in Figure 4, together with the FE validation for maximum deformation, which resulted in a maximum deformation of 0.46 mm, which was below the set threshold of 0.5 mm. The mass of the parts is 0.262 kg also in this case. After this exploration, the HPDC bracket was finally considered eligible for the following process selection phase.

3.1.3. Additive Manufacturing | PBF-LB/M Process Exploration

Although PBF-LB/M systems allow the manufacture of extremely complex shapes [46], some basic limitations should be considered. The range of commercially available materials for PBF-LB/M is still very limited compared to conventional manufacturing processes. The original AA2024 alloy is not commercially available for PBF-LB/M systems, so a similar aluminum alloy had to be considered. A potential challenge in the proposed material substitution is the necessity to maintain the desired product performance. In this case study, the new material must meet the same functional specifications as the original. In particular, the bracket must adhere to the maximum deformation constraint under the working load, as outlined in Table 1. Aluminum alloys are largely used in the aerospace sector due to their lightweight and good mechanical performances [47]. However, there are alternative alloys that offer an excellent strength-to-weight ratio, such as titanium alloys, which are also suitable for use in aerospace applications [48]. Therefore, EOS Aluminium Al2139 AM, a 2000 series aluminum alloy developed specifically for AM [49], was chosen for its excellent mechanical and corrosion resistance properties. In addition to the aluminum alloy, a titanium alloy was also considered to widen the range of materials considered. Ti6Al4V was chosen because of its outstanding mechanical properties and widespread use in the manufacturing and aerospace industries [50].

The volume of commercial PBF-LB/M systems limits the maximum dimensions of the parts that can be manufactured, in order to avoid subsequent assembly operations. However, the part dimensions were well below the PBF-LB/M limits, as shown in Appendix A. Similarly, the minimum wall thicknesses and overall features were considered feasible. As a rule of thumb, thin walls in PBF-LB/M should not be thinner than 1 mm to ensure their structural integrity, although recent studies have pushed the capabilities of commercial systems down to as little as 0.1 mm [51]. Finally, in addition to the simple feasibility of a part, its geometric accuracy and surface finish should also be considered, especially where tight tolerances are required. However, tolerances are not a critical factor when finishing operations follow the main manufacturing stage. In the case study analyzed, the general tolerances are compatible with the AM process, considering that the mating surfaces require the finishing step of machining. Once the main limitations of PBF-LB/M systems have been outlined, the manufacturability of the specific bracket can be asserted. In conclusion, the bracket concept of PBF-LB/M was found to be feasible without the need for design refinements, only a change in material. As previously stated, a change in material necessitates an evaluation of the performance of the product, ensuring that the specific functional requirements are fulfilled. Consequently, both brackets, the PBF-LB/Al2139 bracket and the PBF-LB/Ti6Al4V bracket, were subjected to a static verification process through numerical simulation. The PBF-LB/Al2139 bracket fulfilled the functional specifications with a maximum deformation of 0.43 mm (Figure 4) and a mass of 0.284 kg. The PBF-LB/Ti6Al4V option performed considerably better, with a maximum deformation of only 0.27 mm at a cost of a higher mass, equal to 0.444 kg.

3.2. Process Selection

Once the manufacturability of the part had been successfully stated for all the three process candidates, the MCDM method was applied. The first task was to define criteria weights using the BWM. The considered criteria are here recalled for the sake of simplicity: complexity index, surface finishing, material waste, energy required, time to market, and overall cost. A reduced time to market allows a company to gain a competitive advantage with respect to other competitors. On the other hand, as-built surface roughness was expected to have a minor impact, especially when considering the need of machining operations in all manufacturing scenarios. Thus, for this case study, the time to market was deemed the most important criterion, while the surface finishing was considered the least important. Table 2 reports the Best-to-Others (BO) and Others-to-Worst (OW) vectors, defined by a comparison between touchstones and other criteria. Table 3 reports the final criteria weights computed following the rationale outlined in the Appendix B [51]. The consistency of criteria weights is demonstrated by the computed consistency ratio, equal to 0.052, being significantly close to zero.

Table 2. Best-to-Others and Others-to-Worst vectors.

	Touchstone	Complexity Index	Surface Finishing	Material Waste	Energy Consumption	Time to Market	Overall Cost
BO	Time to market	5	6	2	5	1	2
OW	Surface finish	2	1	4	2	7	5

Table 3. Criteria final weights.

Complexity Index	Surface Finishing	Material Waste	Energy Consumption	Time to Market	Overall Cost
0.083	0.052	0.208	0.083	0.365	0.208

Once the attribute weights were calculated, the decision matrix required by the PIV method was constructed by assigning to each candidate process a score for each attribute, as described in the following subsections.

3.2.1. Complexity Index

The I_C of the refined concept was evaluated for each candidate process by using the three parameters introduced in Section 2, namely volumetric index (I_V), detail index (I_D), and freeform index (I_F). This evaluation is independent of the material. It only concerns the geometry. The AM bracket did not necessitate any alterations of the initial part concept. In this instance, the volume of the bracket was found to be 100,220 mm^3, whereas the volume of the parallelepiped bounding box of the component was 770,100 mm^3, resulting in a final I_V index of 0.130. The slight modification made on the CNC refined concept did not significantly alter the geometrical complexity, resulting approximately in the same I_V index of 0.130. The I_D index yielded for both AM and CNC concepts a relatively low value of only $5 \cdot 10^{-4}$, mostly due to the large number of vertices (110) and edges (80) of the model. Finally, the absence of freeform surfaces set the I_F index to one, which is its maximum value. The sum of the three parameters was therefore rounded to 1.131. Computations conducted on the HPDC bracket concept yielded slightly different indices, reflecting the concept refinement required by the same HPDC process. In particular, the I_V index was equal to 0.128, the I_D index was equal to $8 \cdot 10^{-4}$, while the I_F index remained constant at one. As with previous calculations, the sum of the three indices was 1.129, rounded to the third decimal place.

3.2.2. Surface Finishing

Surface finishing, expressed in terms of average surface roughness, Ra, was estimated at 0.8 µm for machining operations on aluminum alloys, considering the final finishing machining step in the machining cycle. Ra was estimated at 1.5 µm for HPDC, a value that can be easily reached with current HPDC systems [52,53]. The use of aluminum alloys allows for the achievement of a surface roughness of 10 µm Ra for PBF-LB/M, provided that the process parameters and shot peening are properly tuned [54,55]. In comparison, Ti6Al4V exhibits superior performance with an achievable surface roughness of 6 µm Ra.

3.2.3. Material Waste

CNC machining operations usually produce consistent amount of waste materials, typically in the shape of chips, being one of its major drawbacks when machining complex shapes. In the present case study, the volume of the waste material was computed as the difference between the volume of the parallelepiped bounding box surrounding the part and the part itself. Therefore, the resulting mass of waste material was found equal to 1.628 kg, slightly more than six times the mass of the bracket. HPDC usually requires the introduction of local allowances for subsequent finishing operations to achieve the required surface finish and geometric tolerances. In this case, a rule of thumb suggests to consider the allowance equal to the 10% of the mass of the component [56]. Given that the weight of the HPDC bracket was 0.260 kg, the corresponding allowance material was computed as 0.026 kg. PBF-LB/M accessory material consists of the allowances needed for subsequent finishing operations, as for HPDC operations, and the support structures required for the PBF-LB/M. Various approaches have been proposed to estimate the allowances required by AM processes [57,58]. In this work, the approach proposed by Priarone, Ingarao [56] was chosen for computing the machining allowances, mainly due to its immediacy and simplicity, setting the allowance to 10% of the component weight. This resulted in 0.028 kg in the case of PBF-LB/Al2139 and in 0.044 kg in the case of PBF-LB/Ti6Al4V.

Additionally, supports volume was computed using Autodesk Netfabb Premium 2024, by Autodesk (San Francisco, CA, USA). The brackets were oriented and placed on a virtual representation of the building platform of the EOS M 290 system, by EOS GmbH (Krailing, Germany), in accordance with the standard orientation algorithm provided by Netfabb, trying to maximise the volume occupation. A total of 14 brackets were placed on a single platform, arranged as shown in Figure 5.

In accordance with the specified procedure, the fabrication of a single bracket necessitates the utilization of a volume of 38,715 mm^3 of supports, resulting in an estimated mass of accessory material per bracket of 0.027 kg for PBF-LB/Al2139 and 0.043 kg for PBF-LB/Ti6Al4V, considering a support density of 25%. It is important to clarify why the supports were estimated using Netfabb rather than expressed as a simple fraction of the mass of the bracket. The introduction of a second software package is an inherent source of higher costs and longer training times for a company. However, Netfabb, as other commercially available software packages such as Magics by Materialize NV (Leuven, Belgium), allows the accurate definition of the number of parts to be fabricated at the same time, in what is commonly called "job". This piece of information is of utmost importance in the definition of manufacturing time, cost, and energy required, and therefore, cannot be overlooked.

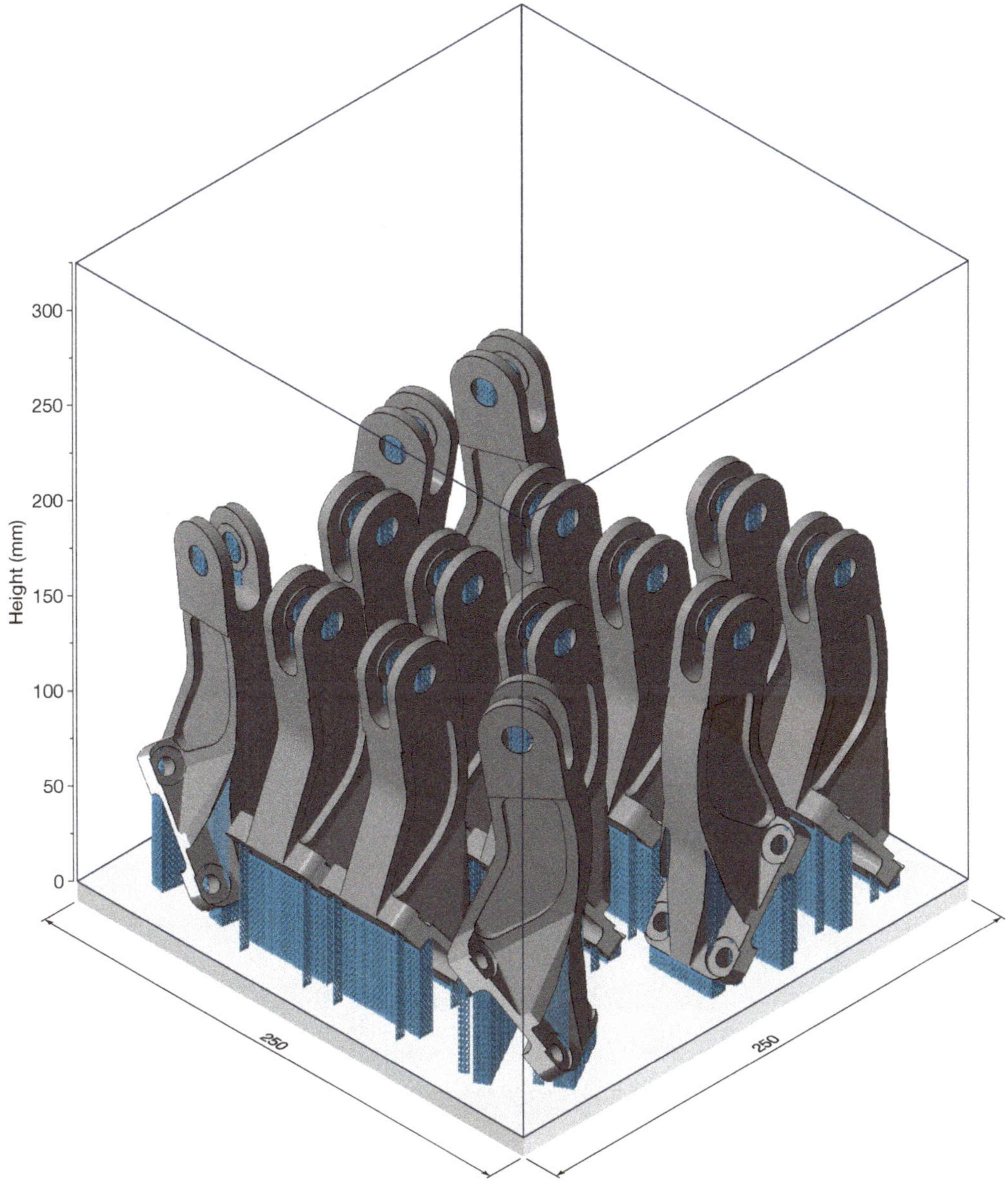

Figure 5. Proposed brackets orientations on the EOS M 290 building platform.

3.2.4. Energy Consumption

The energy consumption of the three candidate processes was estimated considering only the process step and excluding the raw material production. It is important to differentiate the energy required by CNC machining when considering separately the parameters used for roughing and finishing operations. This is because the specific energy consumption (SEC) changes significantly from one condition to the other. Accordingly, the proportions of the total material removed during both machining phases must be established, along with the corresponding specific energy consumption. Priarone et al. [56] suggested that during the machining of aluminum alloys, 85% of the removed material occurs during rough machining, with the remaining 15% occurring during finishing operations. Ingarao et al. [59] also estimated the SECs of both rough machining and finishing operations of aluminum alloys to be $1.9\,\mathrm{MJ}\cdot\mathrm{kg}^{-1}$ and $6.8\,\mathrm{MJ}\cdot\mathrm{kg}^{-1}$ of removed material, respectively. This provides further evidence of the differing energy consumption of the two machining phases. Therefore, the overall energy required to produce the studied bracket by CNC machining was found equal to 4.7 MJ. Similarly, the energy consumption of HPDC was divided in energy used to melt and maintain the aluminum at high temperature, and the energy used by the actuators. Cecchel et al. [60] quantified the former energies using real foundry data, at $7\,\mathrm{MJ}\cdot\mathrm{kg}^{-1}$ and $1.5\,\mathrm{MJ}\cdot\mathrm{kg}^{-1}$, respectively, whereas Liu et al. [61] measured the energy required by all ancillary actuators to be approximately 0.8 MJ per working cycle. Overall, the energy required for the production of the HPDC bracket was found equal to 3.4 MJ. The subsequent finishing by machining of the allowance material, considering the same SEC of $6.8\,\mathrm{MJ}\cdot\mathrm{kg}^{-1}$, accounted for 0.2 MJ. The energy required by the PBF-LB/M process was estimated using the average power consumption of the machine, assumed to be 2.4 kW [62]. The build time, t_{build}, was computed as:

$$t_{build} = \frac{V}{VR} + n \cdot t_{recoat} \tag{4}$$

where V is the aggregate volume of the job on the platform of the EOS M 290, VR is the volume rate allowed by the EOS M 290 machine for the two different materials that were taken into account, t_{recoat} is the time required to recoat a single layer (approximately 10 s on the EOS M 290 machine), and n is the number of layers required to complete the job. The volume rate of PBF-LB/Al2139 production is $7.2\,\mathrm{mm}^3\cdot\mathrm{s}^{-1}$, with a layer thickness of 60 µm [63]. In comparison, the volume rate of PBF-LB/Ti6Al4V is $5\,\mathrm{mm}^3\cdot\mathrm{s}^{-1}$, with a layer thickness of 30 µm [64]. A total of 2927 layers were required for PBF-LB/Al2139, with a total height of 175.6 mm, and 5853 layers were required for PBF-LB/Ti6Al4V. The build time for the PBF-LB/Al2139 job was found to be 83.2 h, while the PBF-LB/Ti6Al4V job required 124.3 h. The total build time for the single PBF-LB/Al2139 bracket was approximately 6 h, while the PBF-LB/Ti6Al4V bracket required 8.9 h. The values of 51.4 MJ and 76.7 MJ were found for the production of the PBF-LB/Al2139 bracket and the PBF-LB/Ti6Al4V bracket, respectively, which is generally in agreement with the high energy density of AM processes [59]. The energy consumption for the finishing operation was deemed negligible.

3.2.5. Time to Market

The time-to-market of the CNC machining bracket was estimated by the Xometry Europe (Ottobrunn, Germany) online service, together with its cost, and was equal to 14 working days. In contrast, the time-to-market for conventional high pressure die casting was estimated to be 30 working days, and to only one week for the PBF-LB/Al2139 and 10 days for the PBF-LB/Ti6Al4V bracket, stressing the different flexibility of these production systems. In fact, it is well known that AM can help reducing the lead time of a part, enabling a quick response from the company, particularly when dealing with small batches [65,66], thus justifying the shortest time-to-market out of the three processes. It is worth noting that the considered time-to-market for HPDC and PBF-LB/M include the consideration of the final finishing.

3.2.6. Overall Cost

The cost of CNC machining operations was estimated using the online free tool offered by Xometry Europe. The online service provided by Xometry carefully considered the 3D CAD model of the bracket, its material and the expected resulting surface roughness, enhancing the accuracy of the final estimate. Therefore, a cost of EUR 95 per bracket was computed this way. As for HPDC, the higher complexity hindered by the process did not allow the use of any online tool for cost estimation, nudging the authors to opt for empirical models to estimate the cost of the bracket. It this scenario, the model developed by Atzeni and Salmi [67] was referenced for the cost evaluation of the HPDC bracket. While reporting the whole breakdown structure of the model would go beyond the scope of this investigation, it is worth noticing some of the assumption made. The overall cost was divided into four items: material cost per part, machine setup cost, machine operation cost, and post-processing costs. Assuming a die cost of roughly EUR 30,000, for a batch of 50 pieces, the price per bracket would be near EUR 659, as reported in the respective column of Table 4. The same study was also considered when estimating the cost of the PBF-LB/M bracket. Also, in this case, the total cost per bracket was divided in the same four cost items: material cost per part, machine setup cost, processing cost, and post-processing costs. The model resulted in a cost of EUR 812 per the PBF-LB/Al2139 bracket and EUR 1348 per the PBF-LB/Ti6Al4V bracket, with the machine cost accounting for over than 85% of the total value. Table 4 presents all data collected in this section and organizes them for an easier implementation of the following hybrid MCDM methodology.

Table 4. Decision matrix.

	Complexity Index (-)	Surface Finishing (μm)	Material Waste (kg)	Energy Consumption (MJ)	Time to Market (Working Days)	Overall Cost (EUR)
CNC Machining	1.131	0.8	1.628	4.7	14	96
HPDC	1.129	1.5	0.026	3.6	30	659
PBF-LB/Al2139	1.131	10	0.055	51.4	7	812
PBF-LB/Ti6Al4V	1.131	6	0.087	76.7	10	1348

The decision matrix was then normalized to enable comparison of different scores. Every element of the matrix was normalized by dividing it by the square root of the sum of squares of the corresponding column, resulting in a dimensionless number. Table 5 presents the normalized data for the batch of 50 pieces. Each column entry was then multiplied by the corresponding weight to generate the weighted normalized decision matrix, as shown in Table 6. From the weighted normalized decision matrix the ideal best, Positive Ideal Solution (PIS), was computed by selecting the smallest options for each attribute in each column, as all attributes were considered costs. PIS components are reported in the last row of the same Table 6.

Table 5. Normalized decision matrix.

	Complexity Index	Surface Finishing	Material Waste	Energy Consumption	Time to Market	Overall Cost
CNC Machining	0.500	0.068	0.998	0.051	0.397	0.056
HPDC	0.499	0.127	0.016	0.039	0.850	0.286
PBF-LB/Al2139	0.500	0.849	0.034	0.556	0.198	0.475
PBF-LB/Ti6Al4V	0.500	0.509	0.053	0.829	0.283	0.789

Table 6. Weighted normalized decision matrix.

	Complexity Index	Surface Finishing	Material Waste	Energy Consumption	Time to Market	Overall Cost
CNC Machining	0.042	0.004	0.208	0.004	0.145	0.012
HPDC	0.042	0.007	0.003	0.003	0.310	0.080
PBF-LB/Al2139	0.042	0.044	0.007	0.046	0.072	0.099
PBF-LB/Ti6Al4V	0.042	0.027	0.011	0.069	0.103	0.164
Ideal best (PIS)	0.042	0.004	0.003	0.003	0.072	0.012

The overall proximity index values, PIV, of the three processes is equal to the Manhattan distance between the ideal best solution and the solutions provided by the same manufacturing processes. PIV is reported in Table 7. It is worth recalling that a lower PIV suggests a closer solution to the ideal best, and therefore, the most suitable solution. Thus, PBF-LB/Al2139 resulted as the most suitable process for the production of the considered bracket. The same procedure deemed less suitable both the CNC machining and the PBF-LB/Ti6Al4V, which both resulted in very close PIVs. Finally, the HPDC was found to be the least adequate option out the investigated four. At this stage, the proposed methodology highlighted the profitability of PBF-LB/M for the production of a bracket for aerospace applications, both in aluminum and titanium alloys, and low production batch.

Table 7. PIV of the explored manufacturing processes.

	PIV	Rank
CNC machining	0.278	2
HPDC	0.310	4
PBF-LB/Al2139	**0.175**	**1**
PBF-LB/Ti6Al4V	0.280	3

3.3. Other Scenarios

It is, therefore, evident that the choice of the right material can severely influence the results of the whole hybrid MCDM method. Ti6Al4V has considerable higher mechanical properties than Al2139, together with a considerably higher density. Using Ti6Al4V as an alternative to aluminum alloys, without coherently change the concept of the same bracket, may partially hinder the potentialities of the material. Therefore, given that the I_C is greater than unity, it might be beneficial to explore the potential of utilizing an inspiring TO to reduce the mass of the titanium bracket, thereby enhancing its suitability for the production by PBF-LB/M and improving its score at the end of the MCDM method.

3.3.1. Topology Optimization

The TO step was completed within the Fusion 360 simulation environment, without the necessity for additional software packages. Figure 6a depicts the outcomes of the TO, highlighting the difference between the initial design and the optimal solution proposed by Fusion 360. The redesigned bracket concept was considerably less bulky than the original one (Figure 6b), with a substantial lower mass that was reduced from the original 0.444 kg to 0.273 kg, marking a 39% reduction. The optimized concept was also positively tested for the initial functional specifications. The maximum displacement computed was equal to 0.40 mm, which is below the threshold of 0.5 mm (Figure 6c), and therefore, considered eligible for process selection.

It was found that the modifications made to the titanium bracket geometry had an appreciable influence on the MCDM analysis. Computations were performed to determine the new I_V and I_D indices, which yielded an I_C of 1.076. The reduction in the allowance, which is directly proportional to the part weight, was offset by the greater necessity for supports, resulting in a final value of 0.090 kg of material waste per bracket. The most consistent changes, which also had the greatest impact on the final process ranking, were related to the overall cost of the bracket and to its energy consumption. The reduction in bracket mass following the TO stage resulted in a decrease in manufacturing time, which in turn led to a reduction in energy consumption, amounting to 72.8 MJ in this scenario. Similarly, the overall cost was reduced to EUR 1155, resulting in savings of EUR 193 per bracket. Table 8 represents the decision matrix updated to consider the PBF-LB/Ti6Al4V bracket after the TO. The incorporation of the novel values in Table 8 resulted in a considerably different final ranking, as reported in Table 9. The PBF-LB/Ti6Al4V process emerged as the second most suitable option, distinguishing itself from the CNC machining process and deepening the distance from the HPDC one.

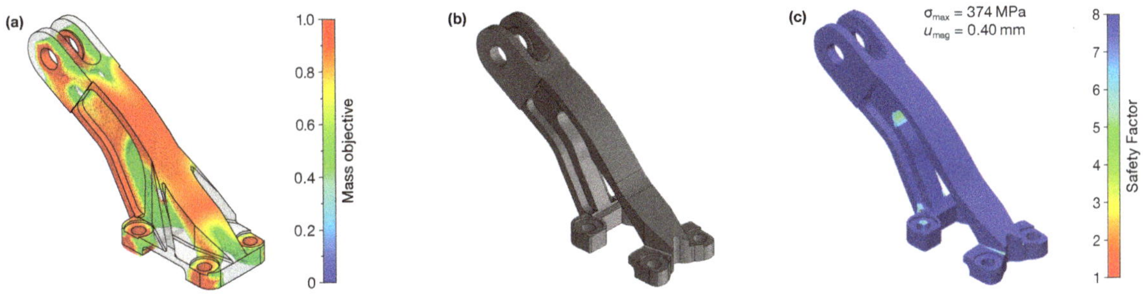

Figure 6. (**a**) TO results. (**b**) Redesigned bracket. (**c**) FE validation of the redesigned bracket.

Table 8. Decision matrix after TO.

	Complexity Index (-)	Surface Finishing (μm)	Material Waste (kg)	Energy Consumption (MJ)	Time to Market (Working Days)	Overall Cost (EUR)
CNC Machining	1.131	0.8	1.628	4.7	14	96
HPDC	1.129	1.5	0.026	3.6	30	659
PBF-LB/Al2139	1.131	10	0.055	51.4	7	812
PBF-LB/Ti6Al4V | After TO	1.076	6	0.090	72.8	10	1155

Table 9. PIVs after TO.

	PIV	Rank
CNC machining	0.278	3
HPDC	0.316	4
PBF-LB/Al2139	**0.184**	**1**
PBF-LB/Ti6Al4V	0.267	2

3.3.2. Production Batch Sensibility

However, the outcomes yielded by the proposed hybrid MCDM method were found to be significantly influenced by the dimensions of the production batch. To assess the impact of varying the batch size, the batch was divided by two, multiplied by two, and multiplied by twenty. A further MCDM analysis was conducted for these scenarios. Although smaller batches do not appear to significantly impact the prioritization of the selected processes (Figure 7), differences were introduced by scenarios of larger batches. In fact, the production batch of 100 pieces was sufficiently large to significantly reduce the cost of a single bracket produced by HPDC, down to EUR 359. This made the HPDC the second-best option, surpassing both the CNC machining solution and the PBF-LB/Ti6Al4V solution. Furthermore, the cost of the HPDC bracket was markedly reduced for the largest production batch considered, comprising 1000 pieces, reaching only EUR 89 per piece. This sharp decline in production costs was reflected in the significantly lower PIV of the HPDC, creating a substantial margin separating the HPDC from the PBF-LB/Ti6Al4V solution. It is evident that this trend would eventually position the HPDC as the most viable option for larger production volumes, even when compared to the PBF-LB/Al2139 solution.

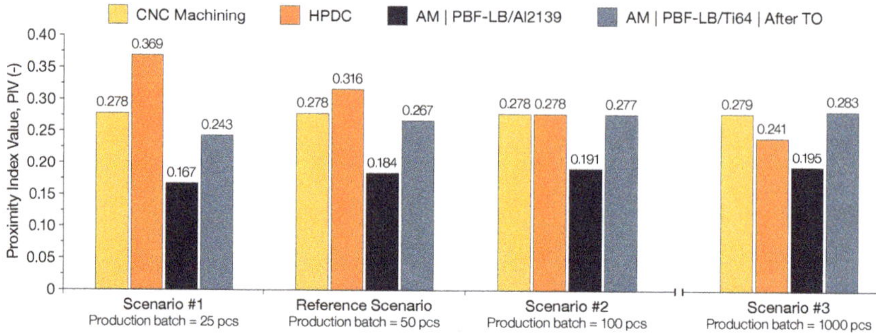

Figure 7. PIV of CNC machining, HPDC, and PBF-LB/M as a function of batch number.

4. Conclusions

The present investigation proposed a methodology aimed at choosing the best manufacturing process for a specific scenario, with special attention on the distinction between AM and conventional processes. The methodology was evaluated on a case study taken from the aeronautical field to show the proficiency of the entire proposed workflow. The main results of the investigation can be summarized as follows:

- The methodology put forth a hybrid MCDM approach to evaluate the relative suitability of AM and CM processes, which can be readily utilized by technical professionals without a strong background in AM.
- AM processes were found to be ideal for the production of small to medium batches, up to 100 pieces, leveraging their higher flexibility due to the absence of initial tooling costs.
- The significance of material selection in the context of AM during the preliminary design phase was emphasized. In fact, the utilization of materials with a high strength-to-weight ratio, such as titanium alloys, necessitated supplementary redesign activities to enhance the suitability of AM techniques in comparison to conventional ones.
- In the context of redesign activities, it was confirmed the positive role that TO may cover. The implementation of TO resulted in a 39% reduction in the weight of the bracket, thereby positively influencing the manufacturing time. The reduction in manufacturing time subsequently resulted in a 10% improvement in terms of cost and 5% improvement for energy consumption, which in turn enhanced the score of AM in the final process ranking.
- The use of CM techniques, such as HPDC, has been demonstrated to offer a highly competitive solution for the production of large batches, larger than 100 pieces, where the initial tooling costs associated with the mold can be distributed across a greater number of components.

In conclusion, the human role in the production planning is still central and high skilled work figures must still rely on their experience while incorporating multiple elements during their decision-making processes. Nonetheless, the methodology proposed can help newcomers, and less skilled workers, to still take a reliable decision thanks to a guided and robust procedure. Future works might go even further in this same direction, trying to use artificial intelligence algorithms in the decision making process.

Author Contributions: Conceptualization, A.S, E.A., and L.I.; methodology, G.V.; validation, G.V.; investigation, G.V.; resources, A.S, E.A., and L.I.; writing—original draft preparation, G.V.; writing—review and editing, A.S, G.V., E.A., and L.I.; visualization, A.S. and G.V. All authors have read and agreed to the published version of the manuscript.

Funding: This study was carried out within the MICS (Made in Italy—Circular and Sustainable) Extended Partnership and received funding from the European Union Next-GenerationEU (PIANO NAZIONALE DI RIPRESA E RESILIENZA (PNRR)—MISSIONE 4 COMPONENTE 2, INVESTIMENTO 1.3—D.D. 1551.11-10-2022, PE00000004). This manuscript reflects only the authors' views and opinions, neither the European Union nor the European Commission can be considered responsible for them.

Data Availability Statement: We are prone to provide the aforementioned data on request.

Acknowledgments: The authors would like to acknowledge the Interdepartmental Centre for Integrated Additive Manufacturing (IAM@PoliTo) at the Politecnico di Torino, Torino, Italy, for the resources to perform the research activities.

Conflicts of Interest: The authors declare no conflict of interest.

Abbreviations

The following abbreviations are used in this manuscript:

AM	Additive Manufacturing
AHP	Analytic Hierarchy Process
BO	Best-to-Others
BWM	Best Worst Method
CAD	Computer-Aided Design
CAE	Computer-Aided Engineering
CM	Conventional Manufacturing
CNC	Computer Numerically Controlled
DfAM	Design for Additive Manufacturing
FDM	Fused Deposition Modeling
FE	Finite Element
FGM	Fuzzy Geometric Mean
GD&T	Geometric Dimensioning and Tolerancing
GHG	Greenhouse Gases
HPDC	High Pressure Die Casting
MCDM	Multi-Criteria Decision-Making
OW	Others-to-Worst
PBF-LB	Powder Bed Fusion with Laser Beam
PIS	Positive Ideal Solution
PIV	Proximity Index Value
RP	Rapid Prototyping
SEC	Specific Energy Consumption
TO	Topology Optimization
TOPSIS	Technique for Order of Preference by Similarity to Ideal Solution
VIKOR	VIekriterijumsko KOmpromisno Rangiranje
WPI	Weighted Proximity Index

Appendix A. Building Volumes and Available Materials of AM Commercial Systems

In this first appendix, the volumes of commercially available AM systems are reported. Ensuring a building volume large enough to accommodate the whole component in production is a key feature of AM systems, avoiding the need of subsequent assembly operations. Table A1 contains the building volume dimensions of some of the most common commercial systems [68]. Similarly, designers must consider the plethora of commercially available AM materials during the initial design phases. Later material changes might require undesired concept changes to respect functional specifications. Table A2 reports some of the most used materials in PBF-LB/M applications.

Table A1. PBF-LB/M commercially available systems.

Company	Model Name	X (mm)	Y (mm)	Z (mm)	Ref.
3D SYSTEMS	DMP Flex 200	140	140	115	[69]
	DMP Factory 350	275	275	420	[70]
	DMP Factory 350 Dual	275	275	420	[70]
	DMP Flex 350	275	275	420	[71]
	DMP Flex 350 Dual	275	275	420	[71]
	DMP Flex 350 Triple	275	275	420	[71]
	DMP Factory 500	500	500	500	[72]
Colibrium Additive	M2 Series 5	245	245	350	[73]
	M Line	500	500	400	[74]
	X Line 2000R	800	400	500	[75]
DMG MORI	Lasertec 12 SLM	125	125	200	[76]
	Lasertec 30 Dual SLM	300	300	300	[77]
EOS	M 290	250	250	325	[62]
	M 300-4	300	300	400	[78]
	M 400	400	400	400	[79]
	M 400-4	400	400	400	[80]
Farsoon Technologies	FS121M	120	120	100	[81]
	FS273M	275	275	355	[82]
	FS200M	425	230	300	[83]
	FS301M	305	305	410	[84]
	FS350M-4	433	358	400	[85]
	FS422M	425	425	550	[86]
	FS721M-CAMS	720	420	390	[87]
	FS721M	720	420	420	[88]
	FS621M	620	620	1100	[89]
Matsuura Machinery	LUMEX Avance-25	256	256	300	[90]
	LUMEX Avance-60	600	600	500	[91]
Prima Additive	Print Sharp 150	Φ150		160	[92]
	Print Genius 150	Φ150		160	[93]
	Print Green	Φ150		160	[94]
	Print Sharp 300	330	330	400	[95]
	Print Genius 300	330	330	400	[95]
	Print Brilliance 300	330	330	400	[95]
	Print Genius 400	430	430	600	[96]
	Print Genius 400 XL	430	430	1000	[96]
Renishaw	RenAm 500 Flex	250	250	350	[97]
	RenAM 500	250	250	350	[97]
	RenAM 500 Ultra	250	250	350	[97]
SLM Solutions	SLM 125	125	125	125	[98]
	SLM 280 PS	280	280	365	[99]
	SLM 280 2.0	280	280	365	[100]
	SLM 500	500	280	365	[101]
	SLM 800	500	280	850	[102]
	SLM NXG XII 600	600	600	600	[103]
Sharebot	metalONE	65	65	100	[104]
TRUMPF	TruPrint 1000	Φ 100		100	[105]
	TruPrint 1000 Basic Edition	Φ 100		100	[106]
	TruPrint 2000	200	200	200	[107]
	TruPrint 3000	Φ 300		400	[108]
	TruPrint 5000	Φ 300		400	[109]
	TruPrint 5000 Green Edition	Φ 300		400	[110]
Velo3D	Sapphire	ϕ 315		400	[111]
	Sapphire 1MZ	315		1000	[111]
	Sapphire XC	600		550	[112]
	Sapphire XC 1MZ	600		1000	[112]

Table A2. PBF-LB/M commercially available materials.

Material Class	Alloy	Providers
Aluminum	Aheadd® CP1	3D SYSTEMS
	Al-HS1®	Höganäs
	AlSi7Mg0.6	3D SYSTEMS, Colibrium Additive, EOS, SLM Solutions
	AlSi10Mg	3D SYSTEMS, Colibrium Additive, EOS, Farsoon Technologies, Prima Additive, SLM Solutions, Höganäs
	AlSi12	3D SYSTEMS
	Al2139	EOS
Cobalt-Chrome	CoCrF75	3D SYSTEMS
	CoCrMo	3D SYSTEMS, Colibrium Additive, EOS, Farsoon Technologies, Prima Additive, Höganäs
	CoCrMoW	Farsoon Technologies, Prima Additive
	SLM MediDent®	SLM Solutions
Copper	Oxygen-Free Copper	3D SYSTEMS, EOS, Prima Additive
	CuCr1Zr	3D SYSTEMS, EOS, SLM Solutions, Höganäs
	GRCop-42	3D SYSTEMS
	CuCr2.4	3D SYSTEMS
	CuNi2SiCr	SLM Solutions
	CuNi30	3D SYSTEMS, EOS
	CuSn10	Farsoon Technologies, Prima Additive
Nickel	HAYNES® 282®	EOS, Höganäs
	GRX-810	3D SYSTEMS
	HX	3D SYSTEMS, EOS, Farsoon Technologies, Prima Additive, SLM Solutions, Höganäs, Oerlikon
	K-500	SLM Solutions
	IN625	3D SYSTEMS, Colibrium Additive, EOS, Farsoon Technologies, Prima Additive, SLM Solutions, Höganäs, Oerlikon
	IN718	3D SYSTEMS, Colibrium Additive, EOS, Farsoon Technologies, Prima Additive, SLM Solutions, Höganäs, Oerlikon
	IN939	EOS, Höganäs
Refractory	C-103	3D SYSTEMS
	Tungsten	3D SYSTEMS
Steel	Invar 36®	SLM Solutions
	M300	3D SYSTEMS, Colibrium Additive, EOS, Farsoon Technologies, Prima Additive
	Tool Steel H11	Höganäs, Oerlikon
	Tool Steel H13	EOS, SLM Solutions, Höganäs, Oerlikon
	316L	3D SYSTEMS, Colibrium Additive, EOS, Farsoon Technologies, Prima Additive, SLM Solutions, Höganäs, Oerlikon
	17-4PH	3D SYSTEMS, Colibrium Additive, EOS, Farsoon Technologies, Prima Additive, SLM Solutions, Höganäs, Oerlikon
	15-5PH	Farsoon Technologies, SLM Solutions, Oerlikon
Titanium	TA15	Farsoon Technologies, SLM Solutions
	CPTi grade 1	3D SYSTEMS, Colibrium Additive
	CPTi grade 2	Colibrium Additive, EOS, SLM Solutions, Höganäs
	Ti6Al4V grade 5	3D SYSTEMS, Colibrium Additive, EOS, Farsoon Technologies, Prima Additive, Höganäs, Oerlikon
	Ti6Al4V grade 23	3D SYSTEMS, Colibrium Additive, EOS, SLM Solutions, Höganäs, Oerlikon
	Ti-6Al-2Sn-4Zr-2Mo	Colibrium Additive
	Ti-5Al-5V-5Mo-3Cr	Colibrium Additive

Appendix B. BWM and PIV Rationales

This appendix presents the rationales behind the BWM and the PIV method used in this investigation. The BWM was used to define the weights of the criteria considered, whereas the PIV method was used to rank the manufacturing processes. As already explained, the BWM was introduced to reduce the number of pair-wise comparisons between different options, improving the consistency of the results obtained [13]. The BWM is carried out as follows:

1. Definition of the set of criteria to compare.
2. Select the best criterion and the worst criterion in the current scenario. Only primary comparisons are carried out, namely between the best criterion and the other options, and between the worst criterion and the other options. This way, all the so-called secondary comparisons can be avoided, drastically reducing the number of comparisons.
3. Define the Best-to-Others vector, whose components quantify how much the best criterion is preferred over the others. The value 1 indicates the same importance between criteria, while the value 9 indicates the utmost importance of the best criterion over the second one:
$$A_B = (a_{B1}, a_{B2}, \ldots, a_{Bn}) \tag{A1}$$
4. Define the Others-to-Worst vector, following the same procedure explained at the previous step. As before, the value 1 indicates the same importance between the criteria, whereas 9 a prominent importance of the others over the worst criterion:
$$A_W = (a_{1W}, a_{2W}, \ldots, a_{nW})^T \tag{A2}$$
5. Defining the vector of the optimal weight w^*, as $w^* = (w_1^*, w_2^*, \ldots, w_n^*)$ for which the differences $|w_B/w_j - a_{Bj}|$ and $|w_j/w_W - a_{jW}|$ are minimized for all j, namely for all the components of the w vector.

The problem can be formulated as finding the minimum value of ξ so that:
$$\begin{cases} \left|\dfrac{w_B}{w_j} - a_{Bj}\right| < \xi \\ \left|\dfrac{w_j}{w_W} - a_{jW}\right| < \xi \\ \sum_j w_j = 1, w_j > 0 \forall j \end{cases} \tag{A3}$$

The smallest ξ granting a nonempty solution space is called ξ^* and defines the optimal weight vector w^*.

The PIV method was firstly introduced to overcome the rank reversal phenomenon often occurring in the TOPSIS method [14]. The rationale behind the PIV method is quite close with the TOPSIS one, with slight differences in the final part of the procedure. The PIV method may be schematically presented as a seven-step procedure:

1. Formulation of the decision problems by defining decision criteria $C_J (j = 1, \ldots, n)$ and alternatives, $A_i (i = 1, \ldots, m)$.
2. Each alternative is evaluated on every criteria, resulting in a score x_{ij}. The x_{ij} scores constitute the decision matrix (DM), as shown in Table A3.
3. The scores x_{ij} are likely to be expressed in various unit of measures, making it difficult to directly compare them. The normalization step solves this problem, bringing all x_{ij} to a common scale. The normalized entry of the decision matrix, r_{ij}, is computed as $r_{ij} = x_{ij} / \sqrt{\sum_{i=1}^{m} x_j^2}$.
4. After the definition of the normalized decision matrix, each r_{ij} must be multiplied by the corresponding w_j weight, defined in advance. Therefore, the weighted entries of the decision matrix are defined as $v_{ij} = w_j \cdot r_i$, as in Table A4.
5. The weighted proximity index (WPI) expresses the distance between each alternative and the ideal best alternative. If the criterion expresses a benefit for the alternatives, the ideal best components is the v_i scoring the highest value along the column. Conversely, if the criterion expresses a cost for the alternative, the ideal best components is represented by the lowest v_i along the column. The components of the WPI, namely u_i, are computed as $u_i = |v_{best} - v_i|$. This step represents the key moment of the whole procedure, distinguishing the PIV method from the TOPSIS one. In fact, the

use of the 1-norm, instead of the Euclidean norm used by the TOPSIS method, should minimize the occurring of the rank reversal.

6. The 1-norm distances between alternative components and ideal best can be summed up into the overall proximity value (d_j), expressing the closeness of the alternative to the ideal best, namely $d_i = \sum_{j=1}^{n} u_j$.
7. In conclusion, the alternatives can be ranked according to their overall proximity value, from the smallest to the highest one.

Table A3. Decision matrix.

	w_1 C_1	w_2 C_2	$\cdots$ $\cdots$	w_n C_n
A_1	x_{11}	x_{12}	$\cdots$	x_{1n}
A_2	x_{21}	x_{22}	$\cdots$	x_{2n}
$\vdots$	$\vdots$	$\vdots$	$\ddots$	$\vdots$
A_m	x_{m1}	x_{m2}	$\cdots$	x_{mn}

Table A4. Weighted normalized decision matrix.

	w_1 C_1	w_2 C_2	$\cdots$ $\cdots$	w_n C_n
A_1	v_{11}	v_{12}	$\cdots$	v_{1n}
A_2	v_{21}	v_{22}	$\cdots$	v_{2n}
$\vdots$	$\vdots$	$\vdots$	$\ddots$	$\vdots$
A_m	v_{m1}	v_{m2}	$\cdots$	v_{mn}

References

1. Dieter, G.E.; Schmidt, L.C. *Engineering Design*, 4th ed.; McGraw-Hill Higher Education: New York, NY, USA, 2009.
2. ASTM International, West Conshohocken, PA. Additive Manufacturing—General Principles—Fundamentals and Vocabulary. 2021. Available online: https://www.astm.org/f3177-21.html (accessed on 5 August 2024).
3. Wohlers, T.; Gornet, T. History of Additive Manufacturing. Wholers Report 2016. 2016. Available online: https://wohlersassociates.com/product/wohlers-report-2016/ (accessed on 1 June 2024).
4. Salmi, A.; Calignano, F.; Galati, M.; Atzeni, E. An integrated design methodology for components produced by laser powder bed fusion (L-PBF) process. *Virtual Phys. Prototyp.* **2018**, *13*, 191–202. [CrossRef]
5. Jared, B.H.; Aguilo, M.A.; Beghini, L.L.; Boyce, B.L.; Clark, B.W.; Cook, A.; Kaehr, B.J.; Robbins, J. Additive manufacturing: Toward holistic design. *Scr. Mater.* **2017**, *135*, 141–147. [CrossRef]
6. Zhai, Y.; Lados, D.; LaGoy, J. Manufacturing: Making Imagination the Major Limitation. *JOM* **2014**, *66*, 808–816. [CrossRef]
7. Fayazfar, H.; Sharifi, J.; Keshavarz, M.K.; Ansari, M. An overview of surface roughness enhancement of additively manufactured metal parts: A path towards removing the post-print bottleneck for complex geometries. *Int. J. Adv. Manuf. Technol.* **2023**, *125*, 1061–1113. [CrossRef]
8. Piscopo, G.; Salmi, A.; Atzeni, E. Investigation of dimensional and geometrical tolerances of laser powder directed energy deposition process. *Precis. Eng.* **2024**, *85*, 217–225. [CrossRef]
9. Yang, S.; Page, T.; Zhang, Y.; Zhao, Y.F. Towards an automated decision support system for the identification of additive manufacturing part candidates. *J. Intell. Manuf.* **2020**, *31*, 1917–1933. [CrossRef]
10. Taherdoost, H.; Madanchian, M. Multi-Criteria Decision Making (MCDM) Methods and Concepts. *Encyclopedia* **2023**, *3*, 77–87. [CrossRef]
11. Velasquez, M.; Hester, P.T. An analysis of multi-criteria decision making methods. *Int. J. Oper. Res.* **2013**, *10*, 56–66.
12. Qin, Y.; Qi, Q.; Shi, P.; Lou, S.; Scott, P.J.; Jiang, X. Multi-Attribute Decision-Making Methods in Additive Manufacturing: The State of the Art. *Processes* **2023**, *11*, 497. [CrossRef]
13. Rezaei, J. Best-worst multi-criteria decision-making method: Some properties and a linear model. *Omega* **2016**, *64*, 126–130. [CrossRef]
14. Mufazzal, S.; Muzakkir, S.M. A new multi-criterion decision making (MCDM) method based on proximity indexed value for minimizing rank reversals. *Comput. Ind. Eng.* **2018**, *119*, 427–438. [CrossRef]
15. Qin, Y.; Qi, Q.; Shi, P.; Scott, P.J.; Jiang, X. Selection of materials in metal additive manufacturing via three-way decision-making. *Int. J. Adv. Manuf. Technol.* **2023**, *126*, 1293–1302. [CrossRef]
16. Jayapal, J.; Kumaraguru, S.; Varadarajan, S. Evaluation of computationally optimized design variants for additive manufacturing using a fuzzy multi-criterion decision-making approach. *Int. J. Adv. Manuf. Technol.* **2023**, *129*, 5199–5218. [CrossRef]

17. Sakthivel Murugan, R.; Vinodh, S. Prioritization and deployment of design for additive manufacturing strategies to an automotive component. *Rapid Prototyp. J.* **2023**, *29*, 2193–2215. [CrossRef]
18. Sartini, M.; Luca, M.; Claudio, F.; Marco, M. A multi-criteria decision-making approach to optimize the part build orientation in additive manufacturing. *Proc. Des. Soc.* **2023**, *3*, 293–302. [CrossRef]
19. Mançanares, C.G.; de S. Zancul, E.; Cavalcante da Silva, J.; Cauchick Miguel, P.A. Additive manufacturing process selection based on parts' selection criteria. *Int. J. Adv. Manuf. Technol.* **2015**, *80*, 1007–1014. [CrossRef]
20. Liu, W.; Zhu, Z.; Ye, S. A decision-making methodology integrated in product design for additive manufacturing process selection. *Rapid Prototyp. J.* **2020**, *26*, 895–909. [CrossRef]
21. Zaman, U.K.u.; Rivette, M.; Siadat, A.; Mousavi, S.M. Integrated product-process design: Material and manufacturing process selection for additive manufacturing using multi-criteria decision making. *Robot. Comput.-Integr. Manuf.* **2018**, *51*, 169–180. [CrossRef]
22. Ghaleb, A.M.; Kaid, H.; Alsamhan, A.; Mian, S.H.; Hidri, L. Assessment and Comparison of Various MCDM Approaches in the Selection of Manufacturing Process. *Adv. Mater. Sci. Eng.* **2020**, *2020*, 4039253. [CrossRef]
23. Wang, Y.; Zhong, R.Y.; Xu, X. A decision support system for additive manufacturing process selection using a hybrid multiple criteria decision-making method. *Rapid Prototyp. J.* **2018**, *24*, 1544–1553. [CrossRef]
24. Wang, Y.C.; Chen, T.; Lin, Y.C. 3D Printer Selection for Aircraft Component Manufacturing Using a Nonlinear FGM and Dependency-Considered Fuzzy VIKOR Approach. *Aerospace* **2023**, *10*, 591. [CrossRef]
25. Grachev, D.I.; Chizhmakov, E.A.; Stepanov, D.Y.; Buslovich, D.G.; Khulaev, I.V.; Deshev, A.V.; Kirakosyan, L.G.; Arutyunov, A.S.; Kardanova, S.Y.; Panin, K.S.; et al. Dental Material Selection for the Additive Manufacturing of Removable Complete Dentures (RCD). *Int. J. Mol. Sci.* **2023**, *24*, 6432. [CrossRef] [PubMed]
26. Raigar, J.; Sharma, V.S.; Srivastava, S.; Chand, R.; Singh, J. A decision support system for the selection of an additive manufacturing process using a new hybrid MCDM technique. *Sādhanā* **2020**, *45*, 101. [CrossRef]
27. Fera, M.; Macchiaroli, R.; Fruggiero, F.; Lambiase, A. A new perspective for production process analysis using additive manufacturing—complexity vs. production volume. *Int. J. Adv. Manuf. Technol.* **2018**, *95*, 673–685. [CrossRef]
28. Kalami, H.; Urbanic, J. Exploration of surface roughness measurement solutions for additive manufactured components built by multi-axis tool paths. *Addit. Manuf.* **2021**, *38*, 101822. [CrossRef]
29. United Nations. Transforming our World: The 2030 Agenda for Sustainable Development. Report, eSocialSciences, 2015. Working Papers. Available online: https://sdgs.un.org/2030agenda (accessed on 12 August 2024).
30. GrabCAD. Bracket. 2023. Available online: https://grabcad.com/library/bracket-493 (accessed on 23 May 2024).
31. Pimenov, D.Y.; Kiran, M.; Khanna, N.; Pintaude, G.; Vasco, M.C.; da Silva, L.R.R.; Giasin, K. Review of improvement of machinability and surface integrity in machining on aluminum alloys. *Int. J. Adv. Manuf. Technol.* **2023**, *129*, 4743–4779. [CrossRef]
32. Carpio, F.; Araújo, D.; Pacheco, F.; Méndez, D.; Garcìa, A.; Villar, M.; Garcìa, R.; Jiménez, D.; Rubio, L. Fatigue behaviour of laser machined 2024 T3 aeronautic aluminium alloy. *Appl. Surf. Sci.* **2003**, *208–209*, 194–198. [CrossRef]
33. Giasin, K.; Hodzic, A.; Phadnis, V.; Ayvar-Soberanis, S. Assessment of cutting forces and hole quality in drilling Al2024 aluminium alloy: Experimental and finite element study. *Int. J. Adv. Manuf. Technol.* **2016**, *87*, 2041–2061. [CrossRef]
34. Ali, R.A.; Mia, M.; Khan, A.M.; Chen, W.; Gupta, M.K.; Pruncu, C.I. Multi-response optimization of face milling performance considering tool path strategies in machining of Al-2024. *Materials* **2019**, *12*, 1013. [CrossRef]
35. Yücel, A.; Yildirim, C.V.; Sarikaya, M.; Şirin, C.; Kivak, T.; Gupta, M.K.; Tomaz, I.V. Influence of MoS2 based nanofluid-MQL on tribological and machining characteristics in turning of AA 2024 T3 aluminum alloy. *J. Mater. Res. Technol.* **2021**, *15*, 1688–1704. [CrossRef]
36. UNI—Ente Italiano di Normazione, Milano, Italy. Tolleranze Generali. Tolleranze per Dimensioni Lineari ed Angolari Prive di Indicazione di Tolleranze Specifiche. 1996. Available online: https://store.uni.com/uni-en-22768-1-1996 (accessed on 10 February 2024).
37. Samuel, A.; Araoyinbo, A.; Elewa, R.; Biodun, M. Effect of machining of aluminium alloys with emphasis on aluminium 6061 alloy—A review. *IOP Conf. Ser. Mater. Sci. Eng.* **2021**, *1107*, 012157. [CrossRef]
38. Okokpujie, I.P.; Tartibu, L.K. A mini-review of the behaviour characteristic of machining processes of aluminium alloys. *Mater. Today Proc.* **2022**, *62*, 4526–4532. [CrossRef]
39. Zimmermann, N.; Müller, E.; Lang, S.; Mayr, J.; Wegener, K. Thermally compensated 5-axis machine tools evaluated with impeller machining tests. *CIRP J. Manuf. Sci. Technol.* **2023**, *46*, 19–35. [CrossRef]
40. Liu, Y.; Xiong, S. Research Progress on Thermal Conductivity of High-Pressure Die-Cast Aluminum Alloys. *Metals* **2024**, *14*, 370. [CrossRef]
41. Ye, W.; WU, S.p.; Xiang, X.; Chen, R.r.; Zhang, J.b.; Xiao, W.f. Formation mechanism and criterion of linear segregation in ZL205A alloy. *Trans. Nonferrous Met. Soc. China* **2014**, *24*, 3632–3638. [CrossRef]
42. Jiang, H.; Zhang, L.; Zhao, B.; Sun, M.; He, M. Microstructure and Mechanical Properties of ZL205A Aluminum Alloy Produced by Squeeze Casting after Heat Treatment. *Metals* **2022**, *12*, 2037. [CrossRef]
43. Li, S.; Yue, X.; Li, Q.; Peng, H.; Dong, B.; Liu, T.; Yang, H.; Fan, J.; Shu, S.; Qiu, F.; et al. Development and applications of aluminum alloys for aerospace industry. *J. Mater. Res. Technol.* **2023**, *27*, 944–983. [CrossRef]
44. Goenka, M.; Nihal, C.; Ramanathan, R.; Gupta, P.; Parashar, A.; Joel, J. Automobile parts casting-methods and materials used: A review. *Mater. Today Proc.* **2020**, *22*, 2525–2531. [CrossRef]

45. MRT Castings. Casting Process. Available online: https://www.mrt-castings.co.uk/pressure-diecasting-methods.html#:~:text=High%20pressure%20die%20casting%20is,little%20as%201%2D2.5mm (accessed on 23 May 2024).
46. Careri, F.; Khan, R.H.; Todd, C.; Attallah, M.M. Additive manufacturing of heat exchangers in aerospace applications: A review. *Appl. Therm. Eng.* **2023**, *235*, 121387. [CrossRef]
47. Martucci, A.; Aversa, A.; Lombardi, M. Ongoing Challenges of Laser-Based Powder Bed Fusion Processing of Al Alloys and Potential Solutions from the Literature—A Review. *Materials* **2023**, *16*, 1084. [CrossRef]
48. Najafizadeh, M.; Yazdi, S.; Bozorg, M.; Ghasempour-Mouziraji, M.; Hosseinzadeh, M.; Zarrabian, M.; Cavaliere, P. Classification and applications of titanium and its alloys: A review. *J. Alloy. Compd. Commun.* **2024**, *3*, 100019. [CrossRef]
49. Rees, D.T.; Leung, C.L.A.; Elambasseril, J.; Marussi, S.; Shah, S.; Marathe, S.; Brandt, M.; Easton, M.; Lee, P.D. In situ X-ray imaging of hot cracking and porosity during LPBF of Al-2139 with TiB2 additions and varied process parameters. *Mater. Des.* **2023**, *231*, 112031. [CrossRef]
50. Peddaiah, P.C.; Dodla, S. Experimental and numerical investigations of aerospace alloys: Effect of machining. *Proc. Inst. Mech. Eng. Part E J. Process. Mech. Eng.* **2024**. [CrossRef]
51. Rezaei, J. BWM Solvers. Available online: https://bestworstmethod.com/software/ (accessed on 23 May 2024).
52. Kittur, J.K.; Manjunath Patel, G.; Parappagoudar, M.B. Modeling of pressure die casting process: An artificial intelligence approach. *Int. J. Met.* **2016**, *10*, 70–87. [CrossRef]
53. Murugarajan, A.; Raghunayagan, P. The impact of pressure die casting process parameters on mechanical properties and its defects of A413 aluminium alloy. *Metalurgija* **2019**, *58*, 55–58.
54. Cao, L.; Li, J.; Hu, J.; Liu, H.; Wu, Y.; Zhou, Q. Optimization of surface roughness and dimensional accuracy in LPBF additive manufacturing. *Opt. Laser Technol.* **2021**, *142*, 107246. [CrossRef]
55. Yang, T.; Liu, T.; Liao, W.; Wei, H.; Zhang, C.; Chen, X.; Zhang, K. Effect of processing parameters on overhanging surface roughness during laser powder bed fusion of AlSi10Mg. *J. Manuf. Process.* **2021**, *61*, 440–453. [CrossRef]
56. Priarone, P.C.; Ingarao, G.; Lunetto, V.; Di Lorenzo, R.; Settineri, L. The Role of re-design for Additive Manufacturing on the Process Environmental Performance. *Procedia CIRP* **2018**, *69*, 124–129. [CrossRef]
57. Chen, N.; Barnawal, P.; Frank, M.C. Automated post machining process planning for a new hybrid manufacturing method of additive manufacturing and rapid machining. *Rapid Prototyp. J.* **2018**, *24*, 1077–1090. [CrossRef]
58. Fuchs, C.; Baier, D.; Semm, T.; Zaeh, M.F. Determining the machining allowance for WAAM parts. *Prod. Eng.* **2020**, *14*, 629–637. [CrossRef]
59. Ingarao, G.; Priarone, P.C.; Deng, Y.; Paraskevas, D. Environmental modelling of aluminium based components manufacturing routes: Additive manufacturing versus machining versus forming. *J. Clean. Prod.* **2018**, *176*, 261–275. [CrossRef]
60. Cecchel, S.; Cornacchia, G.; Panvini, A. Cradle-to-Gate Impact Assessment of a High-Pressure Die-Casting Safety-Relevant Automotive Component. *JOM* **2016**, *68*, 2443–2448. [CrossRef]
61. Liu, W.; Tang, R.; Peng, T. An IoT-enabled Approach for Energy Monitoring and Analysis of Die Casting Machines. *Procedia CIRP* **2018**, *69*, 656–661. [CrossRef]
62. EOS GmbH. EOS M 290. Available online: https://www.eos.info/en-us/metal-solutions/metal-printers/data-sheets/sds-eos-m-290 (accessed on 10 August 2024).
63. EOS GmbH. EOS Aluminium AlSi2139 AM. Available online: https://www.eos.info/en-us/metal-solutions/metal-materials/aluminium#eos-aluminium-alsi10mg (accessed on 23 September 2024).
64. EOS GmbH. EOS Titanium Ti6Al4V. Available online: https://www.eos.info/en-us/metal-solutions/metal-materials/titanium#eos-titanium-ti64 (accessed on 23 September 2024).
65. Leal, R.; Barreiros, F.M.; Alves, L.; Romeiro, F.; Vasco, J.C.; Santos, M.; Marto, C. Additive manufacturing tooling for the automotive industry. *Int. J. Adv. Manuf. Technol.* **2017**, *92*, 1671–1676. [CrossRef]
66. Sgarbossa, F.; Peron, M.; Lolli, F.; Balugani, E. Conventional or additive manufacturing for spare parts management: An extensive comparison for Poisson demand. *Int. J. Prod. Econ.* **2021**, *233*, 107993. [CrossRef]
67. Atzeni, E.; Salmi, A. Economics of additive manufacturing for end-usable metal parts. *Int. J. Adv. Manuf. Technol.* **2012**, *62*, 1147–1155. [CrossRef]
68. 3D Natives. A Comprehensive List of All the Metal 3D Printer Manufacturers. Available online: https://www.3dnatives.com/en/metal-3d-printer-manufacturers/ (accessed on 12 August 2024).
69. 3D SYSTEMS. DMP Flex 200. Available online: https://www.3dsystems.com/3d-printers/dmp-flex-200 (accessed on 12 August 2024).
70. 3D SYSTEMS. DMP Factory 350. Available online: https://www.3dsystems.com/3d-printers/dmp-factory-350 (accessed on 12 August 2024).
71. 3D SYSTEMS. DMP Flex 350. Available online: https://www.3dsystems.com/3d-printers/dmp-flex-350 (accessed on 12 August 2024).
72. 3D SYSTEMS. DMP Factory 500. Available online: https://www.3dsystems.com/3d-printers/dmp-factory-500 (accessed on 12 August 2024).
73. Colibrium Additive. M2 Series 5. Available online: https://www.colibriumadditive.com/printers/l-pbf-printers/m2-series-5 (accessed on 10 August 2024).

74. Colibrium Additive. X Line 2000R. Available online: https://www.colibriumadditive.com/printers/l-pbf-printers/x-line-2000r (accessed on 10 August 2024).
75. Colibrium Additive. M Line. Available online: https://www.colibriumadditive.com/printers/l-pbf-printers/m-line (accessed on 10 August 2024).
76. DMG MORI. LASERTEC 12 SLM. Available online: https://uk.dmgmori.com/products/machines/additive-manufacturing/powder-bed/lasertec-12-slm (accessed on 10 August 2024).
77. DMG MORI. LASERTEC 30 DUAL SLM. Available online: https://uk.dmgmori.com/products/machines/additive-manufacturing/powder-bed/lasertec-30-slm?_gl=1*l463zu*_gcl_au*MTc2NjEwMDMxNi4xNzIyOTMxNDcy*_ga*MTU4NDc1MjA5Ny4xNzIyOTMxNDcz*_ga_XQ3E6CJXX5*MTcyMzU2NjY4Ni4yLjEuMTcyMzU2NzY3OS41MS4wLjA. (accessed on 10 August 2024).
78. EOS GmbH. EOS M 300-4. Available online: https://uk.eos.info/en-gb/industrial-3d-printer/metal/eos-m-300-4 (accessed on 10 August 2024).
79. EOS GmbH. EOS M 400. Available online: https://uk.eos.info/en-gb/industrial-3d-printer/metal/eos-m-400 (accessed on 10 August 2024).
80. EOS GmbH. EOS M 400-4. Available online: https://uk.eos.info/en-gb/industrial-3d-printer/metal/eos-m-400-4 (accessed on 10 August 2024).
81. Farsoon Technologies. FS121M. Available online: https://www.farsoon-gl.com/products/fs121m/ (accessed on 10 August 2024).
82. Farsoon Technologies. FS273M. Available online: https://www.farsoon-gl.com/products/fs273m/ (accessed on 10 August 2024).
83. Farsoon Technologies. FS200M. Available online: https://www.farsoon-gl.com/products/fs200m/ (accessed on 10 August 2024).
84. Farsoon Technologies. FS301M. Available online: https://www.farsoon-gl.com/products/fs301m/ (accessed on 10 August 2024).
85. Farsoon Technologies. FS350M-4. Available online: https://www.farsoon-gl.com/products/fs350m-4/ (accessed on 10 August 2024).
86. Farsoon Technologies. FS422M. Available online: https://www.farsoon-gl.com/products/fs422m/ (accessed on 10 August 2024).
87. Farsoon Technologies. FS721M-CAMS. Available online: https://www.farsoon-gl.com/products/fs721m-cams/ (accessed on 10 August 2024).
88. Farsoon Technologies. FS721M. Available online: https://www.farsoon-gl.com/products/fs721m/ (accessed on 10 August 2024).
89. Farsoon Technologies. FS621M. Available online: https://www.farsoon-gl.com/products/fs621m/ (accessed on 10 August 2024).
90. Matsuura Machinery. LUMEX Avance-25. Available online: https://www.lumex-matsuura.com/english/lumex-avance-25 (accessed on 10 August 2024).
91. Matsuura Machinery. LUMEX Avance-60. Available online: https://www.lumex-matsuura.com/english/lumex-avance-60 (accessed on 10 August 2024).
92. Prima Additive. Print Sharp 150. Available online: https://www.primaadditive.com/en/technologies/powder-bed-fusion/print-sharp-150 (accessed on 10 August 2024).
93. Prima Additive. Print Genius 150. Available online: https://www.primaadditive.com/en/technologies/powder-bed-fusion/print-genius-150 (accessed on 10 August 2024).
94. Prima Additive. Print Green. Available online: https://www.primaadditive.com/en/technologies/powder-bed-fusion/print-green-150 (accessed on 10 August 2024).
95. Prima Additive. 300 Family. Available online: https://www.primaadditive.com/en/technologies/powder-bed-fusion/300-family (accessed on 10 August 2024).
96. Prima Additive. Print Genius 400. Available online: https://www.primaadditive.com/en/technologies/powder-bed-fusion/print-genius-400 (accessed on 10 August 2024).
97. Renishaw. RenAM 500. Available online: https://www.renishaw.com/en/renam-500-metal-additive-manufacturing-3d-printing-systems--37011 (accessed on 10 August 2024).
98. SLM Solutions. SLM 125. Available online: https://slm-solutions.com/products-and-solutions/machines/slm-125/ (accessed on 10 August 2024).
99. SLM Solutions. SLM 280 PS. Available online: https://www.slm-solutions.com/products-and-solutions/machines/slm-280-production-series/ (accessed on 10 August 2024).
100. SLM Solutions. SLM 280 2.0. Available online: https://www.slm-solutions.com/products-and-solutions/machines/slm-280/ (accessed on 10 August 2024).
101. SLM Solutions. SLM 500. Available online: https://www.slm-solutions.com/products-and-solutions/machines/slm-500/ (accessed on 10 August 2024).
102. SLM Solutions. SLM 800. Available online: https://www.slm-solutions.com/products-and-solutions/machines/slm-800/ (accessed on 10 August 2024).
103. SLM Solutions. SLM NXG XII 600. Available online: https://www.slm-pushing-the-limits.com/ (accessed on 10 August 2024).
104. Sharebot. metalONE. Available online: https://sharebot.us/metalone/ (accessed on 10 August 2024).
105. TRUMPF. TruePrint 1000. Available online: https://www.trumpf.com/en_GB/products/machines-systems/additive-production-systems/truprint-1000/ (accessed on 10 August 2024).
106. TRUMPF. TruePrint 1000 Basic Edition. Available online: https://www.trumpf.com/en_GB/products/machines-systems/additive-production-systems/truprint-1000-basic-edition/ (accessed on 10 August 2024).

107. TRUMPF. TruePrint 2000. Available online: https://www.trumpf.com/en_GB/products/machines-systems/additive-production-systems/truprint-2000/ (accessed on 10 August 2024).
108. TRUMPF. TruePrint 3000. Available online: https://www.trumpf.com/en_GB/products/machines-systems/additive-production-systems/truprint-3000/ (accessed on 10 August 2024).
109. TRUMPF. TruePrint 5000. Available online: https://www.trumpf.com/en_GB/products/machines-systems/additive-production-systems/truprint-5000/ (accessed on 10 August 2024).
110. TRUMPF. TruePrint 5000 Green Edition. Available online: https://www.trumpf.com/en_GB/products/machines-systems/additive-production-systems/truprint-5000-green-edition/ (accessed on 10 August 2024).
111. Velo3D. Sapphire and Sapphire 1MZ Printers. Available online: https://velo3d.com/product-brief-sapphire-and-sapphire-1mz-printer/ (accessed on 10 August 2024).
112. Velo3D. Sapphire XC and Sapphire XC 1MZ Printers. Available online: https://velo3d.com/products/#sapphire (accessed on 10 August 2024).

designs

MDPI

Article

Three-Dimensional Printed Auxetic Insole Orthotics for Flat Foot Patients with Quality Function Development/Theory of Inventive Problem Solving/Analytical Hierarchy Process Methods

Tadeus Pantryan Simarmata [1], Marcel Martawidjaja [2], Christian Harito [1,*] and Cokisela C. L. Tobing [3]

1 Industrial Engineering Department, BINUS Graduate Program—Master of Industrial Engineering, Bina Nusantara University, Jakarta 11480, Indonesia; tadeus.simarmata@binus.ac.id
2 Industrial Engineering Department, Faculty of Engineering, Bina Nusantara University, Jakarta 11480, Indonesia; marcel.martawidjaja@binus.ac.id
3 Dare Prosthetic & Orthotic Service, Bekasi 17147, Indonesia; cokiselachristian@gmail.com
* Correspondence: christian.harito@binus.ac.id

Abstract: Foot disorders affect approximately 10% of adults, with plantar heel pain significantly impacting foot-related quality of life and altering walking patterns. Flat feet, characterized by a lack of longitudinal arches, can lead to fatigue during walking. This study aims to develop 3D-printed shoe insoles tailored to the needs of patients. The design process incorporates Quality Function Deployment (QFD), Theory of Inventive Problem Solving (TRIZ), and Analytic Hierarchy Process (AHP) methods to create insoles that alleviate concentrated loads while meeting patient requirements. The AHP analysis indicated that patients prioritize insoles that effectively manage pressure distribution to achieve optimal functionality. QFD and TRIZ facilitated the identification of four product alternatives and production specifications. The analysis indicated that 3D-printed insoles made from TPU filament with 20% auxetic infill best align with patient preferences. This auxetic TPU option emerged as the top choice, achieving a priority value of 0.2506 due to its superior functionality and comfort. Load distribution measurements confirmed that TPU with auxetic infill resulted in the lowest load distribution, with a standard deviation of 0.1434 and a 25.4% reduction in maximum load compared to conditions without the insole.

Keywords: insole; QFD; TRIZ; AHP; auxetic; 3D print

1. Introduction

The foot is a crucial organ for walking and serves as one of the most important sensory organs in the human body. Improper use of the feet can lead to irreversible shape changes, affecting plantar pressure at various levels [1]. The arch of the foot plays a vital role in absorbing impact pressure and is responsible for the functional stabilization of the body during static and dynamic activities such as standing and walking [2]. The longitudinal arch is a key component in the biomechanics of the foot, helping to maintain stability while standing, facilitating weight distribution over a broader area, enhancing speed and agility during movement, and providing both stability and flexibility. The longitudinal arch is formed by the tarsal and metatarsal bones, ligaments, and tendons. Based on the structure of the longitudinal arch, human foot shapes are categorized into three types: normal feet, flat feet, and cavus feet [3].

Foot disease is a common problem that accounts for approximately 1 million patient visits per year, with approximately 60% of these to primary care physicians. This disease is the most common cause of heel pain in adults, with a lifetime incidence of approximately 10% and an increasing incidence in women aged 40 to 60 years. Plantar heel disease is associated with various types of sports but is mostly reported in recreational activities and professional runners (incidence 5% to 10%) [4]. The presence of plantar heel pain affects foot-related quality of life and changes the way people walk. Therefore, to effectively treat plantar heel pain, treatment must be optimized to reduce its burden [5]. Diseases of the soles of the feet can occur due to several factors, such as genes, accidents, and other diseases that cause changes in foot morphology. Examples of foot diseases are Pes Planus or flat foot [6].

Flat feet (Pes Planus) are known to be associated with a high incidence of lower extremity injuries in the population. The occurrence of flat feet can be caused by multifactorial causes. This disease can appear at birth (congenital Pes Planus) or develop later in life (acquired Pes Planus). Flat feet are an anatomical change that can occur in one foot (unilateral Pes Planus) or in both feet (bilateral Pes Planus) [7].

Foot orthoses are used to treat foot pathologies such as plantar fasciopathies. Foot orthoses are still frequently used to support the arch of the foot in individuals with flat feet and alter lower extremity biomechanics during walking, running, and jumping. Therefore, foot orthoses are still used by doctors to treat painful musculoskeletal leg injuries. Compared with traditionally manufactured foot orthoses, 3D-printed foot orthoses decrease plantar pressure under the heel and reduce the sagittal range of motion, dorsiflexion when the heel moves, and maximum eversion of the ankle when walking.

Three-dimensional printing, also referred to as additive manufacturing, is accomplished through layer-by-layer stacking techniques. According to the designed 3D model, complex and diverse physical entities can be produced [8]. Recently, biomedical applications have evolved significantly due to the dedication of scientists, medical practitioners, engineers, and researchers worldwide. With ongoing advancements in technology, researchers are striving to reduce the costs of 3D-printed components and simplify the fabrication process. Examples of 3D printing applications include the production of bones, spinal implants, prosthetics, skin, and organs [9]. Surgeons can develop patient-specific anatomical models for preoperative planning, which enhances surgical precision. Bioprinting has paved the way for the creation of functional tissues and organs, helping to address the shortage of organ transplants. The dental field has also benefited from 3D printing, enabling the production of precise dental crowns, braces, and aligners [8]. Three-dimensional printed foot orthoses were also as effective as traditionally manufactured foot orthoses in supporting arches, demonstrated by similar reductions in arch height. However, other studies reported no differences in peak hindlimb eversion angles and velocities or loading rates during running between 3D-printed and traditionally manufactured foot orthoses [10].

To obtain information effectively and accurately, various plantar pressure measurement systems have been implemented. In general, these systems can be classified into two types, platform systems and shoe insole systems, each of which has advantages in long-term use and mobility. However, shoe insole systems receive more attention than platform systems today due to their wide range of uses. The system maintains its function under repeated and sometimes severe deformations resulting from daily activities but does not cause discomfort when worn [11]. The concept of custom-designed orthotic insoles has gained popularity due to the importance of comfort and preventing injury.

Kuang-Wei Lin conducted research on 3D-printed foot orthoses, which caused a decrease in ankle evertor moment by changing the path of the center of pressure inward [12]. Malia Ho conducted research to determine the biomedical effects, comparing 3D printed insoles and traditional insoles. The results of the study indicated that, with the use of foot orthoses, there is less activation of the plantar flexor muscles. Foot orthoses using 3D printing are more effective in reducing the decrease in foot angle height compared to traditional orthoses [10]. Ramirez also conducted research related to the use of the four TRIZ principles—namely, segmentation, inversion, preservation of new dimensions, and porous materials—as a basis for making foot orthoses with the results that 4 people were uncomfortable, 12 people were less comfortable, 21 people were comfortable, and 8 people were very comfortable with the resulting orthosis soles [13].

The aim of this research is to develop insoles using 3D printing that suit the patient's needs and desires. For this reason, this research uses the QFD, TRIZ, and AHP methods in the process of designing shoe insoles that can reduce the concentrated load on the patient's feet and are in accordance with the patient's needs and desires. QFD is a systematic system used by industry to link consumer needs with product design specifications to be made using House of Quality as the main tool for mapping and analyzing product design requirements and targets. The use of QFD in the product development process can increase the efficiency of time, costs, and techniques required [14].

In selecting the technical methods to be used, this research uses TRIZ. The "Theory of Inventive Problem Solving", also known as TRIZ, was developed by Russian scientist Genrich Altshuller in the 1940s [15]. Spreafico and Russo conducted a critical study of more than 200 case studies from journals on the use of TRIZ in industry. They concluded that TRIZ is one of the most effective and accepted methods for implementing system innovation [16]. TRIZ includes analytical tools for problem solving as well as data-driven tools for system transformation and their theoretical foundations. The TRIZ analysis tool can be used to transform, model, and analyze a problem using all the information about the product problem. The main goal of the TRIZ method is to find the ideal or perfect solution [17].

After finding the required criteria using QFD and alternative methods that will be used using TRIZ, this research uses AHP to determine the priority importance of each criterion in QFD and select alternatives that will be used in the shoe sole design process using 3D printing. AHP is a hierarchical weighted decision analysis method that applies network systems theory and multi-objective comprehensive evaluation methods and combines qualitative and quantitative methods to solve multi-objective complex problems [18]. After identifying the criteria to follow, the alternatives are selected in the process of making shoe soles using 3D printing. Therefore, the next stage is the product manufacturing process. With the combination of the three methods above, it is hoped that the shoe sole products made will be in accordance with consumer needs and can increase the efficiency of the development process.

2. Materials and Methods

2.1. Subject

In this study, we examined 5 patients with flat feet, detailed as follows in Table 1.

Table 1. Research subjects.

Subject	Gender	Age	Weight (Kg)	Shoe Size (UK)
Patient 1	Male	26	70	43
Patient 2	Female	31	60	41
Patient 3	Male	31	85	43
Patient 4	Male	16	60	44
Patient 5	Female	21	45	39

2.2. Methods

The methods used in this study can be seen in Figure 1.

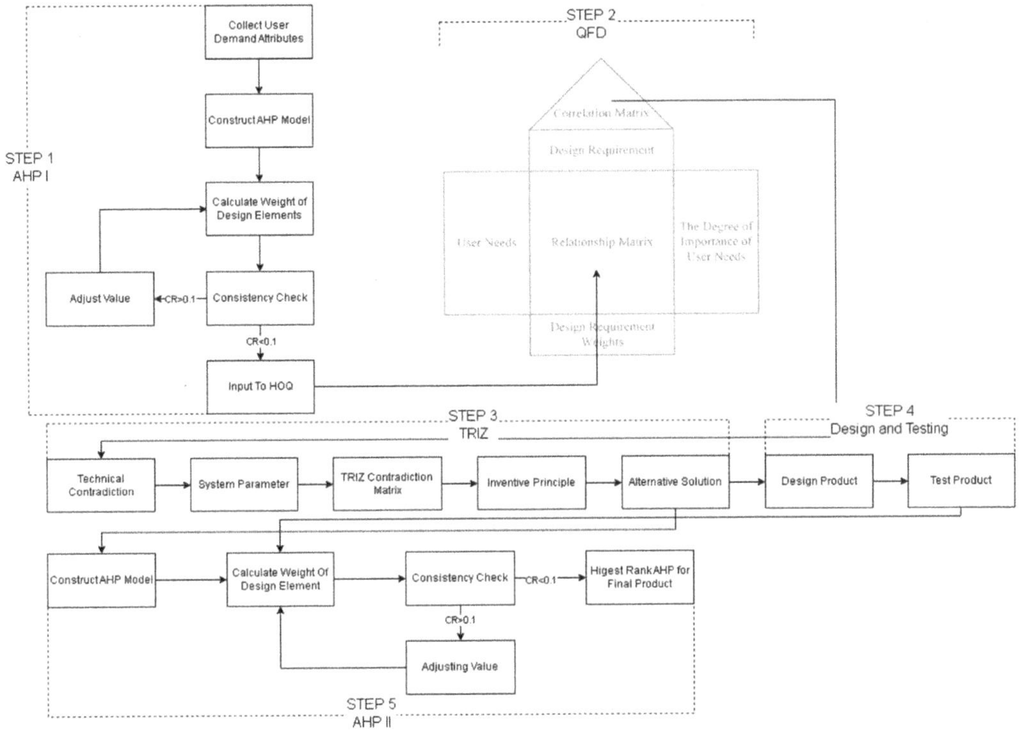

Figure 1. Research flowchart.

2.2.1. AHP I

This stage aims to determine the ranking of the importance of customer needs to be included in the House of Quality under the user needs and relationship matrix. The AHP stages are as follows:

1. Create pairwise comparisons using a scale of 1–9;
2. Establish the comparison matrix;
3. Determine the weight of each element and calculate the eigenvector;
4. Calculate the consistency ratio; if CR > 0.1, the weight values for each pairwise comparison will be recalculated;
5. Rank the consumer needs to be included in the QFD relationship matrix.

Interviews were conducted with consumers to obtain the criteria for consumer needs based on the research conducted by Suchada Rianmora, in her research titled *"Product Characteristics versus Customer Perceptions on a Health-Related Product"*, which identifies six criteria for consumer needs in the development of insoles for individuals with flat feet [19]:

1. Instant pain relief;
2. Odorless;
3. Lightweight;
4. Easy to clean;
5. Durable;

6. Made from various materials.

2.2.2. QFD

This stage aims to determine the correlation between customer needs and the existing technical designs, as well as to identify the correlation matrix among each technical design, which will be resolved using TRIZ. In the study titled *"Product Characteristics versus Customer Perceptions on a Health-Related Product"*, Suchada Rianmora identified five technical designs to produce insoles for individuals with flat feet, which are as follows [19]:

1. Weight
2. Dimension;
3. Material used;
4. Mean lifetime;
5. Shape.

These five technical design criteria served as a reference in this research. The QFD stages are as follows:

1. Enter customer desires into the user needs table;
2. Enter the available technical designs into the design requirement table;
3. Determine the correlation matrix among each technical design;
4. Establish the relationship matrix between customer needs and design requirements. For the customer needs values, use the results from AHP I;
5. Calculate the ranking of the design requirements.

2.2.3. TRIZ

This stage aims to resolve contradictions among technical designs by utilizing the 40 inventive principles obtained from the TRIZ contradiction matrix. The TRIZ stages are as follows:

1. Identify the contradictory technical designs from the QFD results;
2. Select the system parameters for each technical contradiction;
3. Use the TRIZ contradiction matrix to determine the inventive principles that will be applied to resolve the technical contradictions;
4. Create alternative specifications/decisions based on the inventive principles obtained from the contradiction matrix.

2.2.4. Design and Testing

This stage aims to facilitate the product development process in accordance with the alternative specifications proposed by QFD and TRIZ. The resulting product will undergo testing with consumers to determine the weight of design elements that will be used in AHP II as the basis for selecting the final product.

Steps in the design phase:

1. The subject's foot was scanned in three dimensions using a 3D scanner (EinScan Scanner). To prevent the subsidence of the elevated navicular bone during contact with the ground, the scanning was conducted in a non-weight-bearing state, with the subject seated on a chair and the foot suspended in the air. In this non-weight-bearing condition, the inner arch of the foot did not collapse, allowing the height of the arch to be maintained, which further enhanced its shock-absorbing capability [20];
2. A 3D model of the insole was created using the Gensole website and the product was finalized using Blender software (version 4.2);
3. The product was sliced using the Bambu Lab slicer (version 1.9.7.52);
4. The product was 3D printed using the Bambu Lab X1C 3D printer.

The steps in the testing stage occurred as follows:

1. Product testing was conducted with consumers to gather feedback on the insole's performance and comfort;
2. The pressure distribution on the consumer's foot was checked using the **RPPS-2500 array sensor pressure distribution system**. RPPS-2500 has a 370 × 385 sensor size, with 350 mm × 350 mm sensor sensing area size, actuation force 0.1~5 kg, single sensing point 10 mm rubber. The process of checking the distribution of the foot arch using the RPPS-2500 can be seen in Figure 2. The results from the testing phase were used as the basis for the next steps.

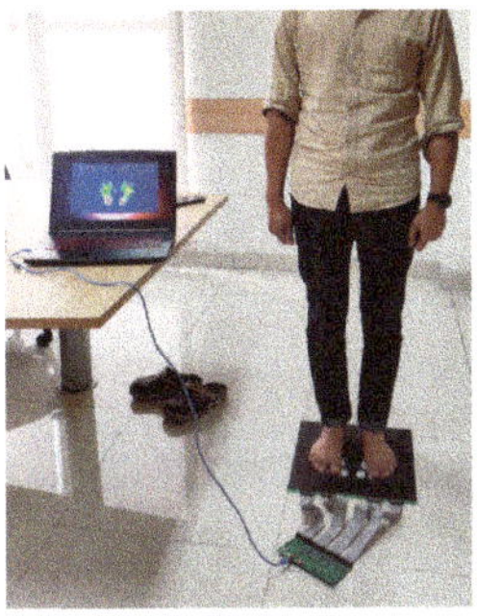

Figure 2. Testing the distribution of load using the RPPS 2500 array sensor distribution system.

2.2.5. AHP II

This stage aims to make the final selection from several alternatives proposed by QFD and TRIZ, where the weights of the elements are derived from direct surveys conducted with consumers who have used all the offered alternatives. The outcome of this AHP process is the final product, which represents the highest value of the priority vector.

3. Results

3.1. STEP 1. AHP I

Suchada Rianmora, in the research titled "*Product Characteristics versus Customer Perceptions on a Health-Related Product*", identified six consumer needs criteria for the production of insoles for individuals with flat feet, which are as follows [19]:

1. Instant pain relief;
2. Odor-free;
3. Lightweight;
4. Easy to clean;
5. Durable;
6. Made from various types of materials.

This development is based on interviews conducted with five individuals suffering from flat feet, which yielded additional criteria to ensure a more comprehensive insole design that meets consumer needs:

1. Good durability;
2. Lightweight;
3. Short production time;
4. Reasonable price;
5. Comfortable to use;
6. Easy to clean;
7. Performs well.

A hierarchical AHP model is created based on these seven primary consumer needs, as illustrated in Figure 3.

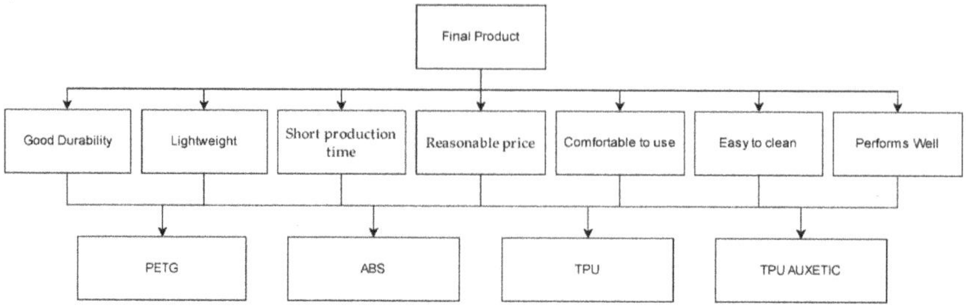

Figure 3. AHP Hierarchy I: comparison between criteria.

From the AHP Hierarchy I: for the comparison between criteria, pairwise comparisons are conducted for each criterion needed by consumers, as shown in Appendix A. The results from these pairwise comparisons will be divided by the total in each column to obtain the AHP normalization matrix. The results from each row in the normalization matrix are presented in Tables A1–A5 at Appendix A. The normalization matrix will then be divided by the total number of criteria to obtain the priority values for each criterion. To determine the eigenvector values, the priority results for each criterion will be divided by the total of the columns in the pairwise comparison matrix.

The AHP was conducted on the five consumers, and the results of the consistency ratios (CRs) are presented in Table 2.

Table 2. The consistency ratio (CR) for each consumer.

Consumer	Consistency Ratio
1	0.091
2	0.076
3	0.095
4	0.070
5	0.053

Based on the consistency ratio results for each consumer, it was found that CR < 0.1, indicating that the AHP results for each consumer are consistent. Once the AHP for each consumer was confirmed to be consistent, the values of each cell in the AHP for each consumer were combined into a consolidated AHP using the geometric mean formula in Excel, as shown in Table 3.

Table 3. Comparison of criteria among the combined five consumers with Geomean.

Criteria	Durability	Lightweight	Production Time	Price	Comfort	Easy to Clean
Durability	1.000	3.178	7.560	5.619	1.644	6.949
Lightweight	0.315	1.000	4.522	3.554	0.392	4.939
Production time	0.132	0.221	1.000	0.415	0.209	1.246
Price	0.178	0.281	2.408	1.000	0.257	2.371
Comfort	0.608	2.551	4.782	3.898	1.000	1.165
Easy to clean	0.144	0.202	0.803	0.422	0.859	1.000
Performs well	2.766	4.573	8.586	7.634	2.862	7.432

The comparison of criteria among the five consumers will involve normalizing the matrix by dividing each cell by the total of its respective column. To obtain the total for each row in the normalized matrix, the sum of the criteria will be used to derive the priority for each criterion. To calculate the eigenvector values, each priority will be multiplied by the total of each criterion in the comparison matrix, as presented in Table 4.

Table 4. Priorities and eigen values for the combined five consumers.

Criteria	Priority	Eigen Value	Rank
Durability	0.232	1.19333	2
Lightweight	0.114	1.37258	4
Production time	0.032	0.95439	7
Price	0.053	1.1923	5
Comfort	0.143	1.03264	3
Easy to clean	0.044	1.10287	6
Performs well	0.382	0.8823	1

The results of AHP I indicate that the comparison among criteria is consistent, as the CR value for each consumer is less than 0.1. The ranking of priorities, from the most important to the least important, is as follows: maximum functionality, high durability, comfort in use, lightweight, affordable price, ease of cleaning, and short production time. The values of each priority for the criteria will be incorporated into the House of Quality in the QFD phase.

3.2. STEP 2. QFD

In our study, we have developed eight technical designs based on broader consumer needs, which include the following:

1. Stable structure;
2. Foot-shape compatibility;
3. Comfort in use;
4. Effective heat absorption;
5. Lightweight;
6. Low production cost;
7. Ease of production;
8. Aesthetic shape.

From the eight technical aspects, a relationship matrix will be created to illustrate the connection between consumer needs and technical aspects, as shown in Table 5. The values will be categorized into three criteria, which are as follows:

1. High relationship between needs and technical aspects: 9;
2. Medium relationship between needs and technical aspects: 3;
3. Low relationship between needs and technical aspects: 1.

Table 5. Relationship matrix between consumer needs and technical aspects.

Relationship Matrix	Stable Structure	Foot-Shape Compatibility	Comfort in Use	Effective Heat Absorption	Lightweight	Low Production Cost	Ease of Production	Aesthetic Shape
Durability	9	1	9	9	1	3	1	1
Lightweight	9	1	9	9	9	9	3	1
Production time	3	9	3	3	3	9	9	1
Price	9	1	1	3	1	9	9	3
Comfort	9	9	1	1	9	1	1	9
Easy to clean	9	1	1	1	9	1	1	1
Performs well	9	1	9	9	3	1	1	1

The values in each relationship matrix, as shown in Table 5, will be multiplied by the respective priority values from AHP I to be incorporated into the relationship matrix in the House of Quality, as illustrated in Figure 4. The results from the House of Quality in this study indicate that having a stable structure is the primary priority among the technical aspects in the production of orthotic insoles.

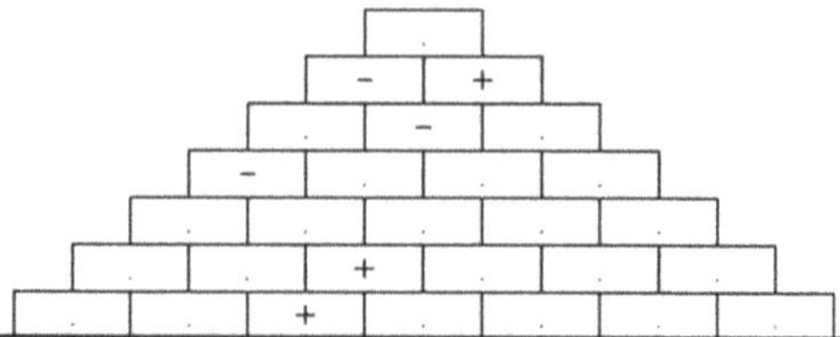

Design Requirements / Customer Requirements	Importance	Stable Structure	Foot Shape Compatibility	Comfort in Use	Effective Heat Absorption	Lightweight	Low Production Cost	Ease of Production	Aesthetic Shape
Durability	0.23	9	1	9	9	1	3	1	1
Lightweight	0.11	9	1	9	9	9	9	3	1
Production Time	0.03	3	9	3	3	3	9	9	1
Proce	0.05	9	1	1	3	1	9	9	3
Comfort	0.14	9	9	1	1	9	1	1	9
Easy to Clean	0.04	9	1	1	1	9	1	1	1
Performs Well	0.38	9	1	9	9	3	1	1	1
Technical Importance: Rating		8.807	2.401	6.888	6.994	4.238	3.059	1.909	2.25
Technical Importance: Rank		1	6	3	2	4	5	8	7

Figure 4. House of Quality structure.

At this stage, the relationships among the technical aspects are also determined, which can be observed in the roof section of the House of Quality. Here, (+) indicates a strong positive relationship between technical aspects; (.) signifies no relationship; and (−) denotes a contradiction among the technical aspects. Any contradictions identified among the technical aspects will be addressed using TRIZ in the subsequent TRIZ phase.

3.3. STEP 3. TRIZ

In the results of the House of Quality, there are three technical parameters that exhibit contradictory relationships. Each technical parameter with a contradictory relationship has its own system parameter from the 39 TRIZ system parameters, which are referenced in Table 6.

Table 6. Technical contradictions and system parameters.

Technical Contradiction	Technical Variable	System Parameter
1	Stable structure Lightweight	Stress or pressure Weight of moving object
2	Ease of production Aesthetic shape	Ease of manufacture Device complexity
3	Stable structure Ease of production	Stress or pressure Ease of manufacture

From each system parameter in Table 6, alternative improvements can be identified using the TRIZ contradiction matrix. The contradiction matrix is a structured arrangement of 39 improvement parameters and 39 worsening parameters (a 39 × 39 matrix). This is organized as a grid with 39 rows and columns, serving as a tool for analyzing the interactions between these features. In the intersection boxes where the parameters of two features converge, inventive principles for addressing the specific problem are organized by frequency, where each cell entry provides the most frequently applied inventive principles to resolve or eliminate contradictions within the technical domain [21].

At this stage, the selected parameters for each technical contradiction from the previous section are cross-referenced in the matrix to identify a set of inventive principles. The contradiction matrix table can present effective solutions to be utilized in addressing technical problems. The results of the matrix table for resolving the technical contradictions in this study can be seen in red frame at Figure 5.

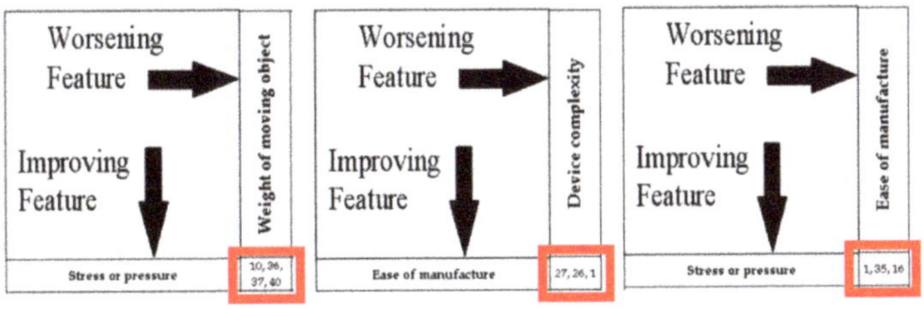

Figure 5. Alternative improvements based on the TRIZ contradiction matrix.

The first contradiction issue can be resolved using the principles of preliminary action (10), phase transitions (36), thermal expansion (37), and composite materials (40). The second technical issue can be addressed through the principles of cheap-short-lived (27), copying (26), and segmentation (1). The third technical issue can be resolved by applying the principles of segmentation (1), parameter change (35), and partial or excessive action (16).

Based on the resolution of the first contradiction, the application of preliminary action, phase transitions, thermal expansion, and composite material ensures that the product manufacturing process is not adversely affected by thermal expansion. In this study, three types of alternative materials were utilized: ABS (Acrylonitrile Butadiene Styrene), PETG (Polyethylene Terephthalate Glycol-Modified), and TPU (Thermoplastic Polyurethane). These materials were selected due to their lightweight properties and adequate pressure resistance, making them suitable to produce orthotic insoles.

For the second technical contradiction, several alternative solutions are available, including cheap-short-lived, copying, and segmentation. This study employs 3D printing as the manufacturing process for insoles. The use of 3D printing technology in this research was chosen due to its ability to create insoles with complex geometries. This technology offers a wider range of material options, along with lower production costs and faster manufacturing times compared to traditional methods [22]. Orthotic insoles can be produced using foot impressions in a foam box; however, this method often suffers from low precision and accuracy when the insoles are fitted to the patient's foot shape, frequently resulting in less comfortable orthotic footwear [23]. A comparison of the production processes for insoles can be seen in Table 7.

Table 7. Comparison of insole production processes.

Categories	Traditional	3D Printing (FDM)	Podograph Insole Machine
Geometric Complexity	Low	High	High
Insole Material Type	Low	High	Low
Additive Material	Low	High	Low
Material Cost per Insole	High	Low	Low
Equipment Cost	Low	Low	High
Customization Ease	Low	High	Low

The type of 3D printing used in this study is Fused Deposition Modeling (FDM). One unique advantage of FDM technology is its ability to adjust the infill density of the object, which significantly reduces weight compared to other 3D printing technologies. The key benefits of FDM technology include its low cost, lightweight, and process simplicity compared to other 3D printing methods [24].

The third technical contradiction has alternative solutions, including segmentation, parameter change, and partial or excessive action. To address this issue, auxetic structures can be utilized as infill in 3D printed products. Auxetic infill is a distinctive porous structure that exhibits lateral expansion under axial tension while contracting laterally under compression. This concept is based on the research conducted by Tong Chen titled "*A Novel Porous Structural Design of the Orthotic Insole for Diabetic Foot*", which employed auxetic infill in flat insoles and regular infill in load-bearing areas, resulting in a lighter product with a stronger structure [25].

The results from the TRIZ analysis in this study identified four types of alternative products that will proceed to the design and testing phases, namely:

1. Using ABS material;
2. Using PETG material;
3. Using TPU material;
4. Using TPU material with an auxetic structure.

Auxetic structures function optimally in the plastic phase of materials [26]. Therefore, in this study, auxetic structures are only applied to elastic materials, specifically TPU (Thermoplastic Polyurethane). Meanwhile, ABS (Acrylonitrile Butadiene Styrene) and PETG (Polyethylene Terephthalate Glycol-Modified) are classified as brittle materials, making them unsuitable for the application of auxetic structures. Thus, the focus of this research is on the use of auxetic structures in materials that possess adequate elasticity properties to support the desired performance and functionality [27,28].

3.4. STEP 4. Design and Testing

The first stage in the product design process involves scanning the consumer's foot using a 3D scanner to obtain the morphological shape of the foot. After the foot morphology is scanned with the 3D scanner, the scan results are exported as an STL file type to be uploaded to the insole creation website, www.gensole.com (accessed on 5 October 2024). Using this website, the insole is designed to follow the contours of the consumer's foot morphology. Once the insole design is completed using Gensole, the design is imported into the Bambuu Lab application for slicing, enabling it to be printed using a 3D printer, as illustrated in Figure 6. The product will be printed at a 0-degree angle following the shape of the product to achieve the highest Young's modulus value [29].

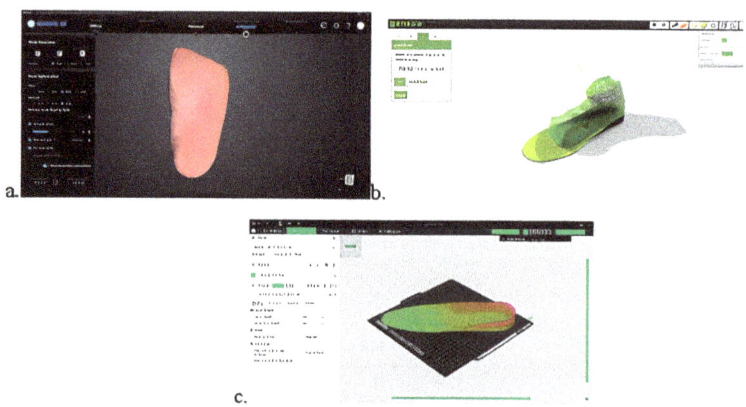

Figure 6. Insole production process: (**a**) 3D scanning of the consumer's foot; (**b**) insole design using the Gensole website; (**c**) slicing process using Bambuu Lab.

Since there is currently no 3D printing slicer that includes an auxetic infill pattern, this study designs the auxetic infill using Blender software (version 4.2). After the auxetic infill design is created with Blender, the infill design is combined with the insole design to achieve the desired insole infill structure, as shown in Figure 7.

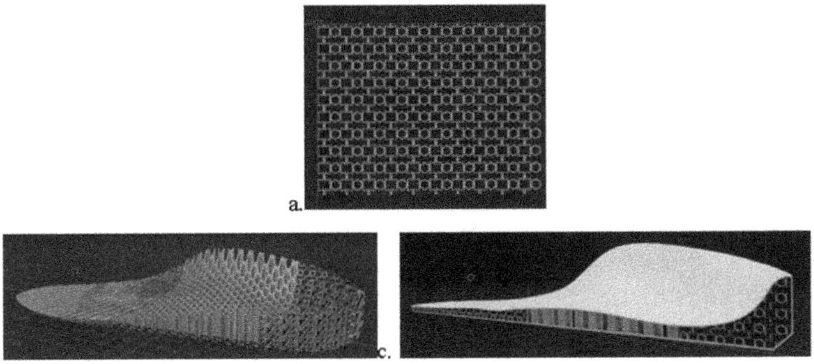

Figure 7. Auxetic insole design process: (**a**) auxetic structure design; (**b**) design structure for insoles; (**c**) combined structure and insole.

The insole is divided into two areas: a soft area located at the forefoot and rearfoot regions; and a support area situated in the midfoot region. The soft area exhibits an 18% greater deformation compared to a common flat structure under pressure. Standing experiments demonstrate that the support area reduces pressure in the forefoot and rearfoot regions by approximately 30% [25]. In this study, a modified auxetic honeycomb structure was utilized, based on research by Aniket Ingrole titled "*Design and Modeling of Auxetic and Hybrid Honeycomb Structures for In-Plane Property Enhancement*". The design selected is Auxetic-Honeycomb 1, as it exhibits the lowest Young's modulus while maintaining a high compressive strength. This makes it suitable for use in soft insoles, ensuring comfort for the user due to the soft material at the front and back of the foot. Additionally, the design provides optimal functionality in the midfoot area as it possesses a higher Young's modulus, ensuring that the contours of the insole retain their shape during use [30]. In this study, a hexagonal infill type is used because the hexagonal infill type has a higher elastic modulus compared to the grid infill type. A 25% infill using the hexagonal type is equivalent to a 50% infill using the grid type [31]. A wall thickness of 2 mm was utilized in this study because a greater wall thickness demonstrated improved performance in terms of stiffness and resistance to deformation under compressive loading [32].

The next step is the production of the products using a 3D printer. All alternatives are printed using the same parameter settings. The ABS, PETG, and TPU filaments used in this study were produced by SUNLU. The results of the printing for each alternative can be seen in Figure 8. After the printing process, all alternatives will undergo testing with consumers to obtain weight values for AHP II. The tests conducted will include comfort testing and functional testing. For the functional test, the Array 2500 sensor pressure distribution system will be used to observe the pressure distribution and the pressure focal points on the consumers' feet, as shown in Figure 9.

The black area in Figure 9 indicates that there is no pressure or very little pressure in that area, making it undetectable by the software. The teal area represents pressure levels between 0 and 0.5 kPa, while the green area indicates pressure levels between 0.5 and 1 kPa. The lavender area shows pressure levels between 1 and 1.5 kPa. The red area indicates that the pressure is above 1.5 kPa. The results without using insoles show a maximum pressure value of 2.167 kPa on the foot sole. In contrast, the maximum pressure value on the foot sole with the ABS insole is 2.020 kPa, with the PETG insole it is 1.806 kPa, with the TPU insole it is 2.088 kPa, and with the TPU insole auxetic infill, the maximum pressure value on the foot sole is 1.620 kPa.

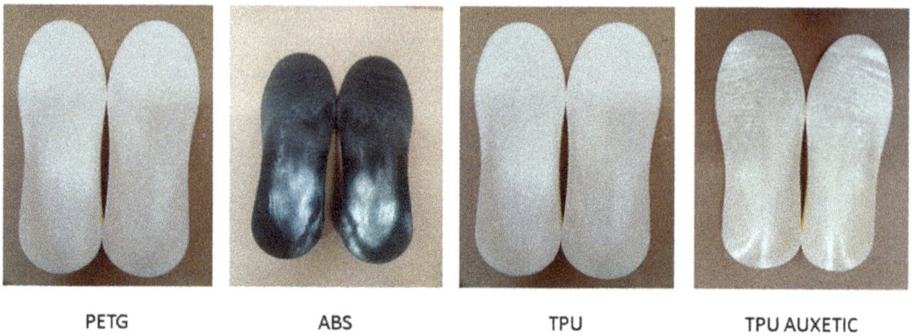

Figure 8. The results of the printing for each alternative.

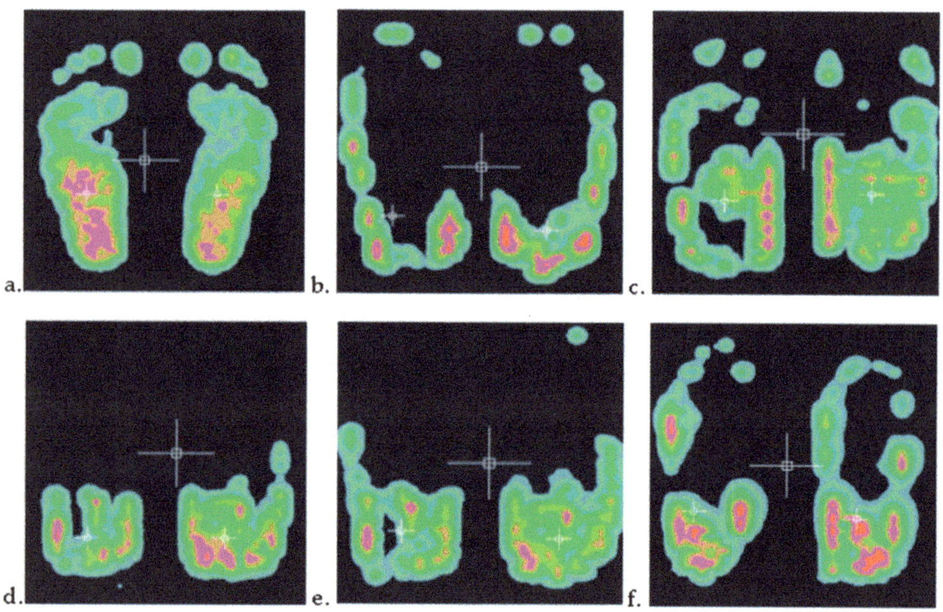

Figure 9. Results of the center-of-mass measurements and load distribution: (**a**) without footwear; (**b**) using ABS material; (**c**) using PETG material; (**d**) using TPU material; (**e**) using TPU material with auxetic structure; (**f**) using traditional insole from marketplace.

The results of this study indicate that the use of insoles leads to a decrease in the maximum load point on the foot. This is consistent with the paper "*A Review of the Plantar Pressure Distribution Effects from Insole Materials and at Different Walking Speeds*" [33].

The results of the pressure distribution testing on the foot were analyzed using an interval plot to assess the load distribution across the foot, as shown in Figure 10. The interval plot results indicate that the TPU auxetic exhibited the most uniform load distribution, with the smallest standard deviation of 0.1434.

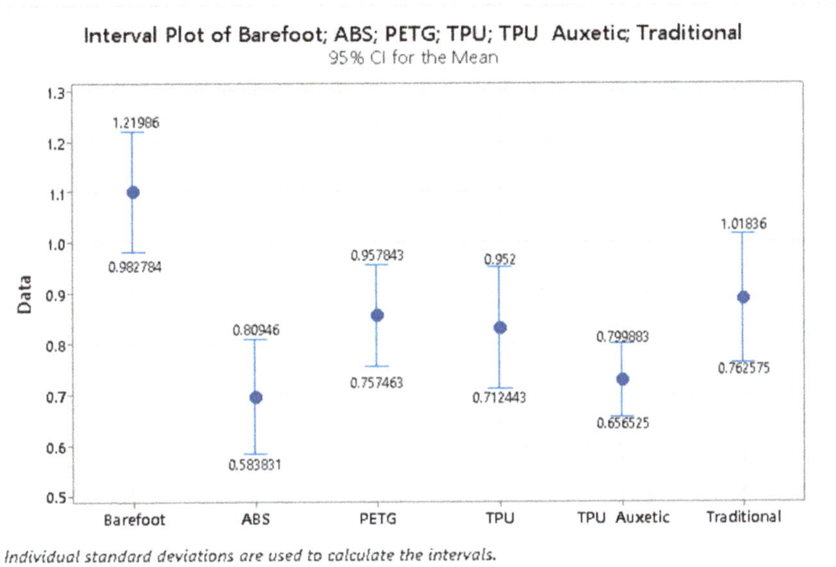

Figure 10. Interval plot of load distribution at each measurement point.

3.5. STEP 5. AHP II

Based on the results of the interviews and the outcomes of QFD and TRIZ, an AHP model was developed to determine the alternatives to be used as the final product, as shown in Figure 11.

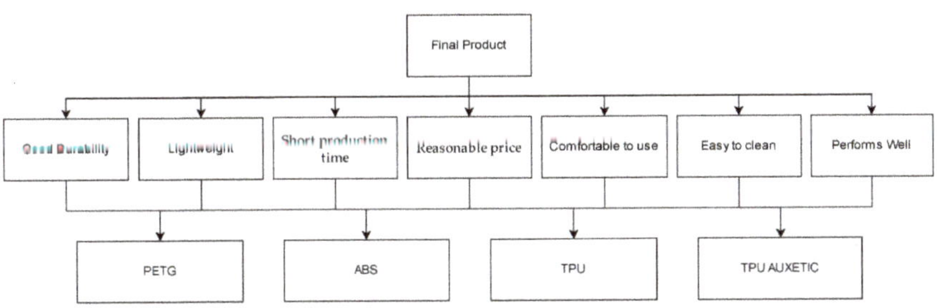

Figure 11. AHP Hierarchy II: comparisons between alternatives.

Using the same method as AHP I, AHP II determined the weights of each alternative for each criterion, with each criterion containing five consumer inputs and a consistency ratio (CR) for each consumer of less than 0.1. By combining the weight values from each consumer for the four materials, Table 8 presents the average weights calculated using the geometric mean for the comfort criterion, based on the results of interviews with patients to determine the weight of each alternative.

Table 8. Average comparison of alternatives for the comfort criterion.

Comfort	ABS	PETG	TPU	TPU Auxetic
ABS	1.000	1.246	0.237	0.117
PETG	0.803	1.000	0.232	0.114
TPU	4.227	4.317	1.000	0.308
TPU auxetic	8.559	8.790	3.245	1.000
Total	14.589	15.354	4.714	1.539

From the pairwise comparison weight matrix of the alternatives, the consistency ratio was calculated for each alternative, and the priority values for each alternative were determined, as shown in Table 9.

Table 9. Priority comparison for each alternative across all criteria.

Alternative	Durability	Lightweight	Production Time	Price	Comfort	Easy to Clean	Performs Well
ABS	0.34	0.09	0.33	0.08	0.068	0.12	0.14
PETG	0.42	0.10	0.45	0.11	0.06	0.17	0.15
TPU	0.13	0.03	0.14	0.04	0.24	0.36	0.19
TPU auxetic	0.11	0.03	0.08	0.02	0.62	0.36	0.53

The comparison data for each criterion across the alternatives will be multiplied by the priority values of each criterion and subsequently ranked according to their priorities, as presented in Table 10.

Table 10. Priority and ranking of each alternative.

Alternative	Total Priority	Rank
ABS	0.165747	3
PETG	0.196854	2
TPU	0.127457	4
TPU auxetic	0.250629	1

4. Discussion

In this study, QFD/TRIZ/AHP were utilized as methods for developing insole orthotics for individuals with flat feet. In the first stage, AHP was employed to translate patient desires into ranked priorities of patient needs, which was incorporated into the House of Quality. The results of AHP I indicate that the maximizing function is the most important criterion in the production of orthotic shoe insoles.

In the QFD stage, a relationship matrix was established between patient needs and existing technical aspects, resulting in the identification of three contradictory problems to be addressed using TRIZ. During the TRIZ phase, solutions for these three contradictory problems were identified as follows:

1. Utilize 3D printing as a production tool to ensure that the resulting product conforms to patient specifications while allowing for rapid production times;
2. Employ three types of alternative materials, ABS, PETG, and TPU, as the base materials due to their high durability and lightweight properties;
3. Implement an auxetic infill type as an alternative infill for TPU material to achieve a more stable structure that is easier to produce.

From these proposed solutions, four types of alternative products were developed for consumer testing to determine the weight of each alternative for use in AHP II. The results of the load distribution and center-of-mass assessment on consumers' feet indicate that using insoles improves load distribution, preventing body weight from concentrating on the central point of the foot, which helps reduce fatigue during walking.

In AHP II, it was determined that the material TPU with an auxetic structure is the primary product choice as it received the highest priority value of 0.2506. This is attributed to the fact that the TPU with auxetic infill offers optimal functionality and high comfort. These findings align with the load distribution measurements, where the TPU with auxetic infill exhibited the lowest load distribution, with a standard deviation of 0.1434, a maximum load of 1620 kPa, and a decrease of 25.4% maximum load.

5. Conclusions

This study aims to develop insole products for orthotics using the QFD/TRIZ/AHP methods. The results of this research identified seven patient needs along with eight technical specifications. The use of auxetic infill effectively resolves technical issues related to the contradiction between stable and lightweight structures. To address the challenge of producing a product that is easy to manufacture while maintaining an aesthetically pleasing design, the use of affordable 3D printing technology is proposed as a solution. Materials such as ABS, PETG, and TPU are recommended to tackle the issue of creating a lightweight product with good pressure resistance. The TPU with auxetic infill is the final product of this research, with a priority value of 0.2506. This product also demonstrates a reduction in value of 25.4% when compared to the condition without the insole. This study also found that structural stability is the most critical technical specification in the production of orthotic insoles. Testing of load distribution on the footbed demonstrated that the use of insoles effectively reduces load distribution on the foot. Overall, the QFD/TRIZ/AHP methodologies can be effectively applied in the product development process of orthotic insoles, prioritizing patient desires while maintaining the primary functions of the insoles.

Author Contributions: Conceptualization: C.H. and C.C.L.T.; methodology: T.P.S., M.M., C.H. and C.C.L.T.; formal analysis: T.P.S. and C.H.; investigation: M.M. and C.C.L.T.; data curation: T.P.S. and M.M.; writing—original draft preparation: T.P.S. and M.M.; writing—review and editing: T.P.S., M.M. and C.H.; visualization: T.P.S. and M.M.; supervision: C.H. and C.C.L.T.; funding acquisition: T.P.S. and C.H. All authors have read and agreed to the published version of the manuscript.

Funding: This study was funded by Direktorat Riset, Teknologi, dan Pengabdian Kepada Masyarakat, Direktorat Jendral Pendidikan Tinggi, Riset, dan Teknologi, Kementerian Pendidikan, Kebudayaan, Riset, dan Teknologi under Penelitian Tesis Magister with contract number 784/LL3/AL.04/2024.

Institutional Review Board Statement: The study was conducted according to the guidelines of the Declaration of Helsinki, and approved by the Institutional Review Board of the Komite Etik Penelitian Kedokteran Fakultas Kedokteran Universitas Jendral Soedirman (010/KEPK/PE/I/2024 and date of approval 20 March 2024).

Informed Consent Statement: Informed consent was obtained from all subjects involved in the study.

Data Availability Statement: The original data presented in the study are openly available in zenodo at https://doi.org/10.5281/zenodo.14752214.

Conflicts of Interest: Author Cokisela C L Tobing was employed by the company Dare Prosthetic & Orthotic Service, Bekasi, Indonesia 17147. The remaining authors declare that the research was conducted in the absence of any commercial or financial relationships that could be construed as a potential conflict of interest.

Appendix A

Table A1. AHP I for consumer 1.

Criteria	Durability	Lightweight	Production Time	Price	Comfort	Easy to Clean
Durability	1.000	3.000	7.000	7.000	2.000	8.000
Lightweight	0.333	1.000	5.000	3.000	0.333	5.000
Production time	0.143	0.200	1.000	0.333	0.200	3.000
Price	0.143	0.333	3.000	1.000	0.250	3.000
Comfort	0.500	3.000	5.000	4.000	1.000	7.000
Easy to clean	0.125	0.200	0.333	0.333	0.143	1.000
Performs well	3.000	5.000	9.000	9.000	3.000	8.000

Table A2. AHP I for consumer 2.

Criteria	Durability	Lightweight	Production Time	Price	Comfort	Easy to Clean
Durability	1.000	3.000	8.000	5.000	2.000	9.000
Lightweight	0.333	1.000	3.000	3.000	0.500	7.000
Production time	0.125	0.333	1.000	0.333	0.200	1.000
Price	0.200	0.333	3.000	1.000	0.333	5.000
Comfort	0.500	2.000	5.000	3.000	1.000	0.143
Easy to clean	0.111	0.143	1.000	0.200	7.000	1.000
Performs well	3.000	5.000	8.000	8.000	4.000	9.000

Table A3. AHP I for consumer 3.

Criteria	Durability	Lightweight	Production Time	Price	Comfort	Easy to Clean
Durability	1.000	4.000	7.000	5.000	3.000	9.000
Lightweight	0.250	1.000	3.000	3.000	0.333	7.000
Production time	0.143	0.333	1.000	0.333	0.200	4.000
Price	0.200	0.333	3.000	1.000	0.333	5.000
Comfort	0.333	3.000	5.000	3.000	1.000	0.143
Easy to clean	0.111	0.143	0.250	0.200	7.000	1.000
Performs well	3.000	5.000	8.000	8.000	4.000	9.000

Table A4. AHP for consumer 4.

Criteria	Durability	Lightweight	Production Time	Price	Comfort	Easy to Clean
Durability	1.000	3.000	9.000	4.000	0.500	5.000
Lightweight	0.333	1.000	7.000	3.000	0.500	4.000
Production time	0.111	0.143	1.000	0.333	0.200	0.500
Price	0.250	0.333	3.000	1.000	0.200	3.000
Comfort	2.000	2.000	5.000	5.000	1.000	5.000
Easy to clean	0.200	0.250	2.000	0.333	0.200	1.000
Performs well	3.000	4.000	9.000	5.000	2.000	7.000

Table A5. AHP for consumer 5.

Criteria	Durability	Lightweight	Production Time	Price	Comfort	Easy to Clean
Durability	1.000	3.000	7.000	8.000	2.000	5.000
Lightweight	0.333	1.000	6.000	7.000	0.333	3.000
Production time	0.143	0.167	1.000	1.000	0.250	0.500
Price	0.125	0.143	1.000	1.000	0.200	0.333
Comfort	0.500	3.000	4.000	5.000	1.000	3.000
Easy to clean	0.200	0.333	2.000	3.000	0.333	1.000
Performs well	2.000	4.000	9.000	9.000	2.000	5.000

References

1. Jiang, Y.; Wang, D.; Ying, J.; Chu, P.; Qian, Y.; Chen, W. Design and preliminary validation of individual customized insole for adults with flexible flatfeet based on the plantar pressure redistribution. *Sensors* **2021**, *21*, 1780. [CrossRef] [PubMed]
2. Wang, Y.-T.; Chen, J.-C.; Lin, Y.-S. Effects of artificial texture insoles and foot arches on improving arch collapse in flat feet. *Sensors* **2020**, *20*, 3667. [CrossRef] [PubMed]
3. Azhagiri, R.; Malar, A.; Hemapriya, J.; Sumathi, G. The cause and frequency of PES Planus (Flat Foot) problems among young adults. *Asian J. Med. Sci.* **2021**, *12*, 107–111. [CrossRef]
4. Trojian, T.; Tucker, A.K. Plantar fasciitis. *Am. Fam. Physician* **2003**, *99*, 744–750. [CrossRef]
5. Bishop, C.; Thewlis, D.; Hillier, S. Custom foot orthoses improve first-step pain in individuals with unilateral plantar fasciopathy: A pragmatic randomised controlled trial. *BMC Musculoskelet. Disord.* **2018**, *19*, 222. [CrossRef]
6. Xu, R.; Wang, Z.; Ma, T.; Ren, Z.; Jin, H. Effect of 3D printing individualized ankle-foot orthosis on plantar biomechanics and pain in patients with plantar fasciitis: A randomized controlled trial. *Med. Sci. Monit.* **2019**, *25*, 1392–1400. [CrossRef]
7. Kodithuwakku Arachchige, S.N.K.; Chander, H.; Knight, A. Flatfeet: Biomechanical implications, assessment and management. *Foot* **2019**, *38*, 81–85. [CrossRef]
8. Pathak, K.; Saikia, R.; Das, A.; Das, D.; Islam, A.; Pramanik, P.; Parasar, A.; Borthakur, P.P.; Sarmah, P.; Saikia, M.; et al. 3D printing in biomedicine: Advancing personalized care through additive manufacturing. *Explor. Med.* **2023**, *4*, 1135–1167. [CrossRef]

9. Kumar, P.; Rajak, D.K.; Abubakar, M.; Ali, S.G.M.; Hussain, M. 3D Printing Technology for Biomedical Practice: A Review. *J. Mater. Eng. Perform.* **2021**, *30*, 5342–5355. [CrossRef]
10. Ho, M.; Nguyen, J.; Heales, L.; Stanton, R.; Kong, P.W.; Kean, C. The biomechanical effects of 3D printed and traditionally made foot orthoses in individuals with unilateral plantar fasciopathy and flat feet. *Gait Posture* **2022**, *96*, 257–264. [CrossRef]
11. Park, J.; Kim, M.; Hong, I.; Kim, T.; Lee, E.; Kim, E.-A.; Ryu, J.-K.; Jo, Y.; Koo, J.; Han, S.; et al. Foot plantar pressure measurement system using highly sensitive crack-based sensor. *Sensors* **2019**, *19*, 5504. [CrossRef] [PubMed]
12. Lin, K.-W.; Hu, C.-J.; Yang, W.-W.; Chou, L.-W.; Wei, S.-H.; Chen, C.-S.; Sun, P.-C. Biomechanical evaluation and strength test of 3d-printed foot orthoses. *Appl. Bionics Biomech.* **2019**, *2019*, 4989534. [CrossRef] [PubMed]
13. Ramírez-Rios, L.Y.; Camargo-Wilson, C.; Olguín-Tiznado, J.E.; López-Barreras, J.A.; Inzunza-González, E.; García-Alcaraz, J.L. Design of a modular plantar orthosis system through the application of triz methodology tools. *Appl. Sci.* **2021**, *11*, 2051. [CrossRef]
14. Al-Dwairi, A.; Al-Araidah, O.; Hamasha, S. An Integrated QFD and TRIZ Methodology for Innovative Product Design. *Designs* **2023**, *7*, 132. [CrossRef]
15. Malhotra, A.; Rajak, S.; Jha, S.K. An eco-innovative green design method by qfd and triz tools-a case study of brass-ware manufacturing. *Pertanika J. Sci. Technol.* **2019**, *27*, 2109–2121.
16. Spreafico, C.; Russo, D. TRIZ Industrial Case Studies: A Critical Survey. *Procedia CIRP* **2016**, *39*, 51–56. [CrossRef]
17. Putri, N.T.; Sutanto, A.; Bifadhlih, N. The improvement of thresher design by using the integration of TRIZ and QFD approach. *Int. J. Prod. Qual. Manag.* **2018**, *25*, 459. [CrossRef]
18. Yuan, G.; Lyu, J.; Wang, Z.; Zhao, H.; Liu, Z. A quality evaluation method for sorting equipment based on AHP and QFD. *Acad. J. Manuf. Eng.* **2019**, *17*, 7–16.
19. Rianmora, S.; Werawatganon, S. Applying quality function deployment in open innovation engineering. *J. Open Innov. Technol. Mark. Complex.* **2021**, *7*, 26. [CrossRef]
20. Joo, J.-Y.; Kim, Y.-K. Effects of Customized 3D-printed Insoles on the Kinematics of Flat-footed Walking and Running. *Korean J. Sport Biomech.* **2018**, *28*, 237–244. [CrossRef]
21. Altshuller, G. *40 Principles: TRIZ Keys to Technical Innovation*; Technical Innovation Center, Inc.: Worcester, MA, USA, 2002; Volume 1.
22. Davia-Aracil, M.; Hinojo-Pérez, J.J.; Jimeno-Morenilla, A.; Mora-Mora, H. 3D printing of functional anatomical insoles. *Comput. Ind.* **2018**, *95*, 38–53. [CrossRef]
23. Anggoro, P.; Bawono, B.; Jamari, J.; Tauviqirrahman, M.; Bayuseno, A. Advanced design and manufacturing of custom orthotics insoles based on hybrid Taguchi-response surface method. *Heliyon* **2021**, *7*, e06481. [CrossRef] [PubMed]
24. Kumar, R.; Sarangi, S.K. 3D-Printed Orthosis: A Review on Design Process and Material Selection for Fused Deposition Modeling Process. In *Advances in Materials Processing and Manufacturing Applications*; Springer: Singapore, 2021; pp. 531–538. [CrossRef]
25. Chen, T.; Tian, M.; Wang, X. A Novel Porous Structural Design of the Orthotic Insole for Diabetic Foot. In Proceedings of the 2021 International Conference on Computer, Control and Robotics (ICCCR), Shanghai, China, 8–10 January 2021; pp. 188–192.
26. Zhang, J.; Lu, G.; You, Z. Large deformation and energy absorption of additively manufactured auxetic materials and structures: A review. *Compos. Part B Eng.* **2020**, *201*, 108340. [CrossRef]
27. Soltanmohammadi, K.; Rahmatabadi, D.; Aberoumand, M.; Soleyman, E.; Ghasemi, I.; Baniassadi, M.; Abrinia, K.; Bodaghi, M.; Baghani, M. Effects of TPU on the mechanical properties, fracture toughness, morphology, and thermal analysis of 3D-printed ABS-TPU blends by FDM. *J. Vinyl Addit. Technol.* **2024**, *30*, 958–968. [CrossRef]
28. Frenkel, D.; Ginsbury, E.; Sharabi, M. The Mechanics of Bioinspired Stiff-to-Compliant Multi-Material 3D-Printed Interfaces. *Biomimetics* **2022**, *7*, 170. [CrossRef]
29. dos Reis, M.Q.; Carbas, R.J.C.; Marques, E.A.S.; da Silva, L.F.M. Effect of the Infill Density on 3D-Printed Geometrically Graded Impact Attenuators. *Polymers* **2024**, *16*, 3193. [CrossRef]
30. Ingrole, A.; Hao, A.; Liang, R. Design and modeling of auxetic and hybrid honeycomb structures for in-plane property enhancement. *Mater. Des.* **2017**, *117*, 72–83. [CrossRef]
31. Żur, P.; Borek, W. Influence of 3D-printing Parameters on Mechanical Properties of PLA defined in the Static Bending Test. *Eur. J. Eng. Sci. Technol.* **2019**, *2*, 65–70. [CrossRef]
32. Cracknell, D.; Battley, M.; Fernandez, J.; Amirpour, M. The mechanical response of polymeric gyroid structures in an optimised orthotic insole. *Biomech. Model. Mechanobiol.* **2024**, 1–19. [CrossRef]
33. Haris, F.; Liau, B.-Y.; Jan, Y.-K.; Akbari, V.B.H.; Primanda, Y.; Lin, K.-H.; Lung, C.-W. A Review of the Plantar Pressure Distribution Effects from Insole Materials and at Different Walking Speeds. *Appl. Sci.* **2021**, *11*, 11851. [CrossRef]

designs

MDPI

Article

Guidelines for Design and Additive Manufacturing Specify the Use of Surgical Templates with Improved Accuracy Using the Masked Stereolithography Technique in the Zygomatic Bone Region

Paweł Turek [1,*], Paweł Kubik [1], Dominika Ruszała [2], Natalia Dudek [2] and Jacek Misiura [1]

1 Faculty of Mechanical Engineering and Aeronautics, Rzeszów University of Technology, 35-959 Rzeszów, Poland; p.kubik@prz.edu.pl (P.K.); jmisiura@prz.edu.pl (J.M.)

2 Faculty of Mathematics and Applied Physics, Rzeszów University of Technology, 35-959 Rzeszów, Poland; dominika.ruszala30@gmail.com (D.R.); n.duudek@gmail.com (N.D.)

* Correspondence: pturek@prz.edu.pl

Abstract: The zygomatic bone area experiences frequent mechanical damage in the middle craniofacial region, including the orbital floor. The orbital floor bone is very thin, ranging from 0.74 mm to 1.5 mm. Enhancing digitization, reconstruction, and CAD modeling procedures is essential to improving the visualization of this structure. Achieving a homogeneous surface with high manufacturing accuracy is crucial for developing precise surgical models and tools for creating titanium mesh implants to reconstruct the orbital floor geometry. This article improved the accuracy of reconstruction and CAD modeling using the example of the development of a prototype implant to replace the zygomatic bone and a tool to form the geometry of the titanium mesh within the geometry of the orbital floor. The masked stereolithography (mSLA) method was used in the model manufacturing process because it is low-cost and highly accurate. Two manufacturing modes (standard and ultra-light) were tested on an Anycubic Photon M3 Premium 3D printer to determine which mode produced a more accurate representation of the geometry. To verify the geometric accuracy of the manufactured models, a GOM Scan1 structured light scanner was used. In the process of evaluating the accuracy of the model preparation, the maximum deviation, mean deviation, range and standard deviation were determined. The maximum deviations for anatomical structures created using the normal mode ranged from ±0.6 mm to ±0.7 mm. In contrast, models produced with the ultra-light mode showed deviations of ±0.5 mm to ±0.6 mm. Furthermore, the results indicate that the ultra-light mode offers better surface accuracy for die and stamp models. More than 70% of the surface of the models is within the deviation range of ±0.3 mm, which is sufficient for planning surgical procedures. However, the guidelines developed in the presented publication need to optimize the CAD process and select 3D-printing parameters to minimize deviations, especially at the edges of manufactured models.

Keywords: masked stereolithography; zygomatic bone; reverse engineering; accuracy; computer aided design; additive manufacturing; surgical template

1. Introduction

The craniofacial bone includes the zygomatic bone. The zygomatic bone is a significant part of the skull structure, giving the face shape and form. It helps protect the orbital region, including the eyeball, from injury. The zygomatic bone joins with other facial bones, such as the temporal, maxillary, and cuneiform bones, to form the complex structure of the facial skeleton. It is one of the bones most frequently subjected to mechanical trauma [1]. Injuries and diseases of the skeletal system in this area result in altered facial appearance, malocclusion, and problems with the essential functions of chewing food, which, in the long term, leads to eating disorders [2,3]. These factors all contribute to deteriorations in the patient's mental and physical state. It is, therefore, essential to undertake treatment and reconstruction of the resulting bone defect. The surgical reconstruction of the zygomatic bone area is very complex, as it requires the implant to be accurately matched to the defect [4,5]. Autogenous grafts are the gold standard in treating zygomatic bone defects [6,7]. They are used in cases of extensive fractures resulting in the loss of a large portion of bone [4]. This procedure is associated with the possibility of additional swelling and complications at the site of bone harvesting for grafting. The geometry of the graft itself also has to be manually adjusted to the size and shape of the defect [8]. This considerably lengthens the procedure, and the final result upon fitting the graft into the defect is not ideal [9]. For this reason, 3D-printing methods are being used more and more frequently to manufacture models of craniofacial anatomical structures, as models made using these methods are individually adapted and prepared for a specific patient [10]. This ensures that they are precisely matched to the defect, significantly reducing the time it takes to perform the procedure and the occurrence of postoperative complications [11].

The entire process of reconstructing the geometry within the region encompassing the zygomatic bone is possible due to the integration of tomographic measurement systems, digital data-editing software, and modern manufacturing methods [12–14]. Despite obtaining a complete geometry of the anatomical structure, it must be borne in mind that at each stage of the measurement process, reconstruction, computer-aided design (CAD) modeling, and manufacturing errors are created that significantly affect the geometric accuracy of the completed model [15–17]. The accuracy of the representation of the geometry of anatomical structures is mainly affected by the digitization stage [16,17] and the processing of volumetric data [18,19]. Multi-detector and cone tomographies are most commonly used in the diagnostic process of the craniofacial region [12,20]. Based on the collected Digital Imaging and Communications in Medicine (DICOM) data, a segmentation and geometry reconstruction is achieved using computer-aided design software. At this stage, the segmentation process plays the most crucial role in the method and the parameters used to extract the anatomical structure from the DICOM data, as well as in the geometry reconstruction methods [16,19]. In developing the geometry of an anatomical structure and a surgical template or implant, it is necessary to use additional CAD modeling software. Most often, hybrid modeling combining solid and surface modeling is used [20,21]. The gold standard uses mirroring and Boolean methods to model zygomatic bone defect geometry in a CAD environment [12]. However, the uncontrolled accuracy of monitoring when using stereolithography (STL) file-processing methods and the CAD-based creation of surgical templates and implant constructs generate additional geometry errors in the final digital model [21].

The choice of manufacturing method also influences the accuracy of reconstructions of the geometry of the anatomical structure [22,23]. A physical model can be obtained using subtractive methods [24] or molding methods [25]. However, due to the complex geometry of the models of anatomical structures within the zygomatic bone area, it is challenging, time-consuming, and thus costly to produce them using, among other things, subtractive methods. Therefore, additive methods are an ideal alternative for creating such models [26,27]. To date, material extrusion (MEX) [12,13,28], vat photopolymerization (VPP) [29], powder bed fusion (PBF) [13,20], and material jetting (MJ) [20] methods are most commonly used to produce models of anatomical structures and surgical templates within the zygomatic bone. Unfortunately, the methods listed also have limitations. These are due to the cost of manufacturing the models and, in some cases, the quality of the manufactured models. Actually, it is becoming increasingly common to manufacture models using masked stereolithography (mSLA) technology, which is a VPP method [30]. SLA technology is becoming increasingly recognized as it has a operating cost than digital-light-processing (DLP) and provides excellent reproduction of details, which is beneficial for high-precision applications [31]. This is a hybrid 3D-printing method combining the advantages of stereolithography (SLA) and DLP. Instead of a projector, as in DLP, mSLA uses an LCD array that masks the UV light emitted by the LED array, allowing the entire resin layer to cure simultaneously [30]. According to the literature review, this method's models are mainly used to produce electronic circuits [32,33]. The application of this method in the medical industry has also been noted. Previous research has primarily evaluated the applicability of the models in terms of assessing the materials available for this technology in the broader sense of bioengineering [34,35]. However, a study on the usefulness of mSLA technology in producing models of anatomical structures and instruments to support surgical procedures is still lacking. One of the possible anatomical areas in which the mSLA method could find a place is the area of the zygomatic bone. This is an area subject to frequent complex mechanical trauma. The bony structure of the zygomatic bone also includes the area of the orbital floor bone. The orbital floor bone is very thin and varies between 0.74 mm and 1.5 mm [36]. To obtain a more accurate visualization of this bony structure, it is necessary not only to develop a more precise digitization, reconstruction, and CAD modeling procedure, but also to obtain a final homogeneous surface with a high degree of manufacturing accuracy, which will allow for the precise development of models for surgery planning [21,37,38] and tools that enable the formation of implants in the form of a titanium mesh for the reconstruction of the orbital floor geometry.

Based on the literature review, guidelines should be established to enhance the accuracy of the reconstruction and CAD modeling of the zygomatic bone. The goal is to create digital models of anatomical structures resembling real geometries. Additionally, due to the significant advantages of the mSLA method in producing precision models, it is worth testing this method in the context of producing models of anatomical structures to verify the obtained geometrical accuracy. The study may significantly extend the applicability of the mSLA method in manufacturing models of anatomical structures, where the high-precision matching of models to each other is required, as is the case within the zygomatic bone area.

2. Materials and Methods

A study was conducted between the F. Chopin University Clinical Regional Hospital in Rzeszów, specifically within the Department of Maxillofacial Surgery, and Rzeszów University of Technology. From 2022 to 2023, 23 patients were involved in the study. This article focuses on two of the most clinically significant cases of patients who experienced injuries from a traffic accident. In the first patient's case, a significant portion of the zygomatic bone was damaged. In the second patient's case, the geometry of the left orbital floor was damaged in a non-standard manner. This was due to significant bone loss of the orbital floor. Diagnostic data were obtained using a Discovery CT750 HD multidetector tomograph from GE Medical Systems. The imaging of the craniofacial region was conducted at the University Clinical Hospital Fryderyk Chopin in Rzeszów. The following scanning protocol is commonly used:

- Scan type: helical;
- Beam collimation: 40 mm;
- Detector configuration: 64 × 0.625 mm;
- Tube settings: 120 kV;

- Slice thickness: 1.25 mm;
- Matrix size: 512 × 512.

2.1. Process of Digital Processing, Segmentation, and 3D Reconstruction of DICOM Data

This protocol achieves high-resolution DICOM data, with voxel dimensions characterized by an isotropic structure of 0.6 mm × 0.6 mm × 0.625 mm. However, a significant influence of the volume-averaging artifact was observed. Its influence significantly impeded the accurate reconstruction of bone structures within the zygomatic bone region. This artifact occurs when a single voxel encompasses multiple tissue types, leading to contrast distortion, particularly in structures with low contrast. While using a thinner CT layer can address this issue, it also increases the radiation dose, posing a danger to the patient. A safer approach is to perform digital processing on the already-collected DICOM data to improve the accuracy of the 3D model of the anatomical structure (Figure 1).

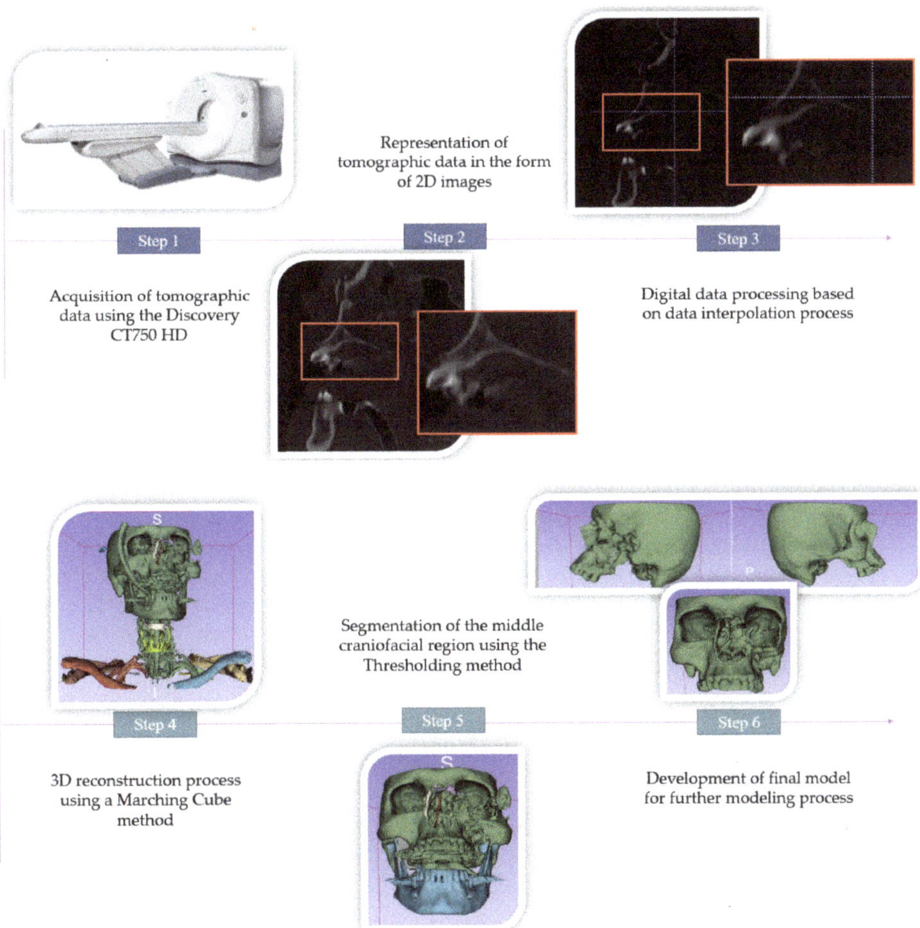

Figure 1. Enhancing the accuracy of anatomical structure reconstruction in the zygomatic bone area using the example of the first patient.

The first step in this process involved applying an interpolation technique to the DICOM data, specifically using a bilinear method [39]. This method analyzes the gray shade values of the four nearest pixels adjacent to the newly calculated pixel. *Bilinear interpolation* determines the gray shade value of the new pixel by taking the arithmetic average of the four neighboring pixels, as described in Equation (1):

$$k(x) = \begin{cases} 1 - |x|;\ |x| < 1 \\ 0; otherwise \end{cases}, \tag{1}$$

The interpolation process significantly enhanced the spatial resolution of the data, reducing the voxel size to 0.1 mm × 0.1 mm × 0.1 mm. Using the reformatted DICOM data, we reconstructed the bone structures in the region, including the zygomatic bone, using the 3D Slicer software. The modeling of the bone structures began by importing a sequence of 2D images into the program's workspace. In the next step, we carried out the segmentation process using a *thresholding* method [16,18]. To improve the accuracy of the segmentation, we focused on determining the thresholds by averaging the gray shade values of pixels only within the zygomatic bone area instead of across the entire craniofacial region. The average pixel gray value within the zygomatic bone area was 248 HU, with a standard deviation of 81 HU. The mean value and standard deviation of the lower segmentation threshold was established in the context of the 23 patients studied. This procedure avoided artificially increasing the volume of segmented anatomical structures due to the incorrect selection of segmentation thresholds. *The marching cubes* algorithm was utilized to visualize the 3D model [40,41]. This algorithm divides space into cubes, each encompassing one or more voxels. The values of the nodes at each corner of the designated cube are then compared against a specified iso-value. Depending on whether a node's value is higher or lower than the iso-value, polygons are generated to represent the iso-surface that intersects the cube. The *marching cubes* method, like any algorithm, has its limitations. These limitations mainly arise from errors that occur during the reconstruction process of the 3D STL model. Common issues include incorrect triangle orientations, duplicated edges and vertices, and duplicated triangles. To prepare models for 3D-printing, it is crucial to eliminate these errors. If these corrections are not made, it can lead to challenges in dividing the facet surface into 3D-print layers, making it difficult or even impossible to create a model of bone structures using the additive method. We used the triangle mesh editing tools in Meshmixer software to correct programming errors. During the generation of the triangle mesh, chord, and angular deviations occur, particularly in areas with rapid changes in the curvature radii. To minimize these errors, we employed an optimization process to enhance the facet structure by compacting the triangle mesh in regions of high surface complexity. This approach significantly improved the mapping quality of the model geometry. The procedure optimization consisted of two steps:

- The surface is smoothed by moving the nodes on which the triangle mesh is spanned. Each node is moved to the average position of its neighbors by applying the Laplace function. The function is the sum of the squares of the lengths of edges sharing a common node (2):

$$f(x,y) = \sum_{i=0}^{k}\left((x - x_i)^2 + (y - y_i)^2\right) \tag{2}$$

- k is the number of neighboring nodes; the position of new nodes is determined using Formula (3):

$$x' = \frac{1}{k}\sum_{i=1}^{k} x_i y' = \frac{1}{k}\sum_{i=1}^{k} y_i \tag{3}$$

- Triangle densities are created in regions of high complexity, and reducing the number of triangles in flatter areas using the isotropic surface remashing algorithm.

2.2. Procedure for Modeling a Defect in the Zygomatic Bone Area

The process of modeling a defect in the zygomatic bone is presented in Figure 2. After verifying programming errors, the plane of symmetry was developed. Based on this, the skull model was divided into two separate parts. Then, the right (undamaged) part was mirrored onto the left (damaged) part. The mirroring operation was carried out by setting the mirror plane's angle and selecting the direction of the mirrored part. The damaged part of the skull was visualized so that a visualization of the common edge was obtained, which formed the basis for the zygomatic bone fragment containing the defect. The edges were smoothed on the selected part of the surface, and then a preliminary separation of the geometry covering the defect area was carried out. After the surfaces were disconnected, the newly created geometry was given a thickness, and its surface was smoothed, obtaining a preliminary version of the defect. After smoothing the edges of the defect, it was reapplied to the rest of the skull to eliminate overlapping surface fragments. This procedure aimed to improve the accuracy of the defect's fit to the skull. The developed model was then examined for any discontinuities arising from the standard edge-extraction process. In the final step, the defect's geometry was refined by modeling holes within the model geometry. This aim of this procedure was to develop a model structure that would allow the defect to grow more quickly into the bone structure.

2.3. Development of a Tool to Form the Geometry of a Mesh Implant to Reconstruct an Orbital Floor Defect

The geometry of the defect in the right orbit was modeled using data from the left, undamaged orbit (Figure 3). Initially, the model generated through the segmentation and reconstruction process from DICOM data was imported into Meshmixer. After verifying the STL file for programming errors, an area of the geometry from the undamaged orbit was selected as a starting point for further modeling.

At this stage, the goal was to adjust the size of the marked area so that it would resemble the size of the damaged area in the right orbit. The selected part of the orbit was then extracted and separated from the rest of the model. This extracted fragment underwent a mirror image operation. To ensure it matched the area of the orbital defect, further editing, including smoothing the edges, was performed on the extracted area. Once the final geometry of the defect was established, the surface modeling process began. The die and stamp model was developed to create the geometry of the implant using CATIA software. After importing the geometry from the previous modeling stage in STL format, it was converted into a surface model using the *Automatic Surface Reconstruction* function. At this stage, the primary surface-matching parameters were defined: *Mean Surface Deviation* was set to 0.025 mm and the *Surface Detail* parameter was specified to include 20,000 elements. Additionally, attention was paid to the *Target Ratio* parameter, which determines the percentage of the STL model surface coverage by the generated surface; this was set to 100%. An edge outline was created using a Spline curve in the following step. This edge outline underwent the Spline drawing process, which produced a surface model. This surface model was then transformed into a solid model in the next step. The entire modeling process resulted in a die model that accurately represents the geometry of the orbital floor. The next step was to create a stamp model based on the geometry of the die model. In the first step, the attitude edge of the earlier model was dropped onto the created plane. Based on the dropped sketch, a *Pad* operation was applied to pull it out in the specified direction, thus creating a new solid. The next step was to trim the newly created solid so that its geometry corresponded to the die model. To achieve this, the *Boolean Remove* tool in the *Part Design* module could be used for logical operations on solids. This procedure allowed for the precise subtraction of one solid from another. The die model was a negative to cut off the corresponding stamp elements. As a result, a stamp model was created that perfectly matched the geometry of the die, which is crucial for the precise formation of the geometry of the orbital floor implant.

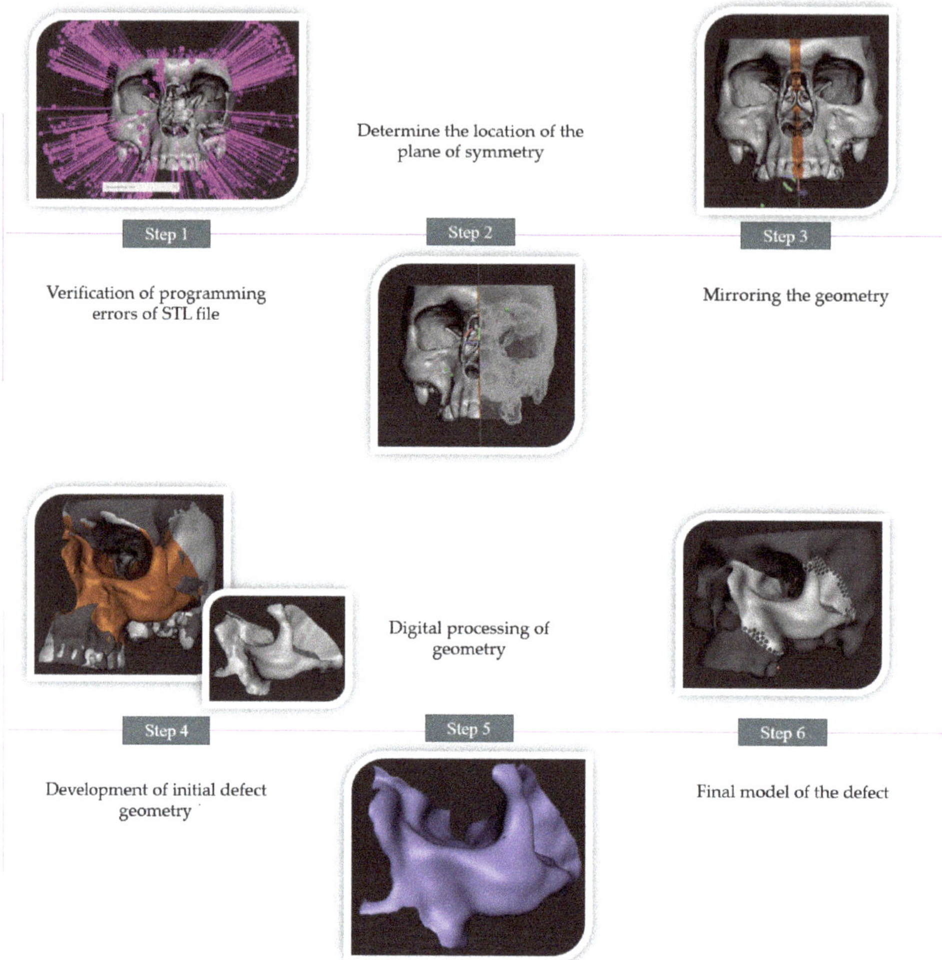

Figure 2. The process of designing defects in the area of the zygomatic bone.

2.4. *Additive Manufacturing of Designed Models Using the mSLA Method*

The models were manufactured using an Anycubic Photon M3 Premium 3D printer using the mSLA method. The process began by filling the 3D printer with Siraya Tech liquid resin. Next, the models' manufacturing parameters were established, focusing on normal and ultra-light modes (Table 1). The Lychee Slicer program generated a support material to ensure the stability of the model and minimize deformations due to its complex geometry.

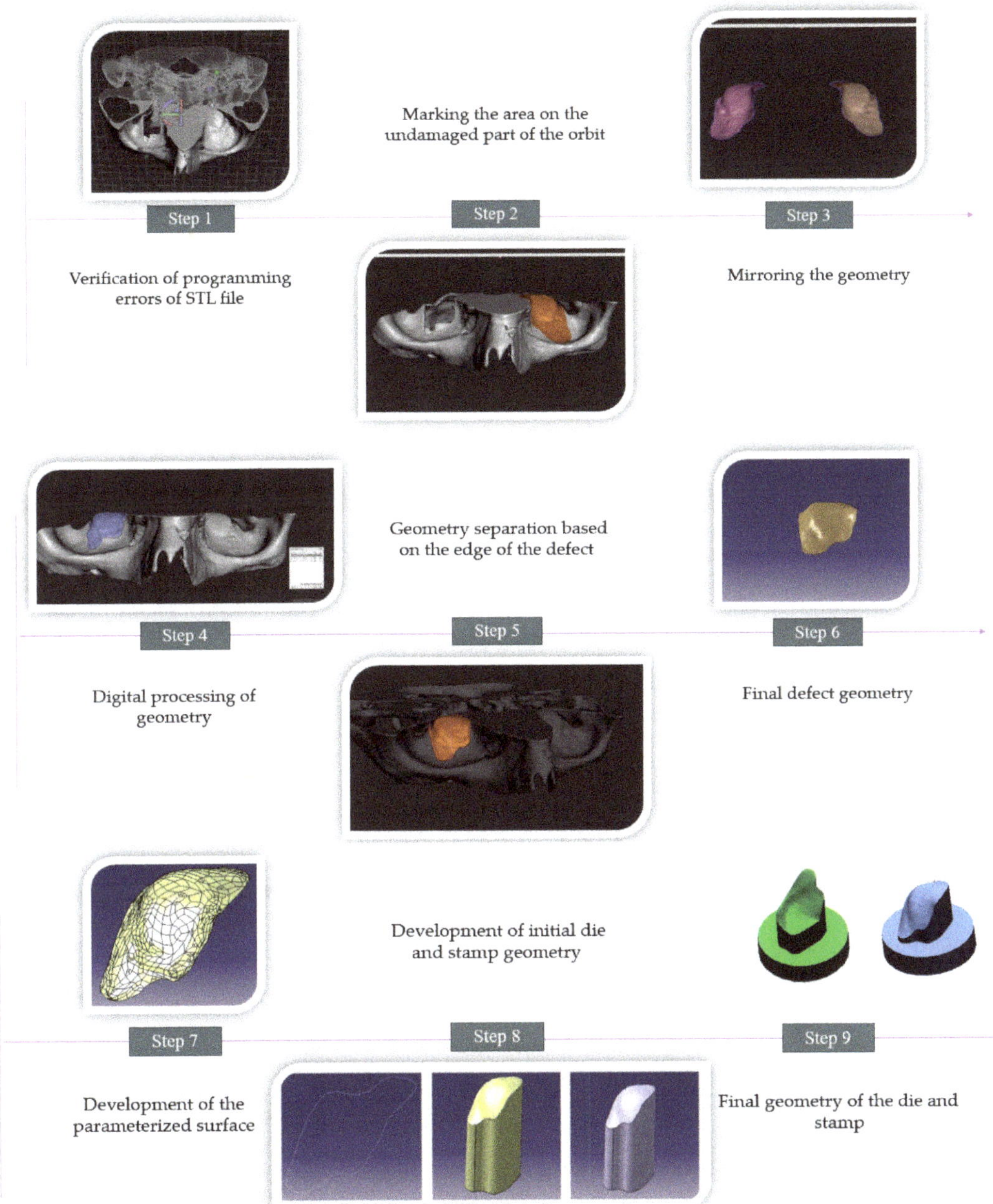

Figure 3. The process of designing the geometry of the die and the stamp used to form the geometry of the implant within the orbital floor area.

After defining the model's layers, the 3D-printing process started (Figure 4). The mSLA method relies on curing photopolymer resin with UV light, which is directed through a mask using an LCD. Once the first layer is cured, the working platform on which the models are being built rises slightly, allowing the next layer to be manufactured. This cycle of exposing the resin and lifting the platform continues until all the models are complete.

Table 1. Applied 3D-printing parameters.

	Parameter	Value
Basic parameters	Layer Thickness	0.050 mm
	Light-Off Delay	2 s
	Exposure Time	2.4 s
	Lift Distance	2.5 mm
	Lift Speed	45 mm/min
	Retract Speed	240 mm/min
Normal mode	Tip Diameter	0.6 mm
	Tip Length	3 mm
	Diameter	1.3 mm
Ultra-light mode	Tip Diameter	0.3 mm
	Tip Length	2 mm
	Diameter	1 mm

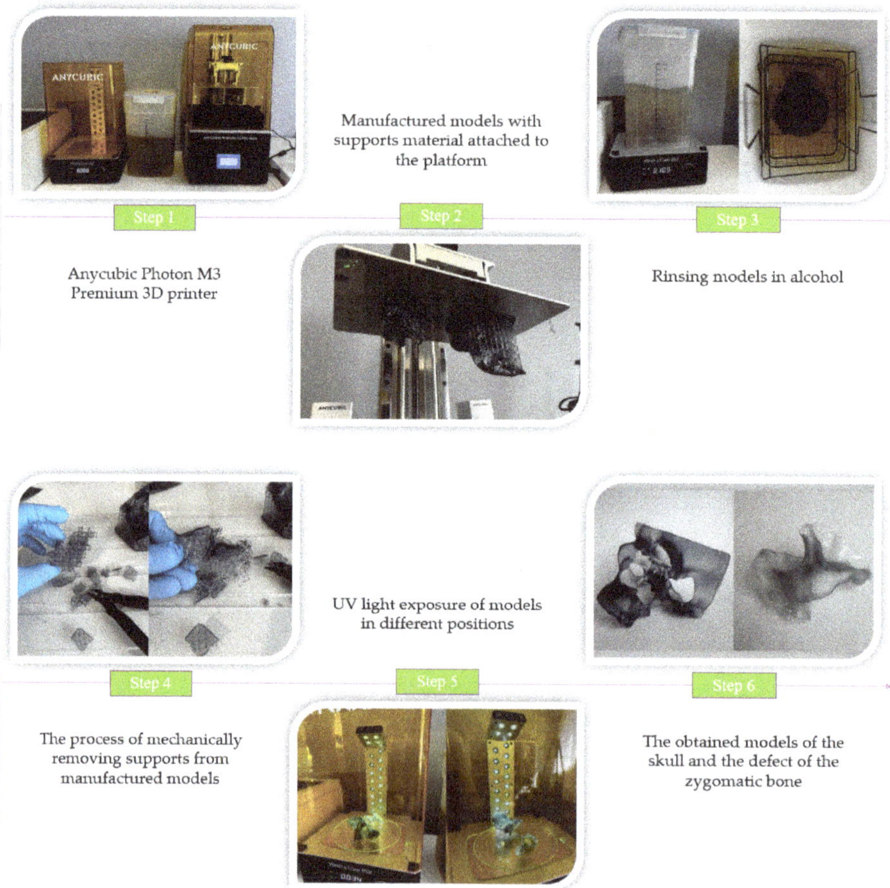

Figure 4. The process of manufacturing models using the example of the first patient.

After the 3D-printing process, the models and supports remained attached to the working platform. The models were then removed from the platform and placed in a special container filled with alcohol. This rinsing process lasted 4 min to eliminate any remaining resin on their surfaces. Next, the models underwent a precision cleaning of the printed supports. This involved using pliers to cut off the supports close to the model's surface and plastic cutters for more precise removal. The final processing step included rinsing the models once again in alcohol for another 4 min and exposing them to ultraviolet light. The completed models were then placed on a rotating platform within a multifunctional device from Anycubic, which was used for drying and curing the manufactured objects. The models were exposed for 2 min in one position, after which they were rotated and exposed again for another 2 min in a second position. As a result of the 3D-printing and post-processing steps, fully processed models were obtained.

3. Results

Before measurements, models made from Siraya Tech resin had smooth and glossy surfaces that could cause reflections and glare, leading to measurement errors when digitizing their geometry. To address this issue, the manufactured models were sprayed with a thin layer of AESUB Blue aerosol to create a matte finish on the surface before data acquisition. After the measurement, the powder layer disappeared within a few minutes, eliminating the need for additional cleaning of the models. In the measurement process, we used an optical measurement system, GOM Scan1 with GOM Professional software using blue structured light, to conduct geometric accuracy tests (Figure 5). This measurement method is based on trigonometric triangulation and projects a light pattern onto an object. An LCD projector emits this pattern. Then, two cameras slightly offset from the projector will examine the shape of the light pattern and calculate the distance from each point in the field of view.

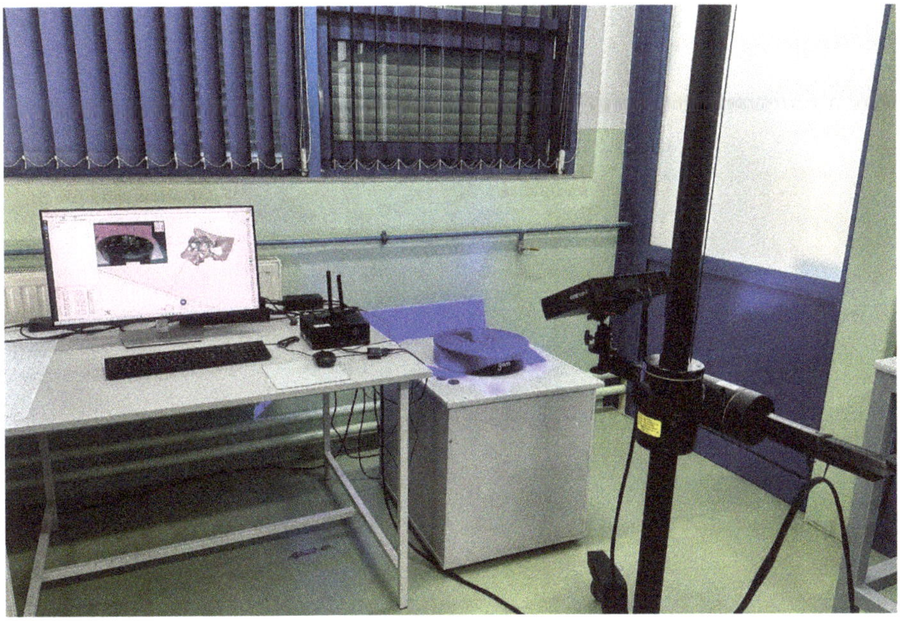

Figure 5. An optical measurement system, GOM Scan1.

Parameters were utilized to acquire the highest measurement resolution with the GOM Scan1 system during the measurement (Table 2). During the measurement process, the orientations of the skull model and the zygomatic bone defect model were adjusted twice to achieve a complete and accurate 3D scan. The models were measured in a single orientation on the measuring table for the geometry measurement of the stamp and die. The measuring table facilitated the measurement process by performing 13 rotations around its axis, allowing for comprehensive geometry data collection for all models.

Table 2. Established measurement parameters for the GOM Scan 1.

Parameters	Value
Pixel-resolution cameras	5,000,000
Measuring area	100 mm × 65 mm × 400 mm
Min. point resolution	0.037 mm
Number of points per scan	5,000,000
Number of rotations of the measuring table	13

The accuracy of the manufactured model was verified using GOM Inspect software. This involved comparing the nominal model created during the RE design stage with the model generated during the measurement stage using the GOM Scan1 optical system. The comparison was conducted using the *best-fit* method, achieving an accuracy of 0.001 mm. As a result of the model adjustments, three-dimensional maps of the geometry, mapping model deviations obtained using two modes, normal and ultra-light (Figures 6–8), were developed. In addition, statistical results were developed for the model of the skull part and the modeled defect of the zygomatic bone (Tables 3 and 4), as well as for the model of the stamp and die (Table 5).

Table 3. Statistical parameters assessing the accuracy of the skull model and the model of the zygomatic bone defect in normal mode.

Parameters	Cranial Model	Defect of the Zygomatic Bone
Maximum deviation [mm]	2.016	2.226
Minimum deviation [mm]	−1.903	−1.092
Range [mm]	3.919	3.318
Mean deviation [mm]	−0.014	−0.047
Standard deviation [mm]	0.277	0.340

Table 4. Statistical parameters assessing the accuracy of the model of the skull part and the defect of the zygomatic bone in ultra-light mode.

Parameters	Cranial Model	Defect of the Zygomatic Bone
Maximum deviation [mm]	1.802	1.197
Minimum deviation [mm]	−1.673	−1.209
Range [mm]	3.475	2.406
Mean deviation [mm]	−0.004	−0.024
Standard deviation [mm]	0.242	0.290

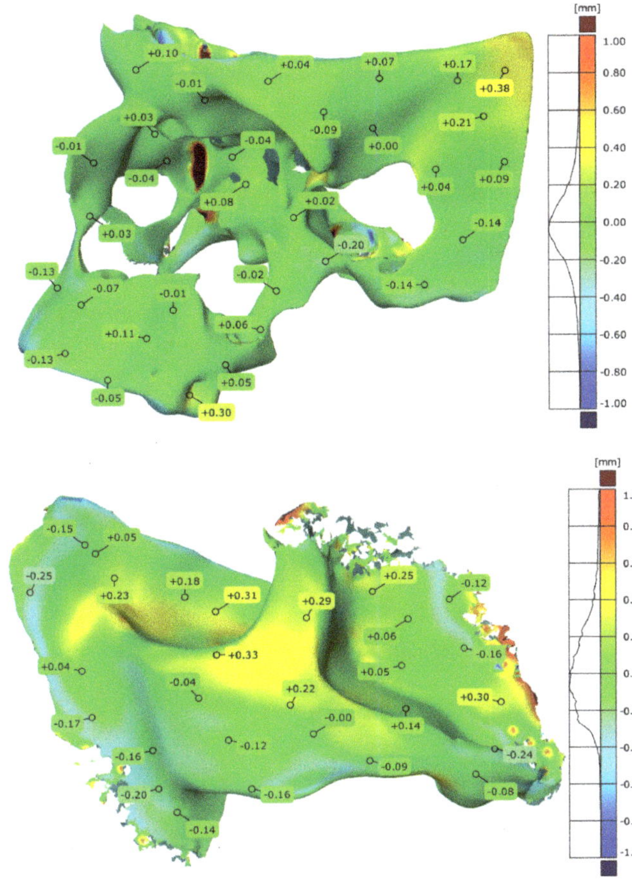

Figure 6. Three-dimensional deviation map of the skull model and the zygomatic bone defect in normal mode.

Based on the statistical results, the skull model was manufactured within a deviation range of ±0.56 mm for the normal mode while the prototype implant of zygomatic bone geometry was within a deviation range of ±0.68 mm. Notably, over 70% of the surfaces of the models fell within the deviation range of ±0.3 mm, which is considered acceptable for planning procedures in the craniofacial region. The most significant positive deviations in the skull model were primarily found in the temporal bone area. Conversely, the most substantial negative deviations were noted mainly at the edges of the completed skull model, particularly in the upper jaw region. Most positive and negative deviations were located at the model's edges for the prototype implant, where the missing zygomatic bone was replaced, especially in the areas corresponding to the designed mesh. Additionally, deviations of approximately 0.3 mm were observed in the frontal process area, lateral surface, and orbital floor geometry. Based on the statistical results obtained for the ultra-light mode, the skull model was manufactured with a deviation range of ±0.48 mm, while the implant prototype of zygomatic bone geometry was within a ±0.58 mm range. Notably, over 70% of the surfaces of both models fell within the deviation ranges of ±0.25 mm for the skull and ±0.29 mm for the implant prototype. Deviations over the ±0.3 mm range occurred at the edges and in the area of the designed mesh, as observed in normal mode. An additional group of negative deviations was observed in the skull model compared to the model created in normal mode. This group of negative deviations was primarily located in the temporal bone area, frontal bone area, and along the edges of the model. Conversely, the upper jaw area noted the most significant positive deviations. For the prototype implant designed to replace the lost zygomatic bone, the most significant positive and negative deviations were also found at the model's edges, particularly in the areas of the designed mesh. Positive deviations of 0.3 mm were observed at similar locations on the implant prototype's surface, particularly in the frontal process area, lateral surface, and orbital floor geometry. Considering the statistical results obtained for the die and stamp model, the positive and negative deviation areas are comparable for both modes. In addition, the die model has a better surface quality for both methods than the stamp model. Higher deviation values were obtained for the normal mode. For normal mode, the deviation range for the stamp model was within ±0.68 mm, and for the ultra-light mode, the deviations were within ±0.52 mm. When using the die model, the maximum deviations were within ±0.46 mm for normal mode and ±0.4 mm for the ultra-light mode. It is worth adding that more than 70% of the surface of the models was within the deviation range of ±0.3 mm, which is sufficient for planning surgical procedures.

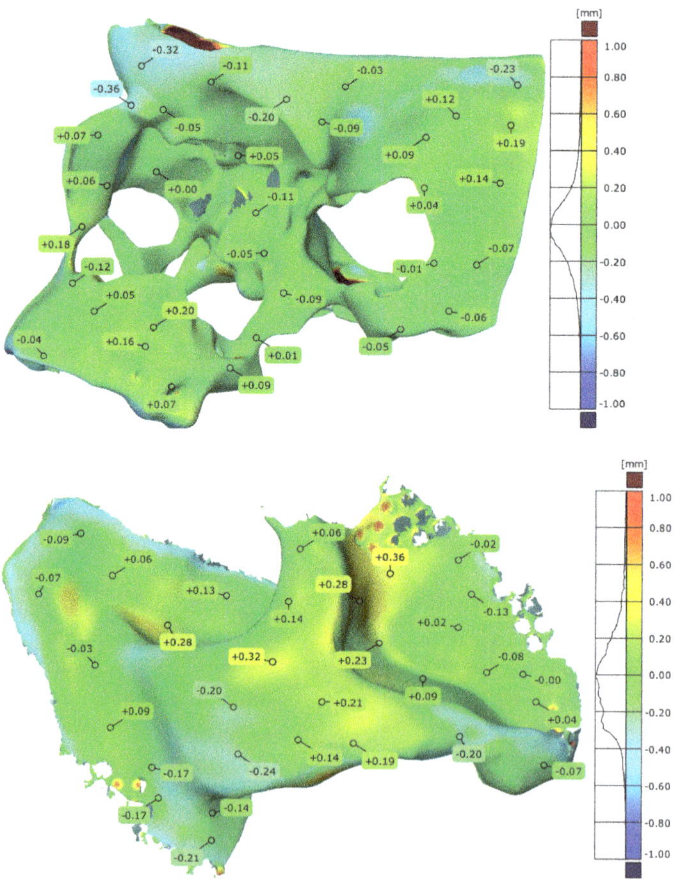

Figure 7. Three-dimensional deviation map of the skull model and the zygomatic bone defect in ultra-light mode.

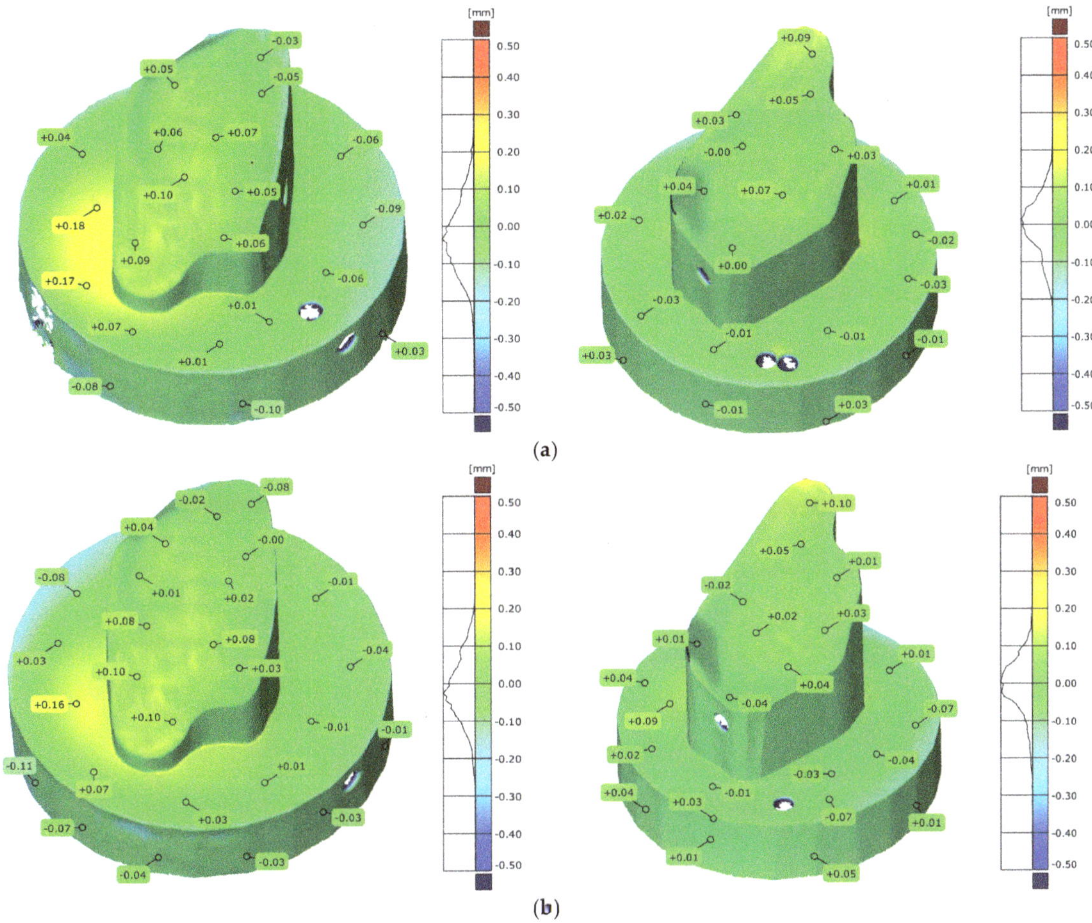

Figure 8. Three-dimensional deviation map of the stamp and the die model: (**a**) in normal mode; (**b**) in the ultra-light mode.

Table 5. Statistical parameters assessing the accuracy of the model of the stamp and the die manufactured in normal mode and in the ultra-light mode.

Parameters	Stamp Model	Die Model	Type of Mode
Maximum deviation [mm]	0.827	0.687	Normal
Minimum deviation [mm]	−1.394	−1.396	
Range [mm]	2.211	2.082	
Mean deviation [mm]	0.009	−0.014	
Standard deviation [mm]	0.341	0.230	
Maximum deviation [mm]	0.547	0.727	Ultra-light
Minimum deviation [mm]	−1.443	−0.696	
Range [mm]	1.990	1.423	
Mean deviation [mm]	0.020	0.045	
Standard deviation [mm]	0.259	0.193	

4. Discussion

Designing and manufacturing a model of an anatomical structure for surgical procedures is a complex task, particularly in the craniofacial area, which contains bony tissues with intricate geometries. Recent advancements in coordinate measuring systems, data processing software, and modern manufacturing techniques have helped to solve this challenge through a process known as reverse engineering. However, errors can occur at various stages—from measuring the patient to creating the final model—that can significantly impact the accuracy of the surgical procedure. Notably, significant errors arise when digitizing geometry. The diagnostic data of the zygomatic bone area most often derive from multidetector CT scanners. High-resolution measurements are necessary to reconstruct the damaged area of the zygomatic bone, including the orbit. This is partly because the orbital floor bone is fragile, and its thickness varies between 0.74 mm and 1.5 mm [36]. High-resolution DICOM data are essential in correctly establishing the diagnosis and especially useful for creating reliable reconstructions and treatment-planning [21,37,38]. Unfortunately, due to the limitations of multidetector tomographic measurement systems, it is often difficult to obtain very high-resolution data. Thus, based on the obtained DICOM data, difficulties arise with the overall segmentation process and reconstruction of the geometry of the zygomatic bone, particularly the orbital floor. It is also necessary to pay attention to the accuracy of manufacturing models using additive methods. MEX methods are most commonly used in manufacturing models of the craniofacial region [14,22]. However, these methods also have their limitations due to, among other things, the use of layer thicknesses close to or higher than 0.1 mm and the anisotropic surface properties obtained during 3D-printing. These factors significantly degrade the geometric accuracy of the obtained models. In addition, using a layer thickness with a value close to the thickness of the orbital floor bone can make it much more difficult or impossible to achieve this part of the geometry at the manufacturing stage. Therefore, when manufacturing models of the craniofacial region, significant detail is often required. Among other dental models, VPP methods are used [42]. These allow for models to be made with a layer thickness of less than 0.1 mm while obtaining a surface with essentially isotropic properties. Therefore, it is essential to develop solutions to ensure the final model closely reflects the anatomical structure.

4.1. Methods to Improve Accuracy in the Numerical Processing of DICOM Data

Most diagnoses use data from multi-detector CT systems to identify injuries or diseases affecting bone structures. The quality of the resulting DICOM data is primarily influenced by spatial and contrast resolution [17,43]. These factors are determined by various parameters, including the design of the diagnostic system and its calibration quality. Low spatial and contrast resolutions can significantly impede the accurate segmentation of bone structures. This issue can be addressed by implementing a measurement protocol that enhances accuracy; however, such procedures may pose risks to the patient's health and, in some cases, their life. For this reason, research to improve accuracy using previously collected DICOM data is ongoing. The authors presented procedures for removing measurement noise, which mainly arises from within the implant area that is to be diagnosed [44–46]. In addition, depending on the quality of the obtained DICOM data, they focus on procedures for smoothing or sharpening the edges of segmented structures [47–49]. The article's authors implemented a digital data-processing technique to enhance the quality of the image and extract essential information. The use of an interpolation method significantly improved the contrast resolution of the DICOM data, which also helped minimize the measurement noise. Additionally, this method increased the spatial resolution of the image by digitally generating extra pixels based on the intensity values of neighboring pixels

(Figure 9). As a result, the issue of volume-averaging that previously complicated the accurate assignment of pixels to segmented bone tissue was significantly reduced.

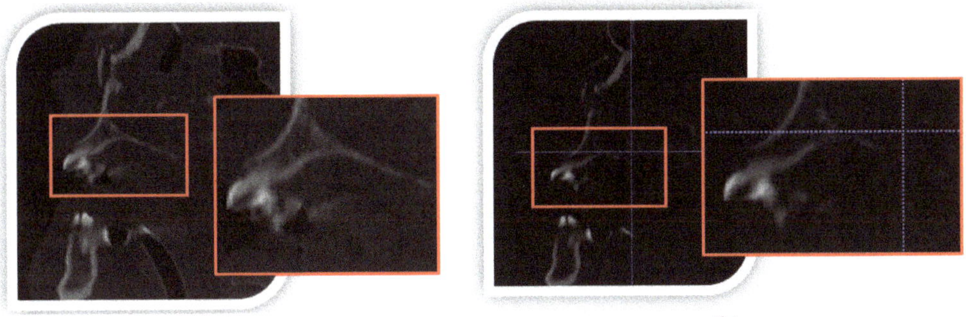

(a) (b)

Figure 9. The effect of applying the data interpolation process: (**a**) DICOM data before processing; (**b**) DICOM data after processing.

A segmentation process was carried out on the digitally processed image. This process involves extracting a selected bone structure from the entire volumetric data set. Several methods are currently used to obtain segmented outlines of anatomical structures. Among the best-known are *Global Thresholding* [50], *Edge Detection* [51], *Region Growing* [52], and machine learning methods [53,54]. However, as with most automatic or semi-automatic methods, it is impossible to accurately determine segmentation thresholds using these methods. This problem of determining accurate HU values is still a significant challenge in modeling anatomical structures [18,50]. The authors of the presented article focused on using the *local thresholding* method. Narrowing the segmentation area and using the interpolation method significantly improved the selection of a more accurate value of the lower threshold against which the middle part of the craniofacial was separated. Thus, the digital model was not artificially increased or decreased in volume during the segmentation process. Various reconstruction methods are used in the literature to depict the three-dimensional model. In the current publications, two main groups, contour-based and Voxel-based, are considered. More often, however, the voxel-based method is used to reconstruct the geometry of anatomical structures. This is because this method does not generate too many programming errors in the structure of the 3D-STL model. For this reason, the authors of the presented publication used the *Marching Cubes* algorithm, one of the voxel-based methods, to develop a 3D representation of the anatomical structure using DICOM data [40,41].

4.2. Methods to Improve Accuracy in the Numerical Processing of 3D-STL Models

The *marching cubes* method, like any algorithm, has its limitations. The quality of the transformation from segmented contours to a faceted surface primarily depends on the layer thickness obtained during tomographic imaging. When the CT layer thickness is significantly larger than the pixel dimensions, gaps can appear in the triangle mesh in some regions of the reconstructed geometry. These gaps result from insufficient diagnostic data, hindering adjacent contours' smooth merging. Also, errors were observed in the triangle mesh structure during geometry reconstruction. The most common issues included incorrectly oriented triangles, duplicated edges and vertices, and duplicated triangles. In certain situations, a triangle mesh transformation process known as remeshing is necessary [55]. This process can involve various quality indicators for the triangle mesh, such as shape modifications, size, diversity, solution error, or a combination of these factors. There are two primary approaches to triangle mesh processing: *parameterization techniques* [56,57] and *mesh adaptation strategies* [58,59]. The most commonly used parameterization techniques include *linear* [60], *non-linear* [61], and *hybrid methods* [62]. In the referenced publication, the authors used the hybrid method. This approach involved optimizing the facet structure by compacting the triangle mesh in areas with high surface complexity, thereby minimizing the errors. (Figure 10). This approach significantly improved the quality of the model's geometry.

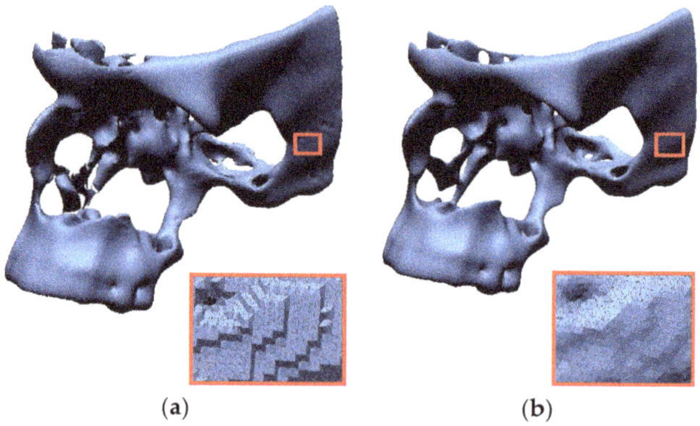

Figure 10. The effect of applying the optimization process: (**a**) 3D-STL model before processing; (**b**) 3D-STL model after processing.

The CAD modeling process focuses on commonly used surface modeling methods and uses functions for mirroring and Boolean functions [63,64]. Appropriate functions are essential, as their configuration can affect the accuracy of the final 3D-CAD model [65]. During the modeling process, it is especially crucial to accurately determine the plane of symmetry on the reconstructed geometries of the anatomic structure models. This procedure significantly shortens the modeling process. One of the crucial steps in preparing data for 3D-printing is converting the developed 3D-CAD models to STL format. During the tessellation process, thousands or even millions of triangles are often created to accurately approximate the curvilinear surface. The resulting differences in mapping accuracy are described by angular and chordal deviations [66,67]. If a smaller value is used for the angular deviation and chord, a 3D-STL model with higher accuracy and a more significant number of triangles will be generated. This generally results in more time being required to carry out the CAD software tessellation process and divide the STL model into print layers. Therefore, it was necessary during the development of the final 3D-STL models to adjust the export parameters in terms of the resolution of the 3D printer in order to not duplicate the errors in the tessellation process when manufacturing the model using the mSLA additive method. In the case of the presented article, the authors assumed that the value of the chord deviation should be 10 times smaller than the resolution of the 3D print. In addition, when selecting the value of the angular deviation, it was determined that it should be no greater than 10 degrees. This ensured sufficient accuracy for the 3D-STL model generation in areas characterized by, among other things, variable radii of curvature.

4.3. Research on Assessing the Accuracy of Models Produced via mSLA Additive Manufacturing

By selecting the appropriate technological parameters during the model's manufacturing, the surface geometry is modified to meet specific technical conditions, ensuring optimal operating performance. The technological evaluation of surface texture is also crucial for objects created using additive methods. Based on research findings in the literature, the accuracy of geometric representation in the mSLA method is significantly influenced by the layer thickness, exposure time, and orientation model in the 3D printer space [68]. Additionally, surface roughness is affected by layer thickness and the object's orientation within the 3D printer space [68]. Additionally, one study [69] observed that exposure time plays an important role in most output measures. However, the studies in the publications were conducted on simple test samples and not on models depicting anatomical structures. In a recent publication [70], attention was drawn to the accuracy of anatomical structure models within the knee joint and the designed reaction templates created using the mSLA method. These models were created using Siraya Tech Fast resin. The bone models and a surgical guide were manufactured with a layer thickness of 0.05 mm, and the exposure times for each layer were consecutively set to 1.8 s. Regarding geometric accuracy, the results for anatomical models and surgical templates were in the ±0.6 mm deviation range. However, the aforementioned publications did not pay attention to yet another research aspect concerning the influence of the method of generating the support material. The use of any 3D-printing method significantly impacts the quality of the surface reproduction. The article presented here examines two modes of generating the support material: normal mode and ultra-light mode. The ultra-light mode produces more accurate models than those created in normal mode. This improvement is attributed to the reduced density and thickness of the supports in the ultra-light mode, which reduces support marks and makes their removal easier (Figure 11).

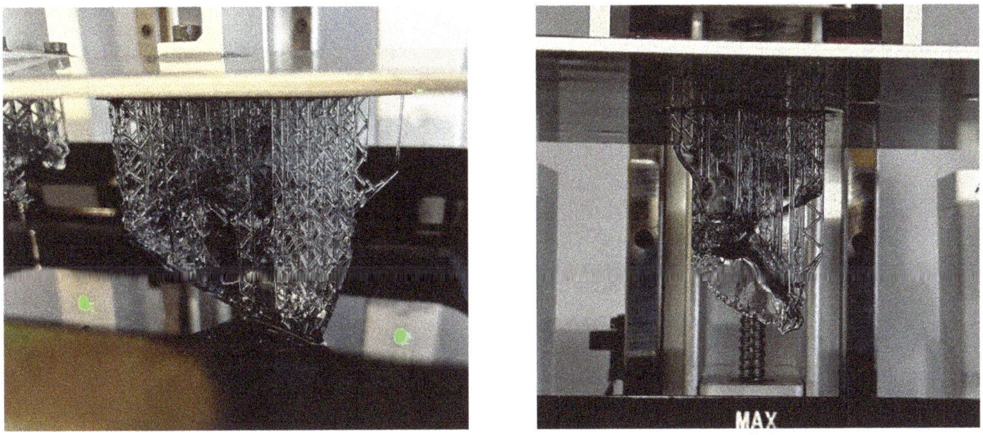

(a) (b)

Figure 11. A view of the completed model with support material for the (**a**) normal mode and (**b**) ultra-light mode.

In both modes, the most significant positive and negative deviation values were found in the areas of the mesh designed for the implant prototype model. Given the established parameters for 3D-printing, producing a CAD-designed mesh using the mSLA method with the recommended geometric accuracy of ±0.3 mm is impossible. If there is considerable negative deviation along the edges of the models, this may be linked to the resin's exposure time. A shorter exposure time means more resin in that area remains uncured, leading to shrinkage of the outer outlines. Additionally, mapping more complex surfaces with the mSLA method can pose challenges. The LCD matrix of the 3D printer produces more accurate images for lines parallel to the X- and Y-axes, whereas lines at an angle are prone to what is known as "step error." Therefore, further research is needed to optimize the parameters used to manufacture models of anatomical structures using the mSLA method.

5. Conclusions

Thanks to the development of coordinate measuring systems, digital data processing methods, and modern manufacturing methods, it is now possible to develop models of anatomical structures, surgical templates, and implants for planning surgical procedures. Designing and manufacturing models for surgical procedures is not a simple task. At each stage of the measurement, reconstruction, CAD modeling, and manufacturing process, geometry errors arise, significantly hindering the development of a final model tailored to a specific patient. In the paper presented here, the following methods were developed to improve geometric accuracy within the zygomatic bone at different stages:

- DICOM data-processing increased spatial and contrast resolution by using a data interpolation process. In addition, the segmentation process used a local thresholding method, which more precisely determined the lower threshold for segmenting bone structures within the zygomatic bone area. Through the use of remeshing methods, the quality of the facet area was significantly increased,
- During CAD modeling, special attention was paid to tessellation, that is, converting the model from CAD to STL format. The values of chordal and angular deviation were adjusted so that errors made during data export were not duplicated in the process of manufacturing the model using the additive method,
- The thinnest layer thickness used in the mSLA method was applied during manufacturing. The recommended model orientation within the 3D printer's workspace was also utilized. The study evaluated two methods for generating the support material. The results indicated that the ultra-light mode produced a more accurate geometrical model. This was attributed to the reduced amount of support material generated during the model's execution, which made the mechanical removal of supports easier during the post-processing stage.

However, the guidelines developed in the presented publication related to the reconstruction process, design, and manufacture using the mSLA method of surgical templates require further improvement. They relate to optimizing the CAD process and selecting 3D-printing parameters to minimize deviations, especially at the edges of models of anatomical structures.

Author Contributions: Conceptualization, P.T. and P.K.; methodology, P.T. and P.K.; software, P.T., P.K., D.R. and N.D.; formal analysis, P.T.; investigation, D.R., N.D. and J.M.; data curation, J.M.; writing—original draft preparation, P.T.; writing—review and editing, P.T. and P.K.; visualization, P.T., P.K., D.R. and N.D.; supervision, P.T. All authors have read and agreed to the published version of the manuscript.

Funding: This research received no external funding.

Data Availability Statement: The raw data supporting the conclusions of this article will be made available by the authors on request.

Conflicts of Interest: The authors declare no conflicts of interest.

References

1. Cortese, A.; Caggiano, M.; Carlino, F.; Pantaleo, G. Zygomatic fractures: Technical modifications for better aesthetic and functional results in older patients. *Int. J. Surg.* **2016**, *33*, S9–S15. [CrossRef] [PubMed]
2. Dechow, P.C.; Wang, Q. Development, structure, and function of the zygomatic bones: What is new and why do we care? *Anat. Rec.* **2016**, *299*, 1611–1615. [CrossRef] [PubMed]
3. Modabber, A.; Rana, M.; Ghassemi, A.; Gerressen, M.; Gellrich, N.C.; Hölzle, F.; Rana, M. Three-dimensional evaluation of postoperative swelling in treatment of zygomatic bone fractures using two different cooling therapy methods: A randomized, observer-blind, prospective study. *Trials* **2013**, *14*, 238. [CrossRef]
4. Markiewicz, M.R.; Gelesko, S.; Bell, R.B. Zygoma reconstruction. *Oral Maxillofac. Surg. Clin.* **2013**, *25*, 167–201. [CrossRef]

5. Litschel, R.; Suárez, G.A. Management of zygomatic fractures: Bone and arch. *Facial Plast. Surg.* **2015**, *31*, 368–375. [CrossRef]
6. Reiss, S.; Kulker, D.; Laure, B.; Paré, A. Reconstruction of the orbitozygomatic framework: State of the art and perspectives. *J. Stomatol. Oral Maxillofac. Surg.* **2024**, *125*, 101788. [CrossRef]
7. Karalashvili, L.; Chichua, N.; Menabde, G.; Atskvereli, L.; Grdzelidze, T.; Machavariani, A.; Chichua, Z. Decellularized bovine bone graft for zygomatic bone reconstruction. *Med. Case Rep.* **2017**, *4*, 10–21767. [CrossRef]
8. Zhu, Z.C.; Yang, Y.F.; Yang, X.; Liu, Y.; Cheng, Y.N.; Sun, Z.Y.; Yang, W.J. Treatment of cryotherapy and orthotopic transplantation following chondromyxoid fibroma of zygomatic bone: A case report. *Medicine* **2018**, *97*, e11707. [CrossRef]
9. Sakkas, A.; Schramm, A.; Karsten, W.; Gellrich, N.C.; Wilde, F. A clinical study of the outcomes and complications associated with zygomatic buttress block bone graft for limited preimplant augmentation procedures. *J. Cranio-Maxillofac. Surg.* **2016**, *44*, 249–256. [CrossRef]
10. Costan, V.V.; Nicolau, A.; Sulea, D.; Ciofu, M.L.; Boișteanu, O.; Popescu, E. The impact of 3D technology in optimizing midface fracture treatment—Focus on the zygomatic bone. *J. Oral Maxillofac. Surg.* **2021**, *79*, 880–891. [CrossRef]
11. Lee, U.L.; Lim, J.Y.; Park, S.N.; Choi, B.H.; Kang, H.; Choi, W.C. A clinical trial to evaluate the efficacy and safety of 3D printed bioceramic implants for the reconstruction of zygomatic bone defects. *Materials* **2020**, *13*, 4515. [CrossRef] [PubMed]
12. Moiduddin, K.; Mian, S.H.; Umer, U.; Alkhalefah, H.; Ahmed, F.; Hashmi, F.H. Design, analysis, and 3D printing of a patient-specific polyetheretherketone implant for the reconstruction of zygomatic deformities. *Polymers* **2023**, *15*, 886. [CrossRef] [PubMed]
13. Moiduddin, K.; Hammad Mian, S.; Umer, U.; Ahmed, N.; Alkhalefah, H.; Ameen, W. Reconstruction of complex zygomatic bone defects using mirroring coupled with EBM fabrication of titanium implant. *Metals* **2019**, *9*, 1250. [CrossRef]
14. Turek, P. The Influence of the Layer Thickness Change on the Accuracy of the Zygomatic Bone Geometry Manufactured Using the FDM Technology. *Eng. Proc.* **2022**, *24*, 26. [CrossRef]
15. Budzik, G.; Turek, P.; Traciak, J. The influence of change in slice thickness on the accuracy of reconstruction of cranium geometry. *Proc. Inst. Mech. Eng. Part H J. Eng. Med.* **2017**, *231*, 197–202. [CrossRef]
16. Van Eijnatten, M.; Berger, F.H.; De Graaf, P.; Koivisto, J.; Forouzanfar, T.; Wolff, J. Influence of CT parameters on STL model accuracy. *Rapid Prototyp. J.* **2017**, *23*, 678–685. [CrossRef]
17. Romans, L. *Computed Tomography for Technologists: A Comprehensive Text*; Wolters Kluwer: Baltimore, MD, USA, 2011.
18. van Eijnatten, M.; Koivisto, J.; Karhu, K.; Forouzanfar, T.; Wolff, J. The impact of manual threshold selection in medical additive manufacturing. *Int. J. Comput. Assist. Radiol. Surg.* **2017**, *12*, 607–615. [CrossRef]
19. Huotilainen, E.; Jaanimets, R.; Valášek, J.; Marcián, P.; Salmi, M.; Tuomi, J.; Wolff, J. Inaccuracies in additive manufactured medical skull models caused by the DICOM to STL conversion process. *J. Cranio-Maxillofac. Surg.* **2014**, *42*, e259–e265. [CrossRef]
20. Moiduddin, K.; Al-Ahmari, A.; Al Kindi, M.; Nasr, E.S.A.; Mohammad, A.; Ramalingam, S. Customized porous implants by additive manufacturing for zygomatic reconstruction. *Biocybern. Biomed. Eng.* **2016**, *36*, 719–730. [CrossRef]
21. Gallo, F.; Zingari, F.; Bolzoni, A.; Barone, S.; Giudice, A. Accuracy of Zygomatic Implant Placement Using a Full Digital Planning and Custom-Made Bone-Supported Guide: A Retrospective Observational Cohort Study. *Dent. J.* **2023**, *11*, 123. [CrossRef]
22. Salmi, M.; Paloheimo, K.S.; Tuomi, J.; Wolff, J.; Mäkitie, A. Accuracy of medical models made by additive manufacturing (rapid manufacturing). *J. Cranio-Maxillofac. Surg.* **2013**, *41*, 603–609. [CrossRef] [PubMed]
23. Turek, P.; Budzik, G. Estimating the Accuracy of Mandible Anatomical Models Manufactured Using Material Extrusion Methods. *Polymers* **2021**, *13*, 2271. [CrossRef] [PubMed]
24. Markopoulos, A.P.; Galanis, N.I.; Karkalos, N.E.; Manolakos, D.E. Precision CNC machining of femoral component of knee implant: A case study. *Machines* **2018**, *6*, 10. [CrossRef]
25. Bagudanch, I.; García-Romeu, M.L.; Ferrer, I.; Ciurana, J. Customized cranial implant manufactured by incremental sheet forming using a biocompatible polymer. *Rapid Prototyp. J.* **2018**, *24*, 120–129. [CrossRef]
26. Sun, B.; Ma, Q.; Wang, X.; Liu, J.; Rejab, M.R.M. Additive manufacturing in medical applications: A brief review. *IOP Conf. Ser. Mater. Sci. Eng.* **2021**, *1*, 012007. [CrossRef]
27. Manmadhachary, A.; Malyala, S.K.; Alwala, A. Medical applications of additive manufacturing. In Proceedings of the International Conference on ISMAC in Computational Vision and Bio-Engineering 2018 (ISMAC-CVB), Palladam, India, 16–17 May 2018; pp. 1643–1653. [CrossRef]
28. Meena, V.K.; Rattan, V.; Luthra, G.; Kalra, P. Development of a custom zygomatic implant using metal sintering. *Rapid Prototyp. J.* **2018**, *24*, 828–831. [CrossRef]
29. Peel, S.; Eggbeer, D.; Sugar, A.; Evans, P.L. Post-traumatic zygomatic osteotomy and orbital floor reconstruction. *Rapid Prototyp. J.* **2016**, *22*, 878–886. [CrossRef]
30. Junk, S.; Bär, F. Design guidelines for Additive Manufacturing using Masked Stereolithography mSLA. *Procedia CIRP* **2023**, *119*, 1122–1127. [CrossRef]
31. Orzeł, B.; Stecuła, K. Comparison of 3D Printout Quality from FDM and MSLA Technology in Unit Production. *Symmetry* **2022**, *14*, 910. [CrossRef]

32. Kumar, S.; Bhushan, P.; Pandey, M.; Bhattacharya, S. Additive manufacturing as an emerging technology for fabrication of microelectromechanical systems (MEMS). *J. Micromanuf.* **2019**, *2*, 175–197. [CrossRef]
33. Sorocki, J.; Piekarz, I. Low-cost microwave components' fabrication in hybrid technology of laminates and additive manufacturing on an example of miniaturized suspended directional coupler. *IEEE Access* **2020**, *8*, 128766–128775. [CrossRef]
34. Mondal, D.; Haghpanah, Z.; Huxman, C.J.; Tanter, S.; Sun, D.; Gorbet, M.; Willett, T.L. mSLA-based 3D printing of acrylated epoxidized soybean oil-nano-hydroxyapatite composites for bone repair. *Mater. Sci. Eng. C* **2021**, *130*, 112456. [CrossRef] [PubMed]
35. Mondal, D.; Willett, T.L. Enhanced mechanical performance of mSLA-printed biopolymer nanocomposites due to phase functionalization. *J. Mech. Behav. Biomed. Mater.* **2022**, *135*, 105450. [CrossRef]
36. Zhou, P.; Chambers, C.B. Orbital Fractures. *Semin. Plast. Surg.* **2021**, *35*, 269–273. [CrossRef]
37. Fan, S.; Sáenz-Ravello, G.; Diaz, L.; Wu, Y.; Davó, R.; Wang, F.; Magic, M.; Al-Nawas, B.; Kämmerer, P.W. The Accuracy of Zygomatic Implant Placement Assisted by Dynamic Computer-Aided Surgery: A Systematic Review and Meta-Analysis. *J. Clin. Med.* **2023**, *12*, 5418. [CrossRef]
38. Lo Russo, L.; Pierluigi, M.; Zhurakivska, K.; Digregorio, C.; Lo Muzio, E.; Laino, L. Three-Dimensional Accuracy of Surgical Guides for Static Computer-Aided Implant Surgery: A Systematic Review. *Prosthesis* **2023**, *5*, 809–825. [CrossRef]
39. Budzik, G.; Turek, P. The impact of use different type of image interpolation methods on the accuracy of the reconstruction of skull anatomical model. *Biomed. Eng. Appl. Basis Commun.* **2020**, *32*, 2050008. [CrossRef]
40. Newman, T.S.; Yi, H. A survey of the marching cubes algorithm. *Comput. Graph.* **2006**, *30*, 854–879. [CrossRef]
41. de Sales Guerra Tsuzuki, M.; Sato, A.K.; Ueda, E.K.; de Castro Martins, T.; Takimoto, R.Y.; Iwao, Y.; Kagei, S. Propagation-based marching cubes algorithm using open boundary loop. *Vis. Comput.* **2018**, *34*, 1339–1355. [CrossRef]
42. Guillaume, O.; Geven, M.A.; Varjas, V.; Varga, P.; Gehweiler, D.; Stadelmann, V.A.; Smidt, T.; Zeiter, S.; Sprecher, C.; Bos, R.R.; et al. Orbital floor repair using patient specific osteoinductive implant made by stereolithography. *Biomaterials* **2020**, *233*, 119721. [CrossRef]
43. Alsleem, H.; Davidson, R. Factors affecting contrast-detail performance in computed tomography: A review. *J. Med. Imaging Radiat. Sci.* **2013**, *44*, 62–70. [CrossRef]
44. Joemai, R.M.; de Bruin, P.W.; Veldkamp, W.J.; Geleijns, J. Metal artifact reduction for CT: Development, implementation, and clinical comparison of a generic and a scanner-specific technique. *Med. Phys.* **2012**, *39*, 1125–1132. [CrossRef] [PubMed]
45. Lubner, M.G.; Pickhardt, P.J.; Tang, J.; Chen, G.H. Reduced image noise at low-dose multidetector CT of the abdomen with prior image constrained compressed sensing algorithm. *Radiology* **2011**, *260*, 248–256. [CrossRef] [PubMed]
46. Siddiqi, A.A. Filter selection for removing noise from ct scan images using digital image processing algorithm. *Biomed. Eng. Appl. Basis Commun.* **2024**, *36*, 2350038. [CrossRef]
47. Zafeiridis, P.; Papamarkos, N.; Goumas, S.; Seimenis, I. A New Sharpening Technique for Medical Images using Wavelets and Image Fusion. *J. Eng. Sci. Technol. Rev.* **2016**, *9*, 187–200. [CrossRef]
48. Hu, Z.; Zheng, H.; Gui, J. A novel interactive image processing approach for DICOM medical image data. In Proceedings of the 2009 2nd International Conference on Biomedical Engineering and Informatics, Tianjin, China, 17–19 October 2009; pp. 1–4. [CrossRef]
49. Kats, L.; Goldman, Y.; Kahn, A. Automatic detection of image sharpening in maxillofacial radiology. *BMC Oral Health* **2021**, *21*, 411. [CrossRef]
50. Van Eijnatten, M.; van Dijk, R.; Dobbe, J.; Streekstra, G.; Koivisto, J.; Wolff, J. CT image segmentation methods for bone used in medical additive manufacturing. *Med. Eng. Phys.* **2018**, *51*, 6–16. [CrossRef]
51. Mahmood, N.; Shah, A.; Waqas, A.; Abubakar, A.; Kamran, S.; Zaidi, S.B. Image segmentation methods and edge detection: An application to knee joint articular cartilage edge detection. *J. Theor. Appl. Inf. Technol.* **2015**, *71*, 87–96.
52. Ren, H.; Zhou, L.; Liu, G.; Peng, X.; Shi, W.; Xu, H.; Liu, L. An unsupervised semi-automated pulmonary nodule segmentation method based on enhanced region growing. *Quant. Imaging Med. Surg.* **2020**, *10*, 233–242. [CrossRef]
53. Morita, D.; Mazen, S.; Tsujiko, S.; Otake, Y.; Sato, Y.; Numajiri, T. Deep-learning-based automatic facial bone segmentation using a two-dimensional U-Net. *Int. J. Oral Maxillofac. Surg.* **2023**, *52*, 787–792. [CrossRef]
54. Tiribilli, E.; Bocchi, L. deep learning-based workflow for bone segmentation and 3D modeling in cone-beam CT orthopedic imaging. *Appl. Sci.* **2024**, *14*, 7557. [CrossRef]
55. Alliez, P.; Ucelli, G.; Gotsman, C.; Attene, M. Recent advances in remeshing of surfaces. In *Mathematics and Visualization*; Springer: Berlin/Heidelberg, Germany, 2008; pp. 53–82. [CrossRef]
56. Laug, P.; Borouchaki, H. Interpolating and meshing 3D surface grids. *Int. J. Numer. Methods Eng.* **2003**, *58*, 209–225. [CrossRef]
57. Attene, M.; Falcidieno, B.; Spagnuolo, M.; Wyvill, G. A mapping-independent primitive for the triangulation of parametric surfaces. *Graph. Models* **2003**, *65*, 260–273. [CrossRef]
58. Wang, D.; Hassan, O.; Morgan, K.; Weatherill, N. Enhanced remeshing from STL files with applications to surface grid generation. *Commun. Numer. Methods Eng.* **2007**, *23*, 227–239. [CrossRef]

59. Lévy, B.; Liu, Y.L. Lp centroidal voronoi tessellation and its applications. *ACM Trans. Graph. (TOG)* **2010**, *29*, 1–11. [CrossRef]
60. Alliez, P.; Meyer, M.; Desbrun, M. Interactive geometry remeshing. *ACM Trans. Graph. (TOG)* **2002**, *21*, 347–354. [CrossRef]
61. Zigelman, G.; Kimmel, R.; Kiryati, N. Texture mapping using surface flattening via multidimensional scaling. *IEEE Trans. Vis. Comput. Graph.* **2002**, *8*, 198–207. [CrossRef]
62. Ben-Chen, M.; Gotsman, C.; Bunin, G. Conformal flattening by curvature prescription and metric scaling. *Comput. Graph. Forum* **2008**, *27*, 449–458. [CrossRef]
63. Mejía Rodríguez, M.; González-Estrada, O.A.; Villegas-Bermúdez, D.F. Finite Element Analysis of Patient-Specific Cranial Implants under Different Design Parameters for Material Selection. *Designs* **2024**, *8*, 31. [CrossRef]
64. Bonda, D.J.; Manjila, S.; Selman, W.R.; Dean, D. The Recent Revolution in the Design and Manufacture of Cranial Implants. *Neurosurgery* **2015**, *77*, 814–824. [CrossRef]
65. Turek, P.; Bezłada, W.; Cierpisz, K.; Dubiel, K.; Frydrych, A.; Misiura, J. Analysis of the Accuracy of CAD Modeling in Engineering and Medical Industries Based on Measurement Data Using Reverse Engineering Methods. *Designs* **2024**, *8*, 50. [CrossRef]
66. Iancu, C.; Iancu, D.; Stăncioiu, A. From CAD model to 3D print via" STL" file format. *Fiability Durab./Fiabil. Si Durabilitate* **2010**, *1*, 73–80.
67. Szilvśi-Nagy, M.; Matyasi, G.Y. Analysis of STL files. *Math. Comput. Model.* **2003**, *38*, 945–960. [CrossRef]
68. Singh, S.; Jain, A.; Chaudhary, P.; Gupta, R.; Mali, H.S. Optimization of dimensional accuracy and surface roughness in m-SLA using response surface methodology. *Rapid Prototyp. J.* **2023**, *29*, 1324–1339. [CrossRef]
69. Borra, N.D.; Neigapula, V.S.N. Parametric optimization for dimensional correctness of 3D printed part using masked stereolithography: Taguchi method. *Rapid Prototyp. J.* **2023**, *29*, 166–184. [CrossRef]
70. Turek, P.; Jakubiec, J. Geometrical precision and surface topography of mSLA-produced surgical guides for the knee joint. *J. Eng. Manag. Syst. Eng.* **2023**, *2*, 150–157. [CrossRef]

designs

MDPI

Article

Comparison of Methods for Reconstructing Irregular Surfaces from Point Clouds of Digital Terrain Models in Developing a Computer-Aided Design Model for Rapid Prototyping Technology

Michał Chlost and Anna Bazan *

Faculty of Mechanical Engineering and Aeronautics, Rzeszów University of Technology, 35-959 Rzeszów, Poland; m.chlost@prz.edu.pl
* Correspondence: abazan@prz.edu.pl; Tel.: +48-178651371

Abstract

This article presents a methodology for developing a three-dimensional terrain model based on numerical data in the form of a point cloud, with an emphasis on reducing mesh surface errors and using a surface smoothing factor. Initial surface generation was based on a point cloud with a square mesh, and an adopted algorithm for mesh conversion to the input form for the computer aided design (CAD) environment was presented. The use of a bilinear interpolation algorithm was proposed to reduce defects in the three-dimensional surface created in the reverse engineering process. The terrain mapping accuracy analyses were performed for three samples of different geometry using two available options in the Siemens NX program. All obtained surfaces were subjected to shape deviation analysis. For each of the analyzed surfaces, changing the smoothing factor from 0% to 15% did not cause significant changes in accuracy depending on the method adopted. For flat regions, in the Uniform Density (UD) method, the size of the area outside the tolerance was 6.16%, and in the Variable Density (VD) method, it was within the range of 5.01–6%. For steep regions, in the UD method, it was 6.25%, and in the VD method, it was within the range of 5.39–6.47%, while for concave–convex regions, in the UD method, it was 6.5% and in the VD method, it was within the range of 4.96–6.36%. For a smoothing factor value of 20%, a sudden increase in the inaccuracy of the shape of the obtained surface was observed. For flat regions, in the Uniform Density (UD) method, the size of the area outside the tolerance was 69.84%, and in the Variable Density (VD) method, it was 71.62%. For steep regions, in the UD method, it was 76.07%, and in the VD method, it was 80.94%, while for concave–convex regions, in the UD method, it was 56.08%, and in the VD method, it was 62.38%. The developed methodology provided high accuracy in the reproduction of numerical data that can be used for further analyses and manufacturing processes, such as 3D printing. Based on the obtained data, three fused deposition model (FDM) prints were made, presenting each of the analyzed types of terrain geometry. Only FDM printing was used, and other technologies were not verified.

Keywords: LiDAR; DTM; reverse engineering; bilinear interpolation; rapid prototyping

1. Introduction

For many years, reverse engineering has been playing a key role as an irreplaceable tool in the processes of obtaining data on existing objects and their digital processing, as well as producing physical copies of these elements. It is used in numerous fields, from the automotive and aerospace industries to biomedical engineering. Especially in combination with additive technologies, such as 3D printing, reconstructive engineering is becoming a basic tool used in the reproduction and production of mechanical parts that may be difficult or impossible to obtain using traditional production methods [1].

The basic method of reconstructing simple elements, such as those presented by Budzik et al. [1], is to map a point cloud by using an approximation through simple solids (cylinder, cone, sphere). In the case of more complex but still regular surfaces, a mapping using free surfaces described by polynomials for both the U and V directions of the surface or a B-spline surface is used [2].

Progress in the field of precise measurement methods, as well as the development of new measurement technologies and data acquisition systems, such as computed tomography, 3D laser scanning, and digital photogrammetry, has allowed for a significant expansion of the possibilities of reconstructive engineering. Thanks to these technologies, it has become possible to accurately reproduce not only simple mechanical elements with regular geometry but also complex biological structures characterized by a high degree of irregularity and complexity of shapes [3–5]. This has opened up new perspectives in fields such as medicine, archaeology, or the protection of cultural heritage, in which the faithful reconstruction of biological or historical structures is of key importance.

A special case of a reverse engineering application is the reconstruction of large-scale objects, which include, among others, topography, extensive geological formations, and large engineering structures [6,7]. The process of digitizing such objects brings with it additional challenges related to their size and the variety of surfaces. For this reason, not all available imaging and measurement technologies are able to meet the requirements for the accuracy and efficiency of data acquisition.

The basic measurement methods currently used in obtaining topographic data include light detection and ranging (LiDAR) technology, photogrammetry, satellite measurements, and ground measurements via the global positioning system (GPS). Each of these methods is characterized by different principles of operation, ranges of applications, and accuracies, which allows for their appropriate selection depending on the specifics of the project and terrain conditions [8].

The LiDAR measurement method is used with great success to generate precise three-dimensional point clouds that represent existing objects and terrain surfaces. Thanks to the use of laser pulses and the precise measurement of their return time, it is possible to obtain very detailed representations, even in difficult environmental conditions, for example, in densely forested areas. Depending on the measurement purpose, the size of the digitized objects, and the possibility of physical access to them by operators, different variants of this technology are used. For data acquisition on a smaller scale, in hard-to-reach places, terrestrial LiDAR systems are used, as described in the literature [7–9]. On the other hand, if it is necessary to cover large areas, systems mounted onboard flying vehicles, such as airplanes or unmanned aerial vehicles [10], are used.

Data from LiDAR are supplemented by elevation measurements performed using GPS receivers, which enable the direct verification and calibration of field measurement results [11]. In particular, high-accuracy geodetic techniques are used, such as the global navigation satellite system (GNSS) or static measurements, which increase the precision of localization of the acquired spatial data.

In parallel to LiDAR technology, photogrammetry is also often used, i.e., a technique for obtaining information about objects and their spatial arrangement based on photo analysis. Photogrammetry, using photos taken from different perspectives, allows for the reconstruction of three-dimensional surface geometry. Due to significantly lower implementation costs compared to LiDAR systems, photogrammetry is widely used in projects in which maintaining the cost-effectiveness ratio is important. Currently, the main carriers of imaging sensors are drones, which, thanks to their mobility and operational flexibility, allow for the capture of high-resolution photos in various terrain conditions [12,13].

In addition, measurement techniques based on satellite observations are used to obtain data on the terrain. Satellite measurement allows for the quick collection of information on very large areas, which is particularly useful in regional and global studies. One of the most popular methods is the use of data from the Advanced Spaceborne Thermal Emission and Reflection Radiometer (ASTER) system, as presented in [14,15]. Alganci et al. [16] present a detailed comparison of various satellite measurement methods, analyzing their accuracy, scope of application, and limitations.

The final effect of using the described measurement methods is the creation of a digital surface model (DSM). This model represents a point cloud that contains information on both the terrain proper and the elements located above the ground surface, such as vegetation, buildings, and other anthropogenic structures, as shown in Figure 1. DSM is the basis for further spatial analyses, the modeling of the natural environment, urban planning, and the monitoring of changes in the landscape.

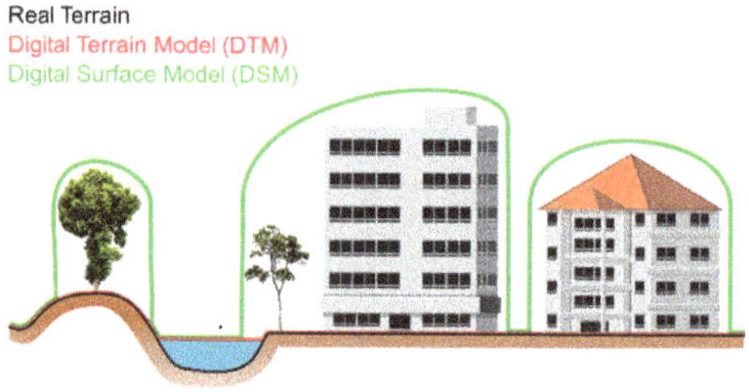

Figure 1. Graphical representation of differences between DTM (red) and DSM (green) terrain data.

A separate challenge is how to filter out objects located above the terrain geometry, i.e., vegetation and buildings, which are an integral part of DSM data. In the process of creating a digital terrain model (DTM), advanced real-time data reduction techniques are increasingly used. One example is the optimum dataset method (OptD) [17], which allows for the generation of DTMs in parallel with data acquisition. Thanks to sequential estimation and dynamic data reduction, this method preserves the most important terrain features while significantly reducing the volume of data by up to 98%, without any significant loss of accuracy. This approach significantly increases computational efficiency and allows for the ongoing creation of 3D models with high precision, which is of great importance in applications that require fast data processing.

In parallel, classical LiDAR point cloud filtering algorithms are being developed. Six popular methods, including the adaptive triangulated irregular network (ATIN), elevation threshold with an expanded window (ETEW), maximum local slope (MLS), progressive morphology (PM), iterative polynomial fitting (IPF), and multiscale curvature classification (MCC), were tested, each showing different levels of effectiveness depending on the terrain characteristics [18,19]. IPF achieved the best results for flat and urban areas, while ETEW was the most effective in difficult, mountainous areas with dense vegetation. Zhang et al. [20] and Lee et al. [21] focused on the challenges associated with filtering photogrammetry data using the digital surface model (DIM) and structure from motion (SfM) methods. Classic LiDAR filters can be partially effective for DIM data but require additional procedures, such as a ranking filter, to minimize systematic errors. In SfM applications, especially for riverine areas, combining vegetation filters, such as normalized difference vegetation index (NDVI) and excess green (ExG), with morphological filters, such as cloth simulation filtering (CSF) and ATIN, has proven effective, thereby improving model quality while maintaining low operating costs.

A separate category of methods is techniques based on object segmentation, as described by Song et al. [22]. The novel approach assumes defining objects as areas enclosed by steep slopes and the ground as smoothly connected spaces, which allows for more consistent filtering over large areas, especially in the urban environment. Additionally, including water bodies and artificial structures, such as bridges, increases the realism of the obtained models. Traditional procedures [23], which are used within the GRASS Geographic Information System (GIS), remain effective in the basic classification of LiDAR data, rejecting buildings and vegetation in order to obtain a precise DTM. Modern filtering approaches increasingly combine several methods to achieve the highest possible model accuracy in diverse terrain conditions.

A separate problem is the noise and irregularity of measurement data, which cause errors in the correct reconstruction of the complex surface. Marton et al., in their article [24], presented a method for filling in missing measurement points using the resampling method. Here, the method of weighted least squares was used to fill in missing data in real time.

The current article presents research indicating the possibility of effective processing of geodetic data and the use of mesh surface creation techniques in the CAD and computer-aided engineering (CAE) environment to prepare models made using 3D printing. The key aim of the publication was to show the entire process of creating a model, from the analysis of input data through the densification of the point grid, which allowed us to reduce errors in creating surfaces, to the use of CAD tools to obtain a digital representation of the surface, ending with obtaining a physical model made via 3D printing. Unlike the method presented by Marton et al. [24], the densification of the point cloud did not consist of filling in the missing areas of the mesh but of creating nodes that facilitate the reconstruction of the surface based on measurement data.

Such models have practical applications, among others, in special education or geo-engineering. The research was carried out on three types of surfaces that represented relatively flat surfaces, which had steep slopes and contained concave–convex forms. An algorithm was developed to increase the quality of the conversion of geodetic data to the mesh surface. The numerical analyses performed allowed for the assessment of the impact of different mesh generation methods and smoothing functions on the accuracy of mapping the terrain topography.

2. Materials and Methods

2.1. Input Data and Their Characteristics

The point cloud, which is the basis of the digital terrain model (DTM), was developed in the form of an ordered grid of regular rectangles with a constant cell size of 1×1 m. This type of spatial data structure allows for the precise analysis of the terrain morphology and further numerical processing, such as surface runoff modeling, slope inclination analysis, or the generation of terrain profiles. The numerical data used to build the model were obtained from a publicly available source: the Geoportal application [25], which is managed by the Head Office of Geodesy and Cartography in Poland.

The applied elevation data refer to the currently applicable national elevation reference system PL-EVRF2007-NH, which is the Polish implementation of the European Height Reference System. This system was officially introduced in Poland by the Regulation of the Council of Ministers Dz.U.2024.0.342 §24 [26] on 1 January 2024, replacing the previous PL-KRON86-NH system. This change aims to unify elevation references on an international scale and ensure compliance with European geodetic and hydrographic standards. Thanks to this change, the elevation data used in spatial analyses are characterized by greater consistency and precision, which is particularly important in the context of data integration at the international level.

The point cloud representing the topography of the analyzed terrain area was obtained in American Standard Code for Information Interchange (ASCII) text file format. This file contains data in the form of a two-dimensional matrix with dimensions of $y \times x$, in which each cell corresponds to the height value of the measured terrain point relative to the adopted reference system. Such an organized data structure allows for the basic visualization of the surface shape but encounters significant limitations in the context of reconstructive engineering applications.

2.2. Transforming Input Data

When using professional reconstructive engineering tools such as Siemens NX in version 23.06.3001 or other CAD/CAE software, the matrix format is not optimal or natively supported as an input database. The main problems resulting from this data structure are as follows:

- The first is the reversal of the orientation of the y-axis corresponding to the north (N) geographic direction. In an ASCII file, data are most often ordered in a way that maps the direction from north to south, which means that the increasing values in the matrix rows correspond to the south (S) direction, not N. This leads to a discrepancy with the spatial orientation expected by engineering software, which interprets the axes in accordance with Cartesian systems (x-width, y-length, and z-height).
- The second is the inappropriate data storage format. Most engineering applications, including NX, expect data in the form of a set of three-dimensional coordinates of points in the format of x; y; z, where each line contains three values: coordinates x and y and the corresponding height, z. A specific storage convention is also required, most often using a field separator in the form of a semicolon (;) or a comma, depending on the regional settings and program requirements.

For this reason, it is necessary to first convert the data from the matrix format to a list of points in the x; y; z format. This process requires taking into account the grid size (e.g., 1×1 m) and the order of rows and columns, as well as shifting or correcting the y-axis orientation in order to adapt the system to the CAD system requirements. Only after such a conversion can the data be effectively imported and used in the process of creating surface or solid models in engineering environments.

In order to adapt the numerical data to CAD software standards, a program was created to convert the $y \times x$ matrix into the x; y; z data format and to reverse the y-axis direction. The schematic diagram of the algorithm is shown in Figure 2.

After the necessary data conversion from the matrix format to the point coordinate format, x; y; z, the point cloud was imported into the design environment of Siemens NX software, which is an advanced CAD tool that is used, among others, in reverse engineering and surface modeling. The aim of the operation was to reconstruct the terrain topography based on the acquired measurement data.

The surface reconstruction process began with the use of tools for creating a surface composed of a triangle mesh based on imported points. Although the base mesh had a regular square layout of 1×1 m, which should theoretically help to obtain a uniform surface, a number of undesirable effects were encountered due to the limitations of the meshing algorithms and the nature of the input data.

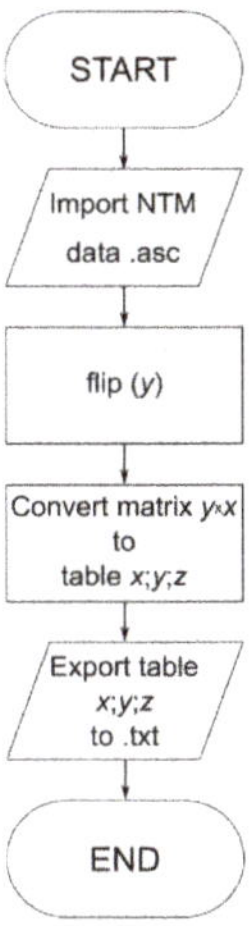

Figure 2. Conversion algorithm for a $y \times x$ matrix into the x; y; z format. The algorithm code is presented in Appendix A.

During surface reconstruction, numerous geometric artifacts appeared in Figure 3, including:

- gaps and discontinuities in the mesh, leading to the creation of empty areas (holes) in the model;
- holes and gaps at the edges, resulting from the lack of clear neighbors for boundary points;
- notches and surface breaks appearing in places where the point data were irregular or insufficiently dense;
- overlapping mesh triangles, i.e., topological errors resulting in surface intersection and generating geometry that is mathematically incorrect and difficult to further process.

Figure 3. Defects created on a surface generated from a square grid of points.

All of the errors listed significantly affect the quality and usability of the obtained terrain reconstruction, thereby limiting the possibilities of using the model in further design stages. The identified problems indicate the need for the prior cleaning and preparation of the point cloud, including smoothing the surface, interpolating missing data, and using dedicated reconstruction algorithms, which are better adapted to work with regular grids with a large number of points.

2.3. Application of the Bilinear Interpolation Algorithm

In order to improve the quality of the generated surface and reduce the number of topological errors that occur during the surfacing process, it was decided to use the bilinear interpolation technique at the central points of each cell of the base grid, as shown in Figure 4a.

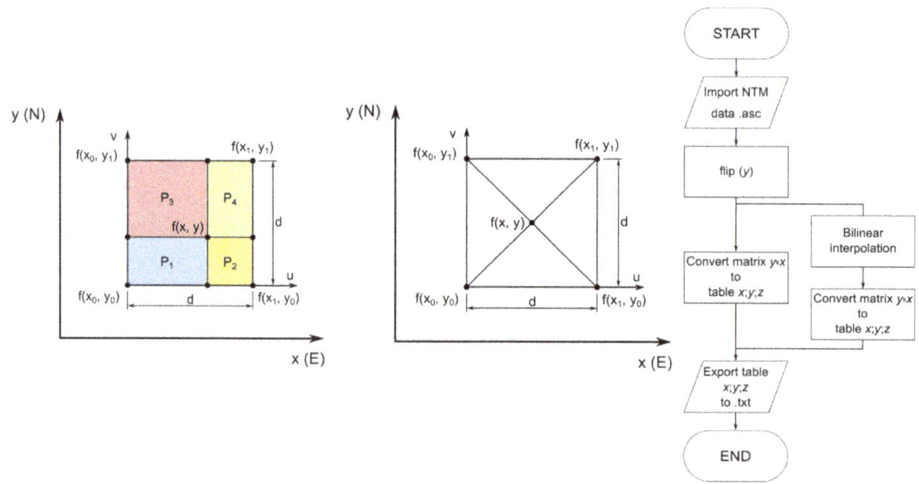

(**a**) (**b**) (**c**)

Figure 4. Graphical representation of bilinear interpolation of a point inside a grid cell: (**a**) idea of bilinear interpolation, (**b**) representation of midpoint interpolation, (**c**) data processing algorithm enriched with a bilinear interpolation term. The algorithm code is presented in Appendix B.

The bilinear interpolation algorithm was based on the input points in ASCII format for a $d \times d$ grid resolution. Therefore, an element performing bilinear interpolation was added to the basic algorithm (Figure 4c). The bilinear interpolation equation is presented in Formula (1) [27,28], where the coefficients, $f(x_i, y_i)$, correspond to the height value, z, at a given grid point (Figure 4a,b).

$$f(x,y) = \frac{f(x_0,y_0)\cdot(d-x)\cdot(d-y) + f(x_1,y_0)\cdot x\cdot(d-y)}{d^2} + \frac{f(x_0,y_1)\cdot(d-x)\cdot y + f(x_1,y_1)\cdot x\cdot y}{d^2} \quad (1)$$

In order to simplify the algorithm, the formula was transformed by replacing the factors, $f(x_i, y_i)$, with the parameter, zi, and the remaining factors with the parameters, Pi (2, 3).

$$\begin{gathered} P_1 = x\cdot y,\;\; P_2 = (d-x)\cdot y,\;\; P_3 = x\cdot(d-y), P_4 = (d-x)\cdot(d-y) \\ z_1 = f(x_0,y_0), z_2 = f(x_1,y_0), z_3 = f(x_0,y_1), z_4 = f(x_1,y_1) \end{gathered} \quad (2)$$

$$f(x,y) = \frac{z_1\cdot P_4 + z_2\cdot P_3 + z_3\cdot P_2\cdot y + z_4\cdot P_1}{d^2} \quad (3)$$

The parameter d^2 was replaced by the sum of the surface areas of the individual segments, P_i (4), thereby obtaining Equation (5).

$$d^2 = P_1 + P_2 + P_3 + P_4 \quad (4)$$

$$f(x,y) = \frac{\sum_{i=1}^{4} z_i\cdot P_i}{\sum_{i=1}^{4} P_i} \quad (5)$$

The original grid, which was made of regular 1×1 m squares, was thus transformed into a system containing additional nodes located exactly at the geometric centers of each square cell.

Bilinear interpolation involves determining the height value, z, at the central point of the grid cell based on the values measured at the four corner nodes of the given mesh. This is an estimation method that assumes a linear change in the values along the x and y axes, which allows for smoother transitions between points and reduces local height jumps. This made it possible to generate a more continuous and coherent terrain surface.

After determining the center points, each square grid cell was divided into four triangles, connecting the corner nodes with the newly determined center point Figure 4b. This approach allowed for the replacement of the square grid with a mesh with an exclusive triangular division, which significantly increased the stability and accuracy of the CAD surface creation algorithms. Triangular elements are more resistant to errors related to the nonplanarity of a surface and better represent terrain irregularities.

It should be noted that bilinear interpolation is only a basic resampling tool. It does not increase the resolution of the mesh itself, and the coordinates of points created during its implementation are burdened with errors, as described by Hu et al. in [26]. However, in the conducted research, bilinear interpolation was used to increase the efficiency of the tool that created surfaces in the CAD environment, which is ultimately based on a triangular grid. The only solution that would exclude the use of bilinear interpolation in the studied case would be to perform an additional measurement of the real terrain, which would fill in the missing grid points.

The use of this method contributed to a significant reduction in the number of reconstruction errors, including a reduction in the number of holes and overlapping surfaces (Figure 5).

Figure 5. Reconstruction of the terrain surface after applying bilinear interpolation.

2.4. Generating Surfaces for Comparison

As part of the experiment, to check the correct operation of the above-described method and to compare the accuracy of surfaces generated with different settings of the Mesh Point Cloud tool, three areas are presented in Table 1.

The surface for each area was generated using the following three methods, which are available in NX software [29]:

- Keep All Points;
- Variable Density (VD);
- Uniform Density (UD).

Table 1. Comparison of visualization methods for three selected research areas.

Type of Surface Area Size x;y	Orthophotomap	Shading	3D Surface
flat region 180; 158 m			

Table 1. *Cont.*

Type of Surface Area Size x;y	Orthophotomap	Shading	3D Surface
steep region 170; 156 m		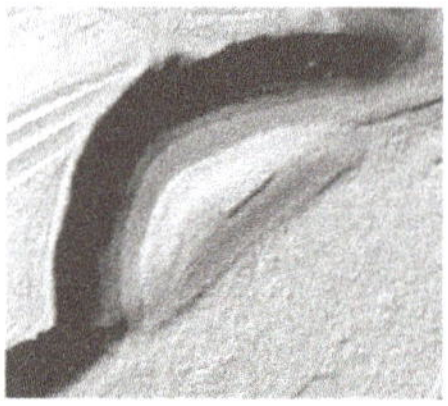	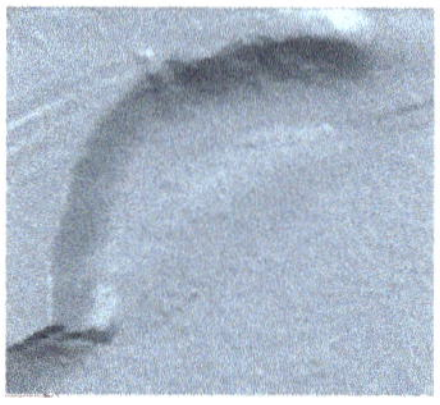
concave-convex region 121; 133 m			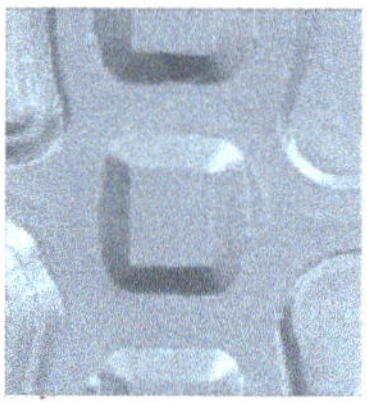

The Keep All Points method is the simplest to operate and involves creating a surface from a triangle mesh between all the points in the cloud. The Variable Density method creates larger segments from the point cloud in regions of low curvature and smaller segments in regions of high curvature. The Uniform Density method creates a surface from the point cloud with a grid of equal-area triangles.

The Keep All Points method was chosen to create a reference surface due to the most accurate rendering of the imported point cloud; however, it does not have a surface smoothing tool. In the other methods, the surface smoothing factor is present, and during the comparative tests, five values were adopted for it: 0, 5, 10, 15, 20%.

3. Results

3.1. Comparative Analysis of Surface Creation Tools

In order to reduce the data size, a 1 × 1 m resolution real grid was imported into the CAD software at a resolution of 1 × 1 mm. The surface shape tolerance was set at ±0.01 mm.

The simulations performed showed slight differences in the accuracy of surface reconstruction depending on the degree of smoothing, ranging from 0% to 15%.

For flat areas, the percentage of areas outside the tolerance range for Variable Density surfaces ranged from 5.01% to 6.17%. For Uniform Density surfaces, the value was 6.16%, as shown in Figure 6.

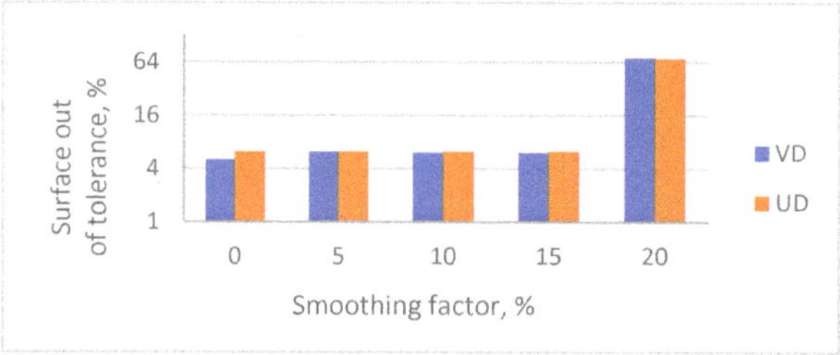

Figure 6. Out of tolerance area range (±0.01 mm) for the flat region sample, expressed in %, in the VD and UD methods.

In the steep regions, the Variable Density surface showed deviations in the range of 5.39–6.47%. In the Uniform Density surface, 6.25% of the surface was out of tolerance, as shown in Figure 7.

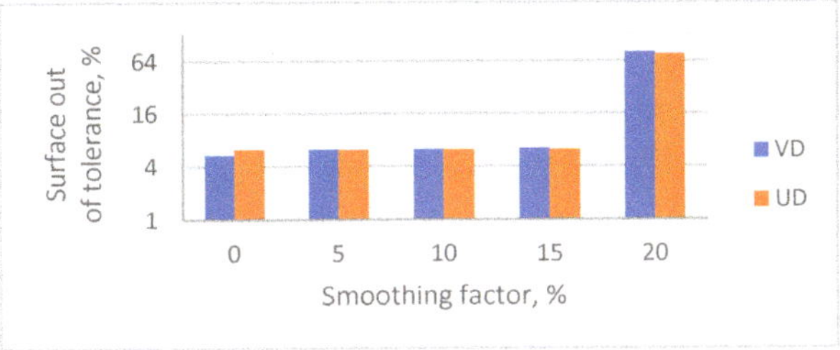

Figure 7. Out of tolerance area range (±0.01 mm) for the steep region sample, expressed in %, in the VD and UD methods.

For the concave–convex regions, the Variable Density surface deviation ranged from 4.96 to 6.44%, while the Uniform Density surface reached 6.5% of the surface out of tolerance (also shown in Figure 8).

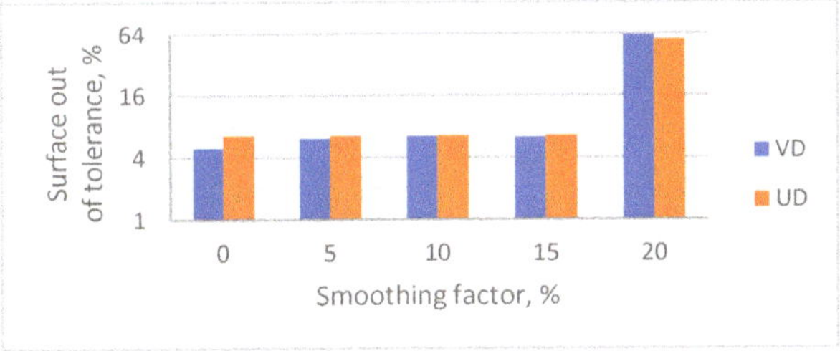

Figure 8. Out of tolerance area range (±0.01 mm) for the concave–convex region sample, expressed in %, in the VD and UD methods.

For all analyzed cases and surface types, the application of 20% smoothing resulted in a significant deterioration in the accuracy of the geometry representation. The highest percentage deviation was recorded for the steep region for the Variable Density surface, while the lowest was recorded for the Uniform Density surface in the concave–convex region (Figure 8).

In general, it has been found that at 20% smoothing, Uniform Density surfaces provide a greater proportion of surface area within tolerance than Variable Density surfaces.

Additionally, surface deviation maps that graphically present areas that exceed the permissible tolerance range were developed (Tables 2–4).

For flat surfaces in both cases—Uniform and Variable—the distribution of out-of-tolerance areas was scattered and random (Table 2).

In the case of steep and concave–convex regions in the Variable Density method, most of the deviations were located in places with a relatively constant curvature (Tables 3 and 4). In the case of the Uniform Density method, the largest clusters of areas outside the tolerance occurred in transitional regions on the boundaries of surfaces with different inclinations.

Table 2. Surface deviation map for the flat region.

Smoothing, %	Variable Density	Uniform Density	Scale
0	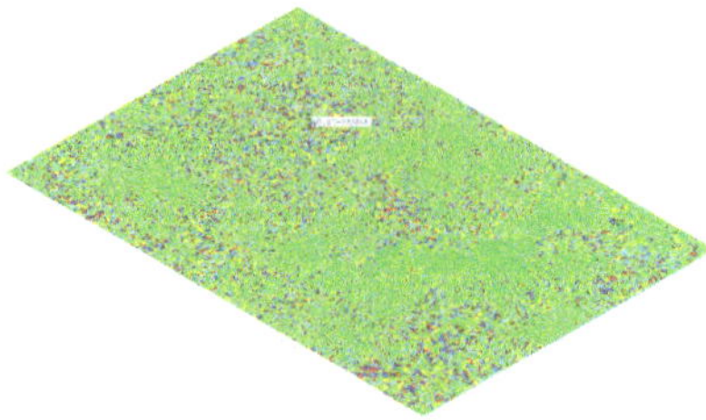	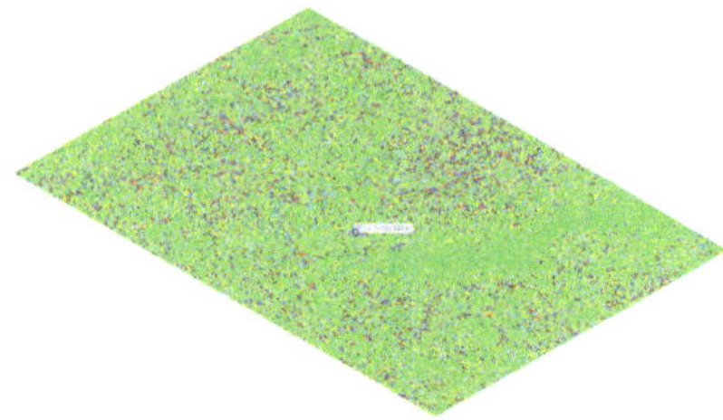	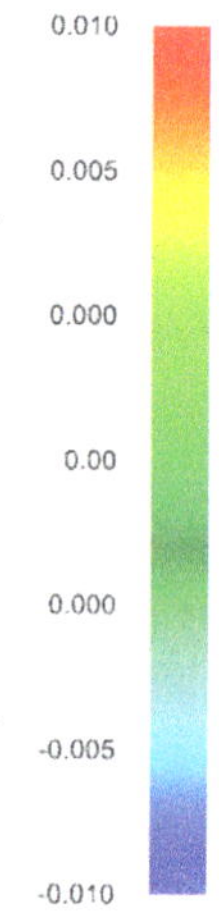 mm
5	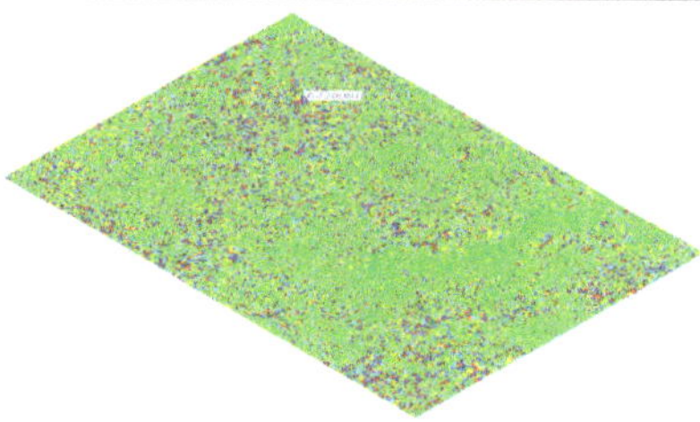	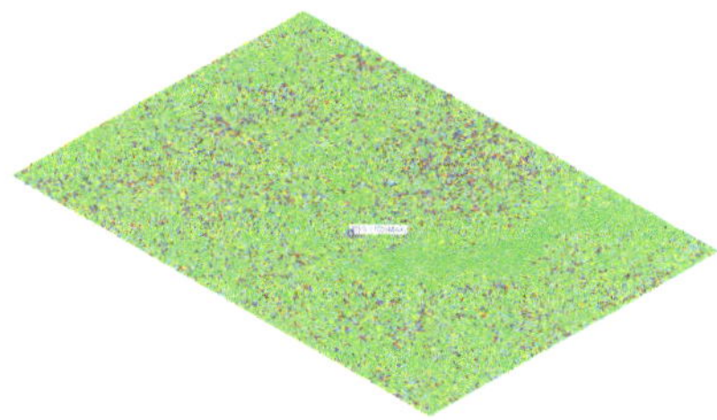	
10	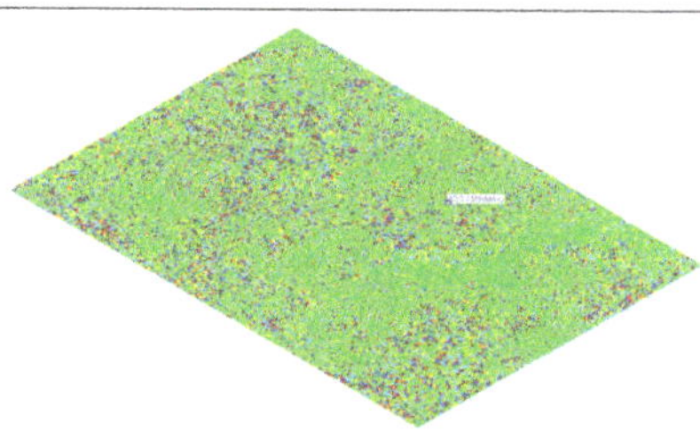	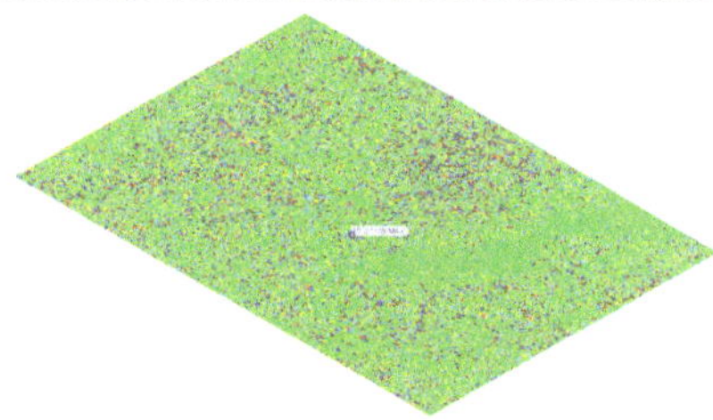	
15	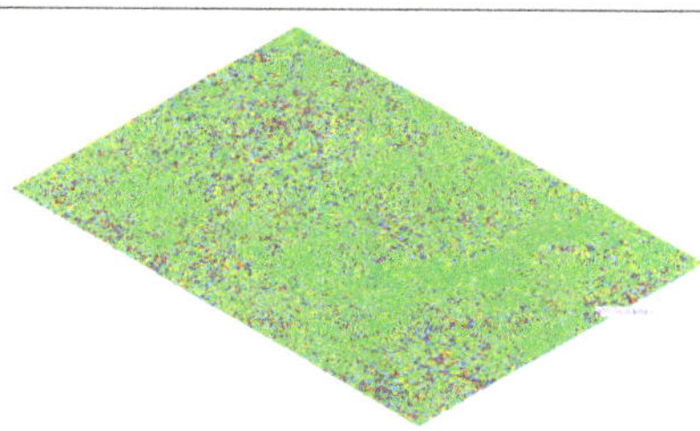	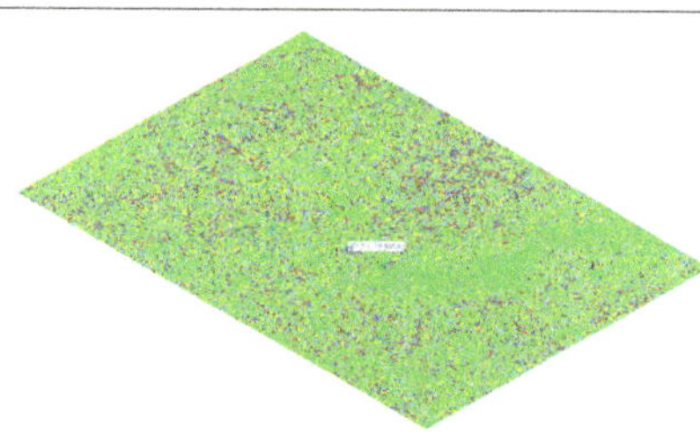	
20	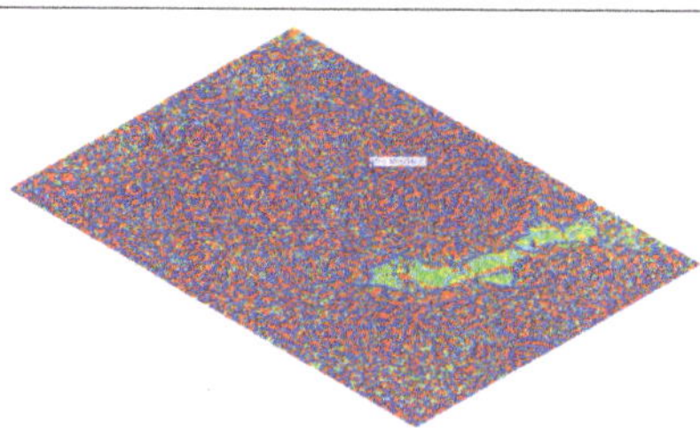	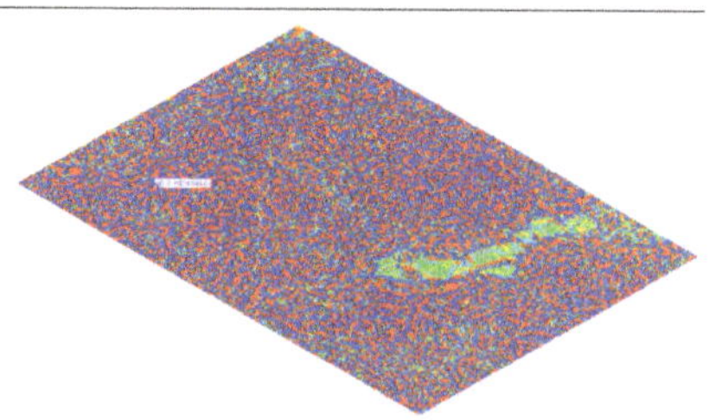	

Table 3. Surface deviation map for the steep region.

Smoothing, %	Variable Density	Uniform Density	Scale
0	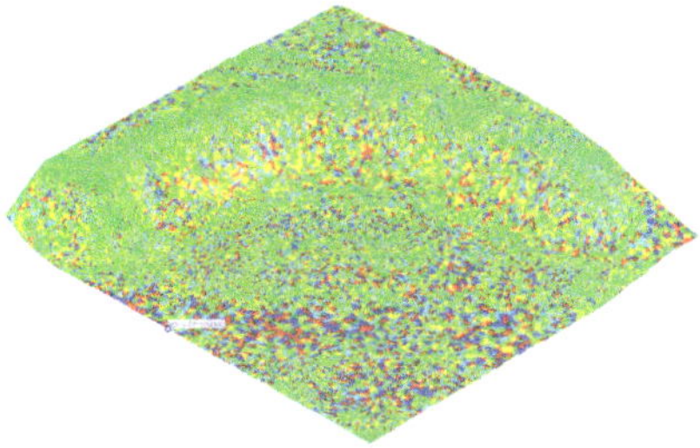	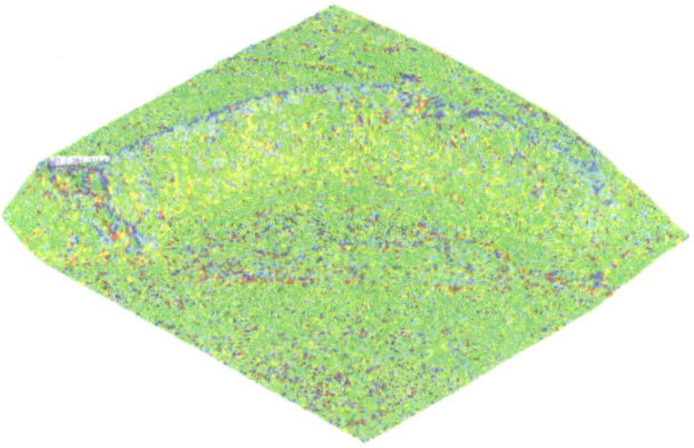	
5	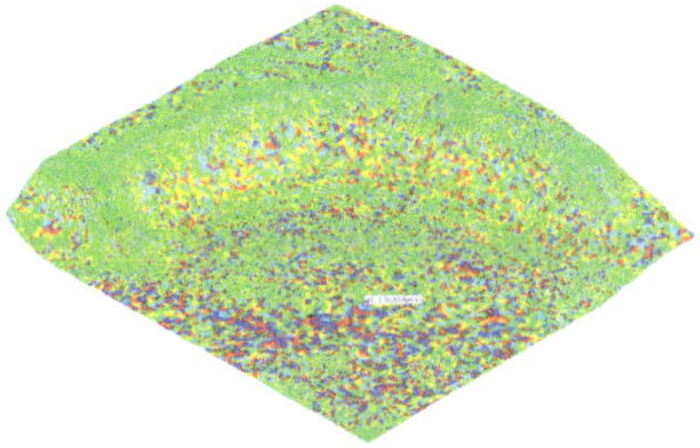	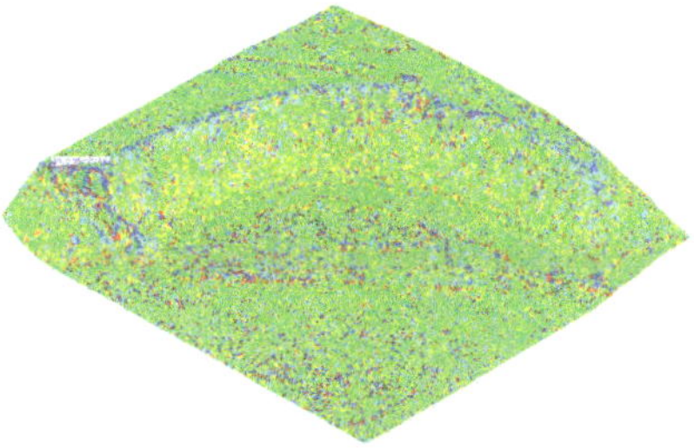	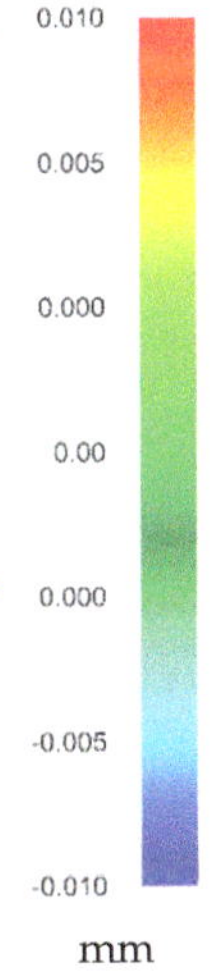
10	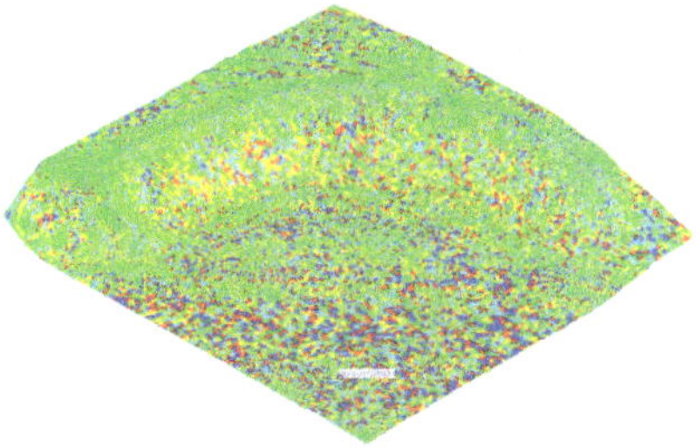	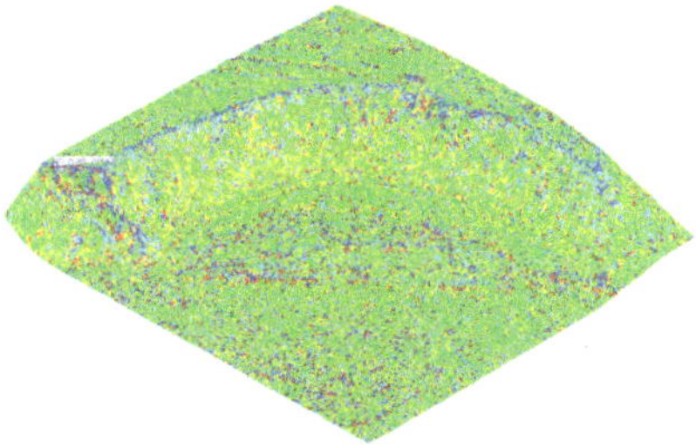	
15	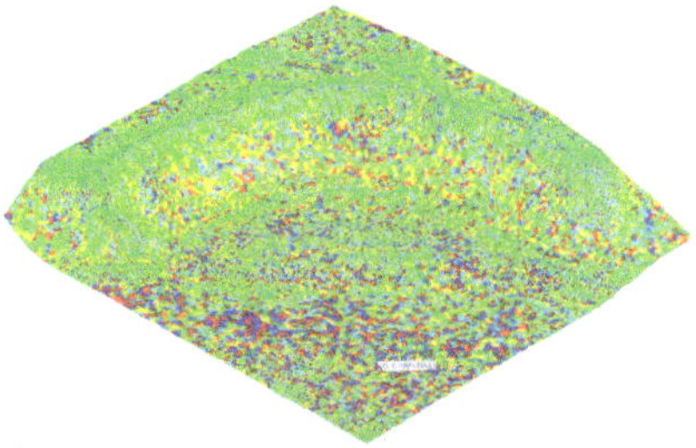	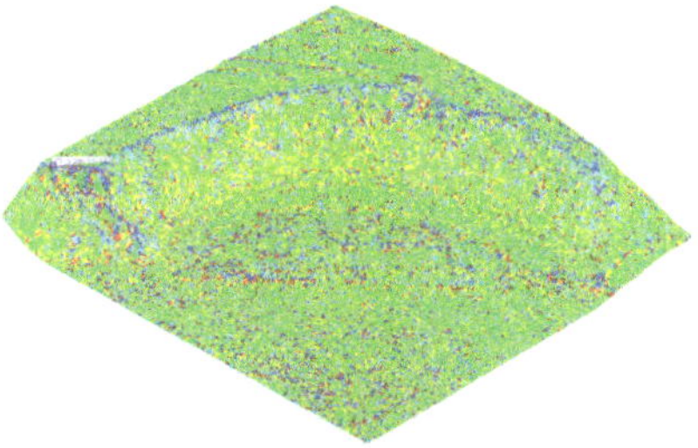	
20	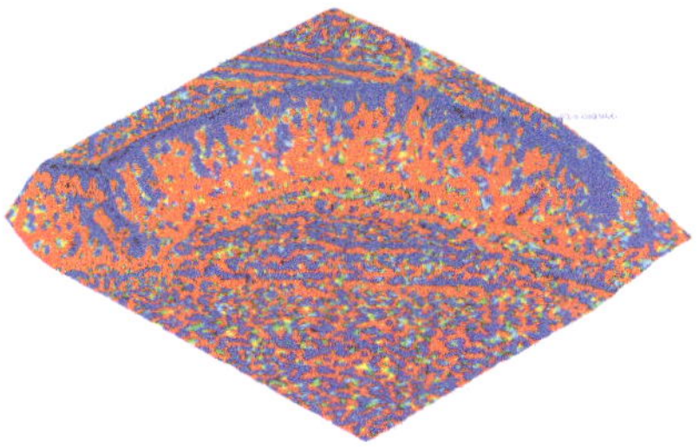	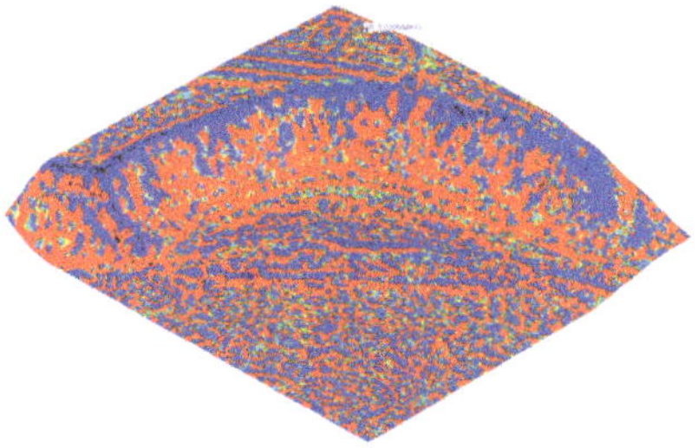	

Table 4. Surface deviation map for the concave–convex region.

Smoothing, %	Variable Density	Uniform Density	Scale
0	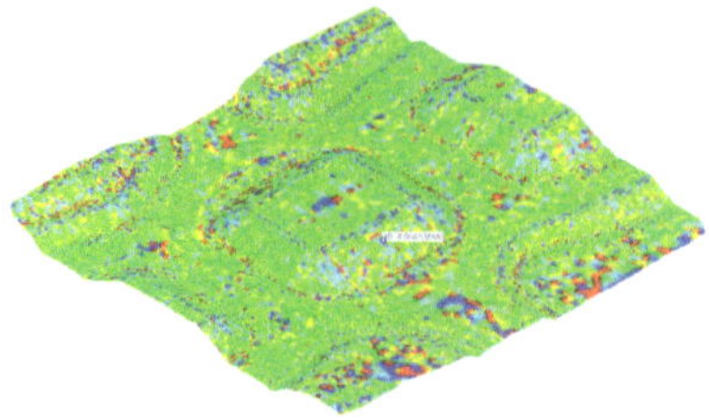	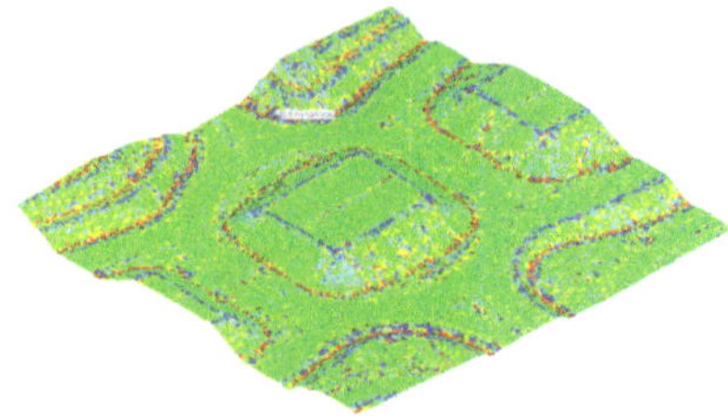	
5	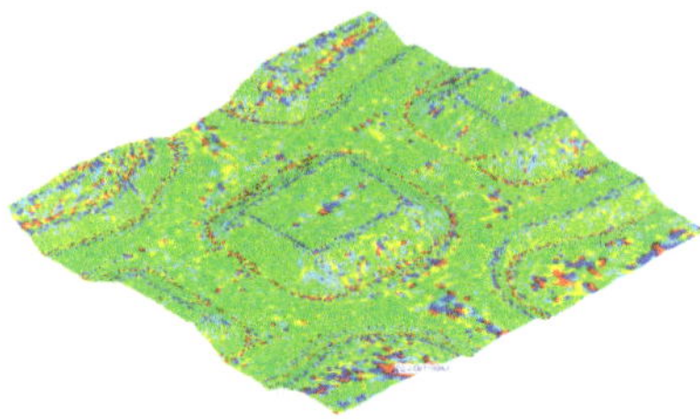	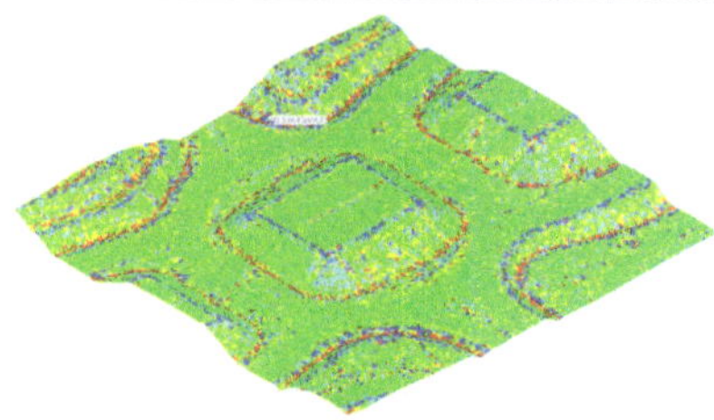	
10	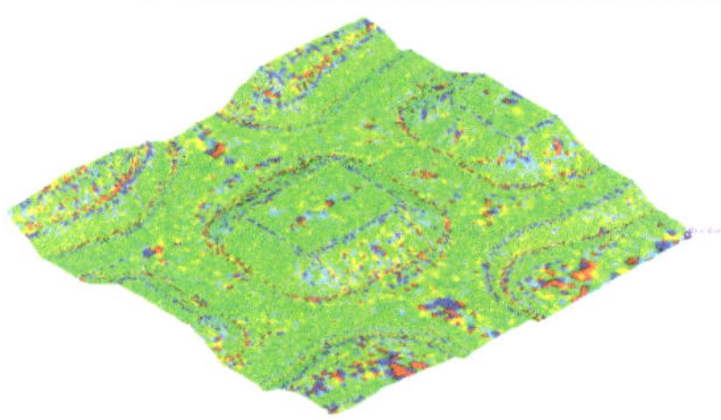	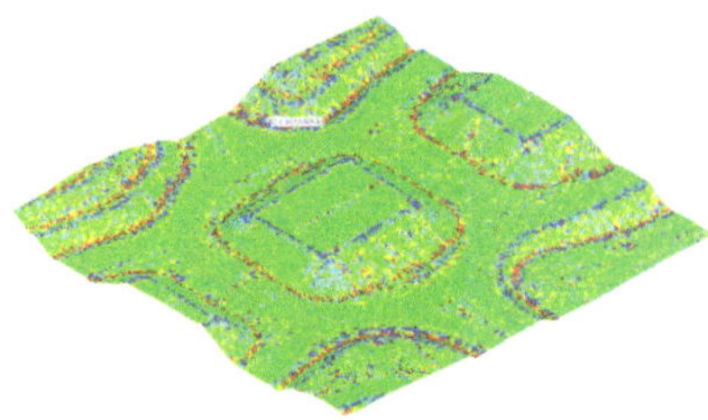	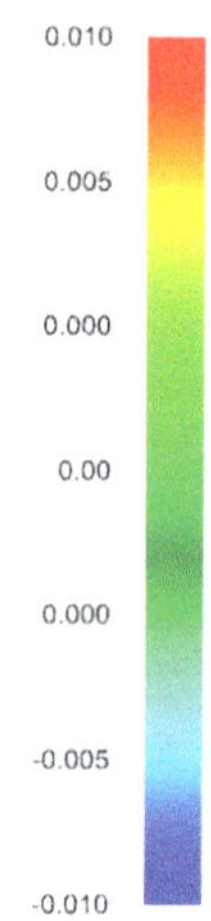
15	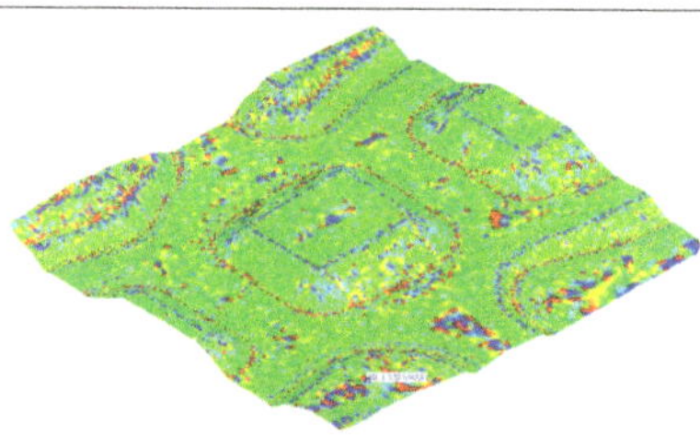	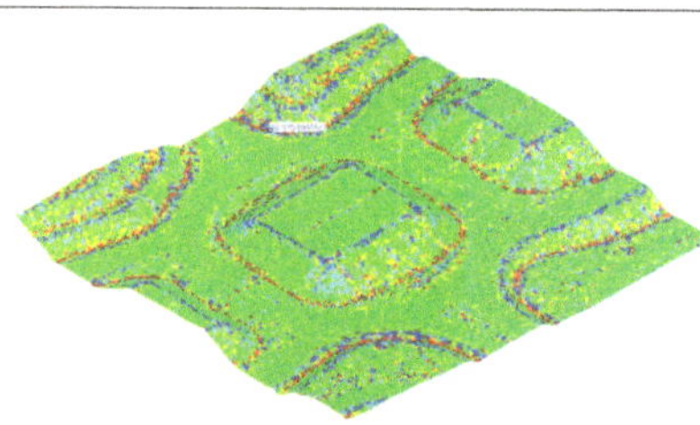	
20	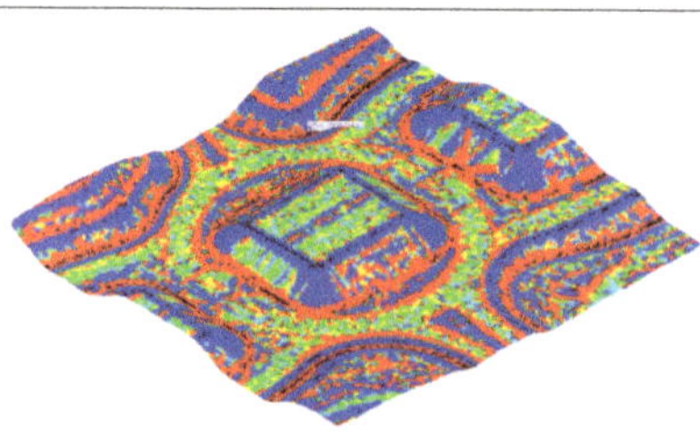	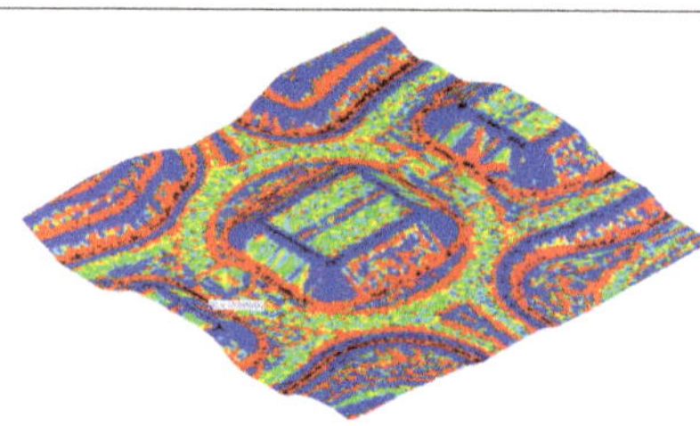	

3.2. Three-Dimensional Printing Based on the Proposed Methodology

The next stage of the research focused on transforming the previously generated surface into a stereolithography (STL) model, which was then used to produce physical models using the additive method, also known as 3D printing. The STL file was exported from NX as a binary file with parameters ensuring the highest accuracy. The Chordal

Tolerance parameter was set at 0.0025 mm, and the Angular Tolerance parameter was set at 1.000 deg. The aim of this stage was to verify the possibility of mapping selected terrain structures in the form of real models using additive technology.

The surfaces that showed the highest percentage of the area within the tolerance range were selected for further analysis and testing, meaning that they were most consistent with the expected topographic profile. In each case, this was a sample created using the Variable Density tool with a smoothing factor of 0% applied, thereby preserving the original detail of the input data.

The models were printed using FDM technology using the Fortus 360mc (Stratasys, Eden Prairie, MN, USA) printer, which has a workspace of 406 × 355 × 406 mm (Figure 9). The model material was PC-10 polycarbonate, which is characterized by good mechanical strength and thermal resistance, making it suitable for mapping complex terrain forms. The support material was PC-S, which allows for the precise printing of even complex geometries, thanks to its easy removal after the process is completed.

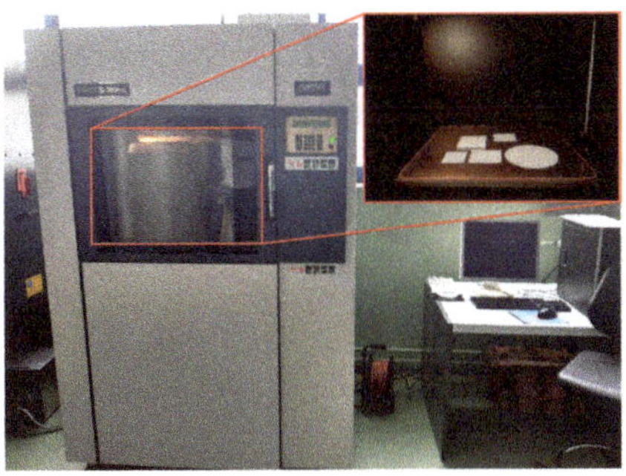

Figure 9. View of the Fortus 360mc FDM printer, including the print chamber.

In order to obtain the appropriate print quality, a nozzle marked with the symbol T12 was used, which allowed us to achieve a single layer of filament thickness of 0.178 mm. As a result of the work carried out, three physical models were obtained, each of which corresponded to one of the distinguished types of terrain morphology (Table 5).

Table 5. Presentation of the STL virtual model and the real body.

Types of Terrain	STL Body	Printed Body
flat region		
steep region		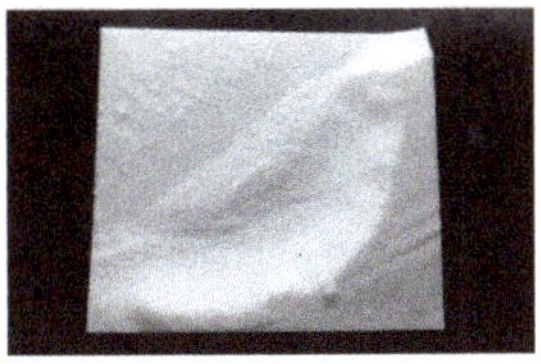

Table 5. *Cont.*

Types of Terrain	STL Body	Printed Body
concave-convex region		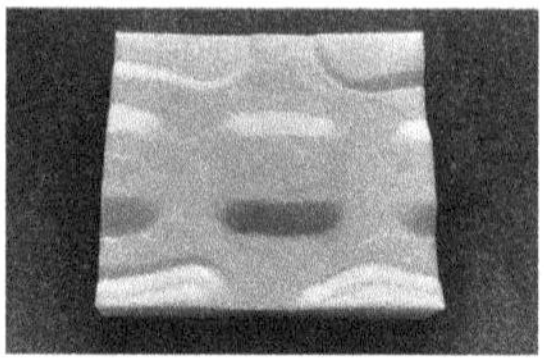

4. Discussion

The numerical analyses carried out have shown that the implementation of the bilinear interpolation algorithm to refine the rectangular grid can be successfully used as an effective method for generating terrain surface models based on available elevation data, which is also shown in the literature [23]. The main advantage of using this technique is a significant reduction in geometric errors that appear during the surface creation process in the case of working without interpolation or with insufficient sampling of input data. Reducing errors results in obtaining a more realistic representation of topography, which is crucial in geoinformatics, engineering, and environmental analyses.

The second important advantage of using bilinear interpolation is the possibility of increasing the resolution of the surface model without the need to acquire additional terrain data. Increasing the resolution leads to obtaining a more accurate representation of local morphological features of the terrain, such as depressions, elevations, and faults, while maintaining the continuity of the geometric grid. However, it should be noted that interpolation does not generate new information; it only estimates the missing values based on the available control points.

In the next part of the analysis, the influence of different meshing options available in the CAD/CAE environment, such as Keep All Points, Variable Density, and Uniform Density, was assessed. The obtained results showed that the choice of the method significantly affects the final geometric character of the model surface. In particular, the use of the Variable Density and Uniform Density options in combination with the smoothing function allows for control of the degree of surface smoothing, which affects the accuracy of the geometry mapping.

In the case of the analyzed model, it was found that the use of the smoothing function with coefficient values up to 20% did not cause any noticeable changes in the accuracy of surface mapping. Only exceeding this value led to a significant deterioration in the quality of the model, which was manifested by the blurring of topographic details and the deformation of the edges of mesh elements. At the same time, it should be noted that due to the abrupt nature of the change in the smoothing coefficient value, it was impossible to precisely determine the limit value at which the quality of mapping deteriorates.

The observed 20% limit may be due to two overlapping factors. First, it is possible that the smoothing function in the Siemens NX environment contains an internal basic smoothing mechanism that automatically applies a value of about 15%, even when the smoothing factor is set to 0%. This would explain the lack of difference in model quality from 0% to 15%. Second, the geometry of the surface, which consists of a triangle mesh, may have a significant effect. When smoothing factors are too high, above 20%, the blurring of edges between triangular surface elements may exceed their size, leading to undesirable geometric distortions.

In Figures 4–6, it can be seen that within a smoothing factor range of 0–15%, the values of the out-of-tolerance area were similar, regardless of the mesh generation method. A distinct difference was visible only for the VD method at a smoothing factor of 0%, for which the out-of-tolerance area was the lowest among all analyzed cases.

To examine statistical differences between the results obtained for the UD and VD methods and various smoothing factor values, a Student's t-test for each pair was performed in the statistical program JMP in version 12.0.1. A total of eight groups were compared: four smoothing factor levels × two methods. The significance level was set at 0.05.

The means for all groups, except for the VD method at a smoothing factor of 0%, were statistically equal ($p > 0.99$). The statistical analysis confirmed that the mean out-of-tolerance area was significantly lower for the VD method at a smoothing factor of 0% than in all other analyzed cases ($p < 0.001$).

The FDM prints made on the basis of the received data were characterized by low vertical resolution, due to the available T12 nozzle. A large simplification of the surface and low detail was observed, especially for the terrain with a flat morphology. In the case of steep and concave–convex areas, the surface detail was sufficient.

During the research, only the FDM technique was used, and other manufacturing methods were not verified. In order to reflect the obtained results as faithfully as possible, the vertical resolution should be increased by changing the printing technology, for example, to the selective laser sintering (SLS), stereolithography (SLA), digital light processing (DLP), or PolyJet methods, some of which were presented in the literature [5,6]. This recommendation can serve as the basis for further research on the accuracy of mapping terrain topography.

5. Conclusions

The described methodology, which is based on bilinear interpolation and advanced surface creation techniques, allows for the quick and efficient generation of a three-dimensional representation of a terrain surface while maintaining a high level of geometric accuracy.

It has been shown that the use of bilinear interpolation significantly supports the process of creating a surface from a point cloud. The application of a smoothing factor in the range of 0–15% did not cause significant changes in the mean out-of-tolerance area value, regardless of the mesh generation method, and was within the range of 6.0–6.5% out-of-tolerance area. Only in the case of the VD method and a smoothing factor value of 0%, the mean out-of-tolerance area value was lower than all other analyzed means: for the flat surface 5.1%, for the steep surface 5.39% and for the concavo-convex surface 4.96%.In the case of using a smoothing factor of 20%, the percentage of the sample surface area outside the tolerance was at a level of 56.89–80.94%.

Solid models generated on the basis of topographic data can be used, among others, for the physical visualization of terrain structures. An example of their practical application is the creation of tactile models dedicated to blind or visually impaired people, who, thanks to such solutions, can gain access to spatial information through the sense of touch. The use of physical models in special education is an important tool that supports the integration and development of spatial thinking in people with disabilities.

In a broader engineering context, the three-dimensional visualization of the terrain surface can be successfully used for advanced analyses in fields such as civil engineering, geoengineering, or urban planning. These models can be used, among others, for:

- designing technical infrastructure, such as roads, bridges, tunnels, and water reservoirs;
- assessing project feasibility through spatial simulations and collision analyses;
- estimating investment costs based on volumetric modeling and ground mass calculations.

From the point of view of natural and environmental sciences, 3D terrain models are a valuable tool in the analysis of natural phenomena. They can be used, among others, for:

- monitoring changes in the terrain topography related to landslides, erosion, floods, and seismic activity;
- forecasting the effects of extreme weather events;
- developing scenarios for environmental risk management.

In addition, this technology is used in space exploration. Processing data from orbital missions allows for the creation of detailed topographic models of the surfaces of celestial bodies, such as the moon or Mars. These models support the planning of future space missions, both unmanned and manned, thereby facilitating the analysis of possible landing sites, the planning of rover routes, and the testing of technical solutions in conditions close to real ones.

Author Contributions: Conceptualization, M.C. and A.B.; data curation, M.C.; investigation, M.C.; methodology, M.C. and A.B.; software, M.C.; supervision, A.B.; writing—original draft, M.C.; writing—review and editing, A.B. All authors have read and agreed to the published version of the manuscript.

Funding: This research received no external funding.

Data Availability Statement: The dataset is available upon request from the authors.

Conflicts of Interest: The authors declare no conflicts of interest.

Abbreviations

The following abbreviations are used in this manuscript:

ASCII	American Standard Code for Information Interchange
ASTER	Advanced Spaceborne Thermal Emission and Reflection Radiometer
ATIN	Adaptive Triangulated Irregular Network
CAD	Computer-Aided Design
CAE	Computer-Aided Engineering
CAM	Computer-Aided Manufacture
CSF	Cloth Simulation Filtering
DIM	Dense Image Matching
DLP	Digital Light Processing
DSM	Digital Surface Model
DTM	Digital Terrain Model
ETEW	Elevation Threshold with an Expand Window
EVRF	European Vertical Reference Frame
ExG	Excess Green
FDM	Fused Deposition Modeling
GIS	Geographic Information System
GNSS	Global Navigation Satellite System
GPS	Global Positioning System
IPF	Iterative Polynomial Fitting
KRON	Kronstadt Elevation System
LiDAR	Light Detection and Ranging
MCC	Multiscale Curvature Classification
MLS	Maximum Local Slope

N	North
NDVI	Normalized Difference Vegetation Index
OptD	Optimum Dataset Method
PM	Progressive Morphology
S	South
SfM	Structure from Motion
SLA	Stereolithography
SLS	Selective Laser Sintering
STL	Stereolithography File Format
UD	Uniform Density
VD	Variable Density

Appendix A

Algorithm A1 Conversion algorithm for $y \times x$ matrix into $x; y; z$ format.

```
clc
clear
z = load('NMT3.asc');
K = size(z);
x = [1:1:K(2)];
y = [1:1:K(1)];
y = flip(y);
X = [];
Y = [];
Z = [];
M = [];
i = 1;
while i < numel(z)
    X(i:i+numel(x)−1) = x(1:numel(x));
    i = i + numel(x);
end
i = 1;
j = 1;
while i < numel(z)
    Y(i:i+numel(x)−1) = ones(1,numel(x)).*j;
    i = i + numel(x);
    j = j + 1;
end
i = 1;
j = 1;
while i < numel(z)
    Z(i:i+numel(x)−1) = z(j,1:end);
    i = i + numel(x);
    j = j + 1;
end
M(:,1) = X;
M(:,2) = flip(Y);
M(:,3) = Z;
writematrix(M,'M_tab.txt','Delimiter',';')
```

Appendix B

Algorithm A2 Data processing algorithm enriched with a bilinear interpolation term

```
clc
clear
N = load('nmt.asc');
K = size(N);
u = K(2);
v = K(1);
j = 1;
z = [];
while j < v
i = 1;
while i < u
    z(j,i:u−1) = (N(j,i) + N(j,i+1) + N(j+1,i) + N(j+1,i+1))/4;
    i = i + 1;
end
j = j + 1;
end
%================================================================
K = size(z);
x = [1:1:K(2)];
y = [1:1:K(1)];
y = flip(y);
X = [];
Y = [];
Z = [];
M = [];
i = 1;
while i < numel(z)
    X(i:i+numel(x)−1) = x(1:numel(x));
    i = i + numel(x);
end
i = 1;
j = 1;
while i < numel(z)
    Y(i:i+numel(x)−1) = ones(1,numel(x)).* j;
    i = i + numel(x);
    j = j + 1;
end
i = 1;
j = 1;
while i < numel(z)
    Z(i:i+numel(x)−1) = z(j,1:end);
    i = i+numel(x);
    j = j+1;
end
X = X+0.5;
Y = Y+0.5;
M(:,1) = X;
M(:,2) = flip(Y);
M(:,3) = Z;
surf(x,y,z);
writematrix(M,'P_sr','Delimiter',';')
```

References

1. Budzik, G.; Dziubek, T.; Kawalec, A.; Turek, P.; Bazan, A.; Dębski, M.; Józwik, J.; Poliński, P.; Kiełbicki, M.; Kochmański, Ł.; et al. Geometrical Accuracy of Threaded Elements Manufacture by 3D Printing Process. *Adv. Sci. Technol. Res. J.* **2023**, *1*, 35–45. [CrossRef] [PubMed]
2. Moola, A.; Corpuz, A.; Burkhart, M.; Ross, C.; Mir, A.; Burkhart, H.; Lee, C.H.; Hsu, M.C.; Pawar, A. VALVEFIT: An analysis-suitable B-spline-based surface fitting framework for patient-specific modeling of tricuspid valves. *arXiv* **2025**, arXiv:2505.09790.
3. Turek, P.; Budzik, G. Development of a procedure for increasing the accuracy of the reconstruction and triangulation process of the cranial vault geometry for additive manufacturing. *Facta Univ. Mech. Eng.* **2025**, *23*, 95–108. [CrossRef]
4. Turek, P.; Dudek, E.; Grzywa, M.; Więcek, K. The Process of Digital Data Flow in RE/CAD/RP/CAI Systems Concerning Planning Surgical Procedures in the Craniofacial Area. *Knowledge* **2024**, *4*, 265–279. [CrossRef]
5. Dong, Q.; Wei, T.; Zhang, Q.; Jia, X.; Pan, B. The texture of Chinese garden rockery stones: Based on 3D point cloud and 3D printing technology. *NPJ Herit. Sci.* **2025**, *13*, 47. [CrossRef]
6. Wabiński, J.; Mościcka, A. Natural Heritage Reconstruction Using Full-Color 3D Printing: A Case Study of the Valley of Five Polish Ponds. *Sustainability* **2019**, *11*, 5907. [CrossRef]
7. Ji, H.; Luo, X. 3D scene reconstruction of landslide topography based on data fusion between laser point cloud and UAV image. *Environ. Earth Sci.* **2019**, *78*, 534. [CrossRef]
8. Sastras, P.; Badea, G.; Badea, A.C.; Salagean, T.; Roșca, S.; Kader, S.; Remondino, F. Land surveying with UAV photogrammetry and LiDAR for optimal building planning. *Autom. Constr.* **2025**, *173*, 106092. [CrossRef]
9. Kovanič, Ľ.; Blišťan, P.; Topitzer, B.; Peťovský, P.; Tokarčík, O. Experiences of UAS photogrammetric rockslide monitoring in the alpine terrain in high Tatras. *GIS Odyssey J.* **2024**, *4*, 51–71.
10. Abate, N.; Roubis, D.; Aggeli, A.; Sileo, M.; Amodio, A.M.; Vitale, V.; Frisetti, A.; Danese, M.; Arzu, P.; Sogliani, F.; et al. An Open-Source Machine Learning–Based Methodological Approach for Processing High-Resolution UAS LiDAR Data in Archaeological Contexts: A Case Study from Epirus, Greece. *J. Archaeol. Method Theory* **2025**, *32*, 38. [CrossRef]
11. Zhang, J.; Lin, X. Advances in fusion of optical imagery and LiDAR point cloud applied to photogrammetry and remote sensing. *Int. J. Image Data Fusion* **2016**, *8*, 1–31. [CrossRef]
12. Sestras, P.; Badea, G.; Badea, A.C.; Salagean, T.; Oniga, V.E.; Roșca, S.; Bilașco, Ș.; Bruma, S.; Spalević, V.; Kader, S.; et al. A novel method for landslide deformation monitoring by fusing UAV photogrammetry and LiDAR data based on each sensor's mapping advantage in regards to terrain feature. *Eng. Geol.* **2025**, *346*, 107890. [CrossRef]
13. Jiménez-Jiménez, S.I.; Ojeda-Bustamante, W.; Marcial-Pablo, M.D.J.; Enciso, J. Digital Terrain Models Generated with Low-Cost UAV Photogrammetry: Methodology and Accuracy. *ISPRS Int. J. Geo-Inf.* **2021**, *10*, 285. [CrossRef]
14. Hirano, A.; Welch, R.; Lang, H. Mapping from ASTER stereo image data: DEM validation and accuracy assessment. *ISPRS J. Photogramm. Remote Sens.* **2003**, *57*, 356–370. [CrossRef]
15. Li, B.; Xie, H.; Liu, S.; Ye, Z.; Hong, Z.; Weng, Q.; Xu, Q.; Tong, X. Global digital elevation model (GDEM) product generation by correcting ASTER GDEM elevation with ICESat-2 altimeter data. *Earth Syst. Sci. Data* **2025**, *17*, 205–220. [CrossRef]
16. Alganci, U.; Besol, B.; Sertel, E. Accuracy Assessment of Different Digital Surface Models. *ISPRS Int. J. Geo-Inf.* **2018**, *7*, 114. [CrossRef]
17. Błaszczak-Bąk, W.; Kamiński, W.; Bednarczyk, M.; Suchocki, C.; Masiero, A. Real-Time DTM Generation with Sequential Estimation and OptD Method. *Appl. Sci.* **2025**, *15*, 4068. [CrossRef]
18. Suleymanoglu, B.; Soycan, M. Comparison of filtering algorithms used for DTM production from airborne lidar data: A case study in Bergama, Turkey. *Geod. Vestn.* **2019**, *63*, 395–414. [CrossRef]
19. Sulung, P.P.; Soeksmantono, B. Comparative Analysis of DTM Extraction from Airborne LiDAR Point Cloud Data with Adaptive TIN Filter, Cloth Simulation Filter, and Progressive Morphological Filter Methods. *J. Geogr. Gea* **2025**, *25*, 1–14.
20. Zhang, Z.; Gerke, M.; Vosselman, G.; Yang, M.Y. Filtering photogrammetric point clouds using standard lidar filters towards DTM generation. *ISPRS Ann. Photogramm. Remote Sens. Spat. Inf. Sci.* **2018**, *4*, 319–326. [CrossRef]
21. Lee, H.; Gou, J.; Park, J.; Jang, S.; Song, I. Development of a river DTM generation algorithm based on SfM point clouds using vegetation and morphological filters. *Res. Sq.* **2024**, *15*, 11613.
22. Song, H.; Jung, J. An Object-Based Ground Filtering of Airborne LiDAR Data for Large-Area DTM Generation. *Remote Sens.* **2023**, *15*, 4105. [CrossRef]
23. Brovelli, M.A.; Cannata, M.; Longoni, U.M. LIDAR Data Filtering and DTM Interpolation Within GRASS. *Trans. GIS* **2004**, *8*, 155–174. [CrossRef]
24. Marton, Z.C.; Rusu, R.B.; Beetz, M. On fast surface reconstruction methods for large and noisy point clouds. In Proceedings of the IEEE International Conference on Robotics and Automation, Kobe, Japan, 12–17 May 2009; pp. 3218–3223.
25. DTM data source. Available online: https://www.geoportal.gov.pl/pl/dane/numeryczny-model-terenu-nmt/ (accessed on 23 June 2025).
26. Dz.U.2024.0.342 §24. Available online: https://dziennikustaw.gov.pl/D2024000034201.pdf (accessed on 23 June 2025).

27. Hu, P.; Liu, X.; Hu, H. Accuracy Assessment of Digital Elevation Models based on Approximation Theory. *Photogramm. Eng. Remote Sens.* **2009**, *75*, 49–56. [CrossRef]
28. Li, Z.; Zhu, Q.; Gold, C. Interpolation Techniques for Terrain Surface Modeling. In *Digital Terrain Modeling Principles and Methodology*; Taylor and Francis: New York, NY, USA, 2005; pp. 115–131.
29. SIEMENS NX 23.06.3001. Available online: https://docs.sw.siemens.com/en-US/doc/209349590/PL20221117716122093.modeling/xid1754092 (accessed on 23 June 2025).

designs

MDPI

Article

Assessment of Accuracy in Geometry Reconstruction, CAD Modeling, and MEX Additive Manufacturing for Models Characterized by Axisymmetry and Primitive Geometries

Paweł Turek *, Piotr Bielarski, Alicja Czapla, Hubert Futoma, Tomasz Hajder and Jacek Misiura

Faculty of Mechanical Engineering and Aeronautics, Rzeszów University of Technology, 35-959 Rzeszów, Poland; 166930@stud.prz.edu.pl (P.B.); 166940@stud.prz.edu.pl (A.C.); 168610@stud.prz.edu.pl (H.F.); 166955@stud.prz.edu.pl (T.H.); jmisiura@prz.edu.pl (J.M.)

* Correspondence: pturek@prz.edu.pl

Abstract

Due to the rapid advancements in coordinate measuring systems, data processing software, and additive manufacturing (AM) techniques, it has become possible to create copies of existing models through the reverse engineering (RE) process. However, the lack of precise estimates regarding the accuracy of the RE process—particularly at the measurement, reconstruction, and computer-aided design (CAD) modeling stages—poses significant challenges. Additionally, the assessment of dimensional and geometrical errors during the manufacturing stage using AM techniques limits the practical implementation of product replicas in the industry. This paper provides an estimation of the errors encountered in the RE process and the AM stage of various models. It includes examples of an electrical box, a lampshade for a standing lamp, a cover for a vacuum unit, and a battery cover. The geometry of these models was measured using a GOM Scan 1 (Carl Zeiss AG, Jena, Germany). Following the measurement process, data processing was performed, along with CAD modeling, which involved primitive detection, profile extraction, and auto-surface methods using Siemens NX 2406 software (Siemens Digital Industries, Plano, TX, USA). The models were produced using a Fortus 360-mc 3D printer (Stratasys, Eden Prairie, MN, USA) with ABS-M30 material. After fabrication, the models were scanned using a GOM Scan 1 scanner to identify any manufacturing errors. The research findings indicated that overall, 95% of the points representing reconstruction errors are within the maximum deviation range of ±0.6 mm to ±1 mm. The highest errors in CAD modeling were attributed to the auto-surfacing method, overall, 95% of the points are within the average range of ±0.9 mm. In contrast, the lowest errors occurred with the detect primitives method, averaging ±0.6 mm. Overall, 95% of the points representing the surface of a model made using the additive manufacturing technology fall within the deviation range ±0.2 mm on average. The findings provide crucial insights for designers utilizing RE and AM techniques in creating functional model replicas.

Keywords: reverse engineering; additive manufacturing; accuracy; structured light system; fused deposition modeling; computer-aided design

1. Introduction

The traditional approach to modeling machine components relies on computer-aided design (CAD) systems widely used in industrial product design. A challenge arises when there is a physical model but no design documentation. Fortunately, advancements in coordinate measuring systems and data processing have led to the reverse engineering (RE) process [1,2], which reconstructs geometries from data gained through contact [3,4] or optical [5,6] measurement methods. The accuracy of the reconstructed model depends on factors such as the quality of measurement data [5,7], triangulation methods [8], and CAD modeling techniques [9,10]. The manufacturing stage is also crucial, with subtractive methods traditionally used [11], although additive manufacturing is gaining traction for cost reduction and efficiency [12].

The acquisition of data in the RE process involves various measuring instruments, from coordinate measuring machines [13] and measuring arms [14] to 2D [15] and 3D scanners [16], as well as tomographic systems [17]. Selecting the right measurement system is crucial, considering factors like resolution, repeatability, measurement range, non-invasiveness, and speed. Measurement data is usually expressed as coordinates in either a global or local coordinate system. After obtaining the point cloud data, it is edited by filtering, merging, and assembling into a model [9,18], typically saved in STL format, which approximates the geometry using a triangle mesh [19]. However, errors can arise during this conversion due to tessellation algorithms, making it essential to balance geometry accuracy with manufacturing machine resolution. Information on common errors when exporting to STL format is provided in ISO/ASTM standards [20–23]. Few studies have explored this topic in depth. With a complete 3D STL model, it can be converted into a parameterized 3D CAD model using methods like characteristic geometry detection [24], section profiles [25], and auto-surfaces [26], all of which rely on approximations from the point cloud [15]. The data processing workflow is complex due to factors such as the acquisition process [27], incomplete data collection, and inherent noise [28]. As a result, new methods are being explored to enhance the data reconstruction process. A review of the literature indicates that deep learning and hybrid analytical-neural approaches are currently in use, significantly improving the accuracy and automation of reconstruction. In particular, the article [29,30] is crucial because it identifies the shortcomings of existing solutions and proposes an innovative approach to address them. In the case of 3D printers, each model has specific characteristics and operating conditions requirements. These factors include the 3D printing parameters, the materials used, and environmental conditions [31]. Consequently, differences may arise between the nominal 3D-CAD model and the finished product. These differences can affect dimensional and geometrical accuracy [32,33]. Among the 3D printing parameters, layer thickness has the most significant impact on the accuracy of the geometry [34,35]. This influence is determined by several factors that depend on the specific 3D printing technique used. The effect of layer thickness on geometric accuracy is particularly pronounced when the model is built in different directions [36,37]. Additionally, the presence of a supporting structure during the model manufacturing process can also impact geometric representation accuracy [38]. Often, this involves post-processing treatments to remove the supporting material through mechanical or chemical methods. As a result, the surface of the model, after the support material has been cleaned off, may differ significantly from the designer's original assumptions, particularly in terms of dimensional and geometrical accuracy. Research in this field focuses on optimizing processes and materials, particularly using machine learning to predict the quality and mechanical properties of parts [39,40].

Functional models are often made using reverse engineering (RE) and additive manufacturing (AM) technologies. To ensure quality, it is crucial to meet standards related to dimensional and geometric accuracy. Common standards used during design include ISO 8015 [41], ISO 1101 [42], ISO 286-2 [43], ISO 22081 [44], and ASME Y14.5 [45]. However, there are no clear design criteria to estimate accuracy during the RE process, especially at the measurement, reconstruction, and CAD modeling stages. Additionally, measuring dimensional and geometric errors during AM production is often unclear, which slows down the ability to bring products to market. It is essential to focus on axisymmetric models and models representing regular shapes, as these are often used in machine parts. Their simple designs make the manufacturing process more efficient and accurate. This publication aims to estimate the errors that occur during both the RE process and the AM stage of production. It will use test models, such as an electrical box, a lampshade for a standing lamp, a cover for a vacuum unit, and a battery cover, as examples.

2. Materials and Methods

The research process focused on three test models: an electrical box, a lampshade for a standing lamp, a cover for a vacuum unit, and a battery cover (Figure 1).

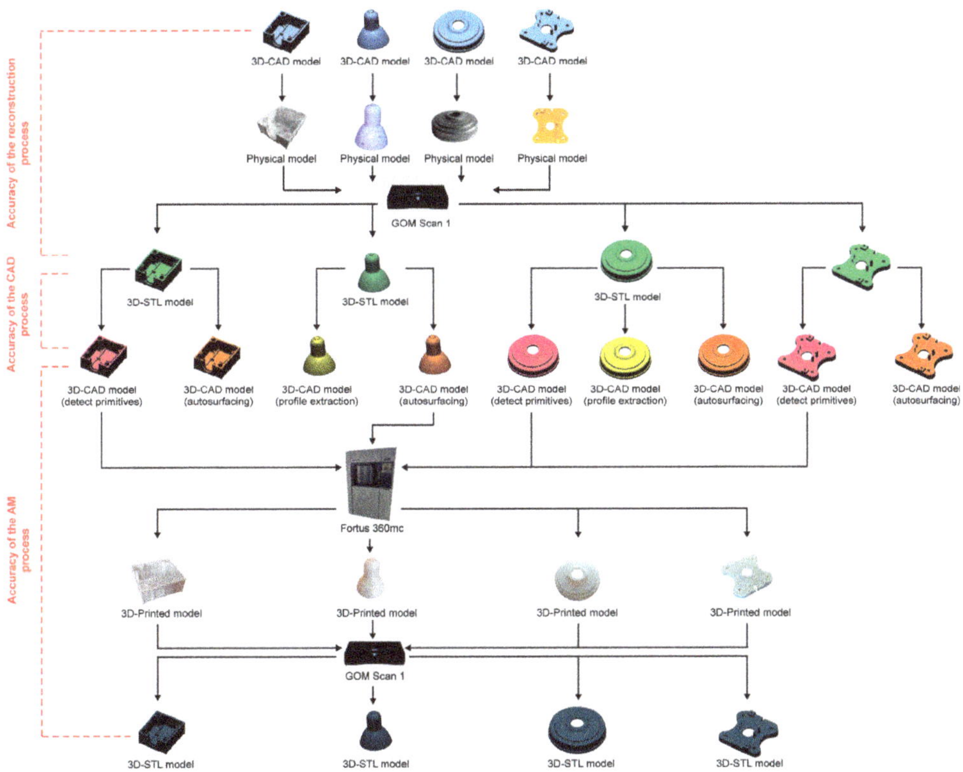

Figure 1. The flowchart of the adopted research procedure.

The research process incorporated axisymmetric models along with models featuring regular geometric shapes, both of which play a vital role in industrial applications. These models are essential due to their considerable influence on key factors such as precision, operational efficiency, and overall production costs. The inherent geometric simplicity of these models offers a range of specific advantages that enhance performance and streamline processes across the entire production chain, ultimately leading to improved outcomes in manufacturing and resource management. To measure these objects, a measurement system utilizing structured light from the GOM Scan 1 was used. After acquiring the measurement data, it was imported into the Siemens NX program. During this phase, a parametric modeling process was implemented using various CAD modeling techniques. By conducting accuracy analyses during the reconstruction and CAD modeling stages, the path that minimized geometry errors was identified. The resulting model, developed through this optimized path, was then used in the AM process. For the additive manufacturing process, a Fortus 360mc 3D printer was used. After 3D printing, the produced model underwent a geometry measurement process using the GOM Scan 1 system to evaluate any errors introduced during the AM production process.

2.1. The Process of Measuring and Reconstructing Geometries Using the GOM Scan 1 System

The complexity of 3D measurement affects the measurement uncertainty of optical coordinate measuring machines (CMMs), necessitating the establishment of a standardized procedure for defining the accuracy of these systems. Currently, the calibration of optical systems that utilize structured light is primarily conducted according to the German VDI/VDE 2634 standard [46]. This standard provides recommendations for the acceptance and re-verification of measuring systems. It also specifies the conditions necessary for properly calibrating optical systems, such as temperature, mechanical vibration, and lighting conditions. The guidelines set forth by the standard include the assessment of the following types of errors:

- Probing error—test performed on a single ball;
- Sphere—spacing error—test performed on a 'ball bar' standard;
- Flatness measurement error—test performed on a flat rectangular plate.

Based on the calibration process carried out, the results are presented in Table 1.

Table 1. The results of verification of optical systems using the standard procedure.

Acceptance Test	Measured Value/Maximum Permission Error (2σ)
Probing error	±0.003 mm/±0.006 mm
Sphere—spacing error	±0.007 mm/±0.020 mm
Flatness measurement error	±0.020 mm

A GOM Scan 1 (100) head was used during the calibration and measuring process, enabling the digitization of geometry with a resolution of 0.037 mm. In evaluating the number of rotations of the measuring table for selected models, four options were tested: 5, 10, 15, and 20 rotations. The first two options did not provide complete digitization of the models' geometry, resulting in significant gaps in the three-dimensional point cloud. When using 20 rotations, there was a notable increase in the size of the measurement file, which was accompanied by the occurrence of overscans. However, with 15 rotations, a nearly complete representation of the scanned model surfaces was achieved, and no overscans were observed. Ultimately, 15 rotations of the measuring table were used in the research process. Detailed information regarding the measurement parameters can be found in Table 2.

Table 2. Established measurement parameters for the GOM Scan 1 [15].

Parameters	Value
Pixel resolution cameras	5,000,000
Measuring area	100 mm × 65 mm × 400 mm
Min. point resolution	0.037 mm
Number of points per scan	5,000,000
Number of rotations of the measuring table	15

During the process of measuring the electric box model, it was not necessary to coat the model with a matting substance but attach reference points. Measurements were taken for two configurations of the object: first, the internal geometry was measured (Figure 2a), and second, the external geometry was measured (Figure 2b). For each configuration, a fixed number of rotations of the measuring table, totaling 15, was used. The two scans were then combined using the best-fit option, resulting in a deviation of 0.010 mm during the merging process. This led to a complete reconstruction of the model's geometry, which was then exported in STL format (Figure 2c).

(a) (b) (c)

Figure 2. Process of measuring electrical box: (**a**) measurement of internal part; (**b**) measurement of external part; (**c**) Model 3D-STL.

The lampshade for a standing lamp was scanned in two positions: external (Figure 3a) and internal (Figure 3b). Each position involved 15 rotations, resulting in two sets of measurement points that were later merged. During scanning, it is generally recommended that the surface of the object be as non-reflective as possible, as reflectivity can negatively affect the final measurement results. In such cases, the object is typically covered with a suitable material (e.g., chalk). However, during the scanning of the lampshade, this procedure was not necessary, even though the object was partly reflective. The attached reference points were sufficient to achieve a final scan with a fitting accuracy of 0.011 mm. The resulting model was then exported to an STL file format (Figure 3c).

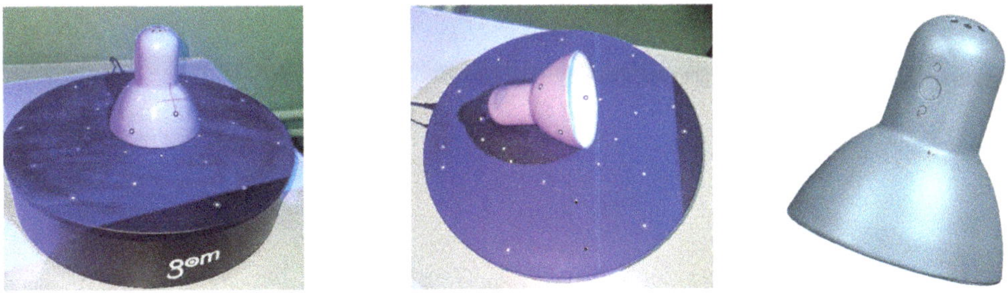

(a) (b) (c)

Figure 3. Process of measuring lampshade for standing lamp: (**a**) measurement of external part; (**b**) measurement of internal part; (**c**) Model 3D-STL.

The process of reconstructing the cover for a vacuum unit began with applying reference points and matte spray to eliminate the reflective surface. According to the manufacturer, the thickness of the spray layer should not exceed 8–15 μm. After this preparation, the element was mounted on a rotary table. Measurements were conducted in two stages, focusing on both the external and internal geometry of the unit. The transformation method chosen to merge the two scans used reference points, with a deviation from these points measured at 0.05 mm. During both measurement processes, the measuring table made a total of 15 rotations. As a result of the measurements and subsequent point cloud conversion, an STL model was generated (Figure 4a).

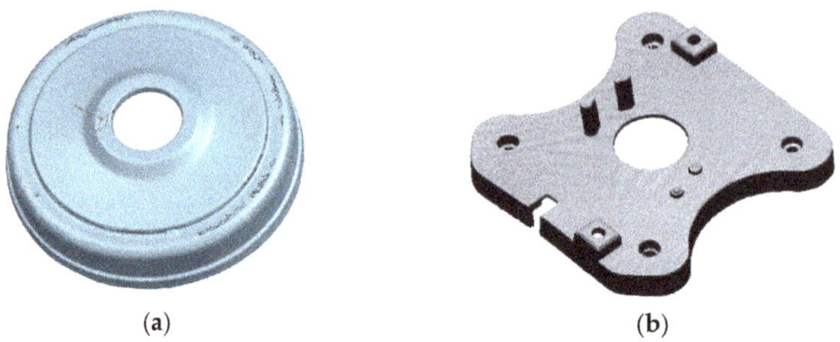

Figure 4. The 3D-STL models obtained: (a) the cover for a vacuum unit; (b) the battery cover.

The process of reconstructing the geometry of the battery covers began by applying reference points. The model was mounted on a rotary table, and a total of 15 complete rotations were performed during scanning. Measurements were taken in two stages, examining both the external and internal geometry. The transformation method selected to combine the two scans used the best-fit option, with a deviation of 0.025 mm. The resulting model was then exported to an STL file format (Figure 4b).

2.2. The Process of CAD Modeling

The 3D-STL model of the electrical box was reconstructed in Siemens NX using the reverse engineering module. The process began by importing a scan of the physical model and aligning it with the global coordinate system by creating three reference surfaces with the Fit Surface command. Surfaces for each face of the object were first reconstructed individually, using Fit Plane for planar surfaces and Fit Cylinder for cylindrical surfaces, while minimizing the Average Error for high accuracy. Once all faces were generated, they were extended with the Extend Sheet command, trimmed using the Trim Sheet function, and stitched with the Sew command to create a continuous, watertight surface model (Figure 5a). To utilize the auto-surface option, a curve mesh was created and projected onto the scanned model with the Project Curve option. The Rapid Surfacing command was then used to create a parameterized surface based on these projected curves (Figure 5b).

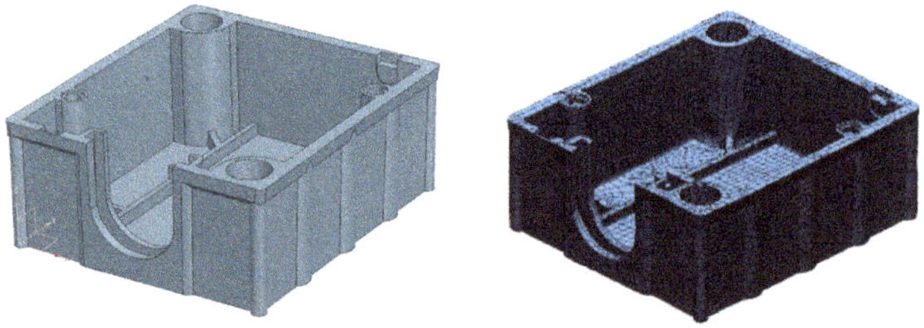

Figure 5. The process of parametric modeling of the electric box: (a) Model 3D-CAD (detect primitives); (b) Model 3D-CAD (auto-surfacing).

To reconstruct the 3D-STL model of the lampshade, a new coordinate system was created and aligned with the primary system, simplifying the process. A plane was positioned at the lampshade's midpoint for contour projection using the Section Curve command. Corrections were made to the projected outer contour shape using a sketch, which was then rotated appropriately. Next, holes were created on the lampshade's rear side. A new plane was set up parallel to the holes for projecting their outlines. The circular hole reconstruction involved creating points along its outline with variable density for accuracy. The process included sketching, projecting outlines, adding points, and using the Fit Curve function for precision. Finally, to generate a parameterized model (Figure 6a), a curve mesh was projected onto the lampshade. To use the auto-surface option in NX, a curve mesh was created and projected onto the lampshade's scanned model using the Project Curve option. The Rapid Surfacing command was then used to create the surface, selecting the "Import Curves" operation to reconstruct it based on the projected curves. This process resulted in a parameterized surface model of the lampshade (Figure 6b).

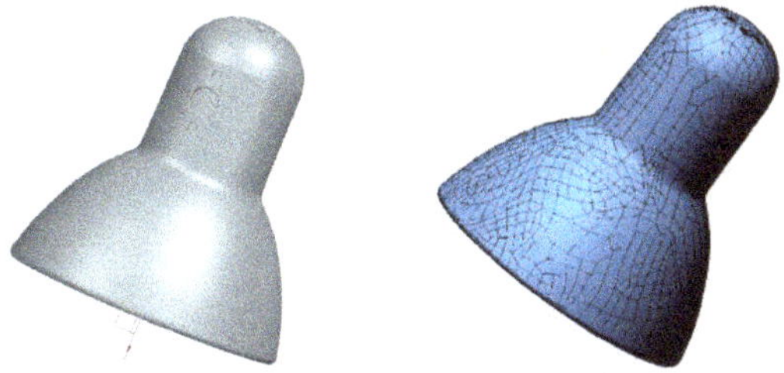

(a) (b)

Figure 6. The process of parametric modeling of the lampshade for a standing lamp model: (**a**) Model 3D-CAD (profile extraction); (**b**) Model 3D-CAD (auto-surfacing).

During the measurement process, a 3D STL model of a vacuum unit cover was created, which led to a subsequent CAD modeling process. The initial method detected primitive geometries for quick approximations using simple shapes like cylinders and cones, with adjustments for structural continuity. Surfaces were modified with extended and trimmed functions for accuracy, resulting in a refined solid model with filets (Figure 7a). The second method involved rotating the STL object's cross-section in the Z-Y plane for an outline projection. After cutting and correcting imperfections with a spline curve, the outline was rotated around the Z-axis to form the model. In NX, we used the Project Curve option to mesh curves onto the scanned model (Figure 7b) and applied the Rapid Surfacing command to create a parameterized surface model of the vacuum unit cover (Figure 7c).

(a) (b) (c)

Figure 7. The process of parametric modeling of the cover for a vacuum unit: (**a**) Model 3D-CAD (detect primitives); (**b**) Model 3D-CAD (profile extraction); (**c**) Model 3D-CAD (auto-surfacing).

The STL mesh from the battery cover scan was imported into Siemens NX's reverse engineering module as the starting dataset. The raw model exhibited common scanning artifacts, so instead of extensive preprocessing, it was divided into basic geometric shapes like cylinders and flat surfaces for effective surface reconstruction. The Fit Surface functionality approximated these shapes with parametric surfaces, primarily using planar and cylindrical patches. A dedicated coordinate system was established for accurate alignment, resulting in a solid model with high fidelity to the original cover (Figure 8a). To utilize the auto-surface option, I created a detailed curve mesh representing the desired contours and projected it onto the scanned model using the Project Curve option for precise alignment. Then, I employed the Rapid Surfacing command, selecting Import Curves to incorporate the projected mesh into the surface definition. This process generated a highly parameterized surface model, improving accuracy and streamlining future design iterations (Figure 8b).

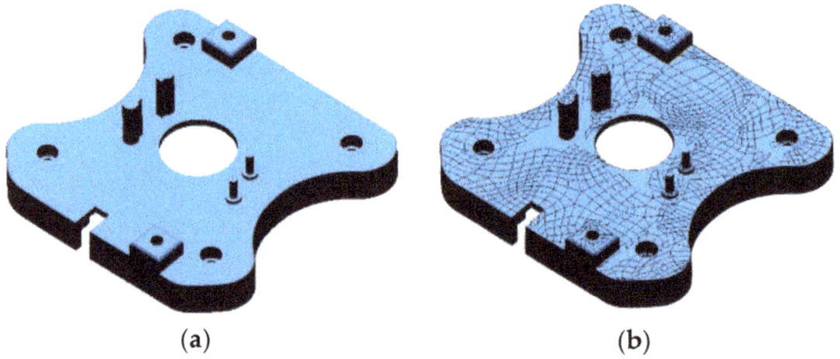

Figure 8. The process of parametric modeling of the battery cover model: (**a**) Model 3D-CAD (detect primitives); (**b**) Model 3D-CAD (auto-surfacing).

After completing the CAD modeling process, models with the smallest geometrical errors were selected. A tessellation process was then conducted on these models to convert from CAD into STL format. Given the resolution capabilities of modern 3D printers, specific parameters for exporting data from CAD to STL format have been selected in Siemens NX software:

- Select the binary format to save the STL file, as the ASCII format (American Standard Code for Information Interchange) results in larger file sizes;
- Set the chordal deviation to less than 0.01 mm;

Set the angle deviation to a value of less than 10°.
These settings will help ensure optimal performance and quality in 3D printing.

2.3. The Process of Additive Manufacturing and Measuring 3D Printed Models

The first preparatory step before beginning the 3D printing process of the electrical box, a lampshade for a standing lamp, a cover for a vacuum unit model, and a battery cover was importing its reconstructed geometry into the Insight V1980-6633 software (Stratasys, Eden Prairie, MN, USA). The initial operation involved orienting the model using the Automatic Orientation command, with the additional option to minimize support structures. This approach positively affects both 3D printing time and material consumption. The selected infill type was Sparse, which results in denser material deposition near the model's outer walls, while the inner volume remains more sparsely filled. This significantly reduces 3D printing time and conserves building material. For the surface finish quality of the external surfaces, the Enhanced option was selected, while for the support structures, the Basic option was used. The next stage involved slicing the model into layers and generating support structures, both of which were performed automatically by the software. Additionally, a machine control program was generated along with a corresponding Coordinate Machine Binary (CMB) file. The generated file was then imported into the Control Center 7.0 software (Stratasys, Eden Prairie, Minnesota, USA), which serves as an interface between the PC and the 3D printer. In this program, the model was positioned within the printer's workspace, and the Build Job command was executed. This action transferred the model along with its control program to the Fortus 360mc 3D printer. In the process of manufacturing the models, ABS-M30 material and a layer thickness of 0.127 mm were used. This is the highest resolution at which models can be produced on the Fortus 360-mc 3D printer. The manufactured models are presented in Figure 9.

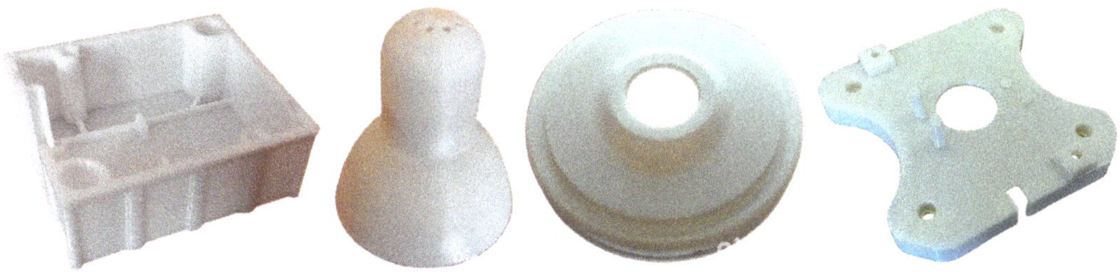

(**a**) (**b**) (**c**) (**d**)

Figure 9. Additive manufacturing models: (**a**) electrical box; (**b**) lampshade; (**c**) cover for a vacuum unit; (**d**) battery cover.

The final stage of the research involved measuring the geometry of the models after they were 3D printed (Figure 10). For this purpose, we used the GOM Scan 1 scanner, which had been used previously. We used the same measurement parameters as before.

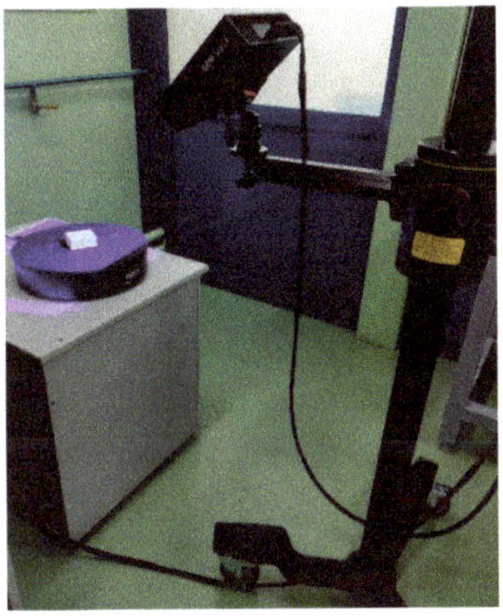

Figure 10. Measurement of electrical box model made using MEX additive manufacturing technique.

3. Results

The accuracy of the model geometry was verified during the stages of reconstruction, CAD modeling, and additive manufacturing using Zeiss Inspect 2024 software (Carl Zeiss AG, Jena, Germany). The model fitting process used the best-fit method, achieving an accuracy of 0.001 mm. The results are presented at the following stages:

- Geometry reconstruction in the form of three-dimensional deviation maps (Figure 11) and statistical parameters (Table 3);
- CAD modeling in the form of three-dimensional deviation maps (Figure 12) and statistical parameters (Table 4);
- Additive manufacturing in the form of three-dimensional deviation maps (Figure 13) and statistical parameters (Table 5);
- RE + CAD + AM (the nominal model with a scan of the 3D printed model)—total error in the form of three-dimensional deviation maps (Figure 14) and statistical parameters (Table 6);

Table 3. Statistical parameters representing reconstruction geometry errors.

Parameters	Electrical Box	Lampshade	Cover for a Vacuum Unit	Battery Cover
Maximum deviation [mm]	1.414	0.618	2.210	1.325
Minimum deviation [mm]	−1.025	−0.713	−1.069	−1.403
Range [mm]	2.440	1.332	3.279	2.728
Mean deviation [mm]	−0.020	0.105	0.119	−0.231
Standard deviation [mm]	0.364	0.467	0.559	0.387

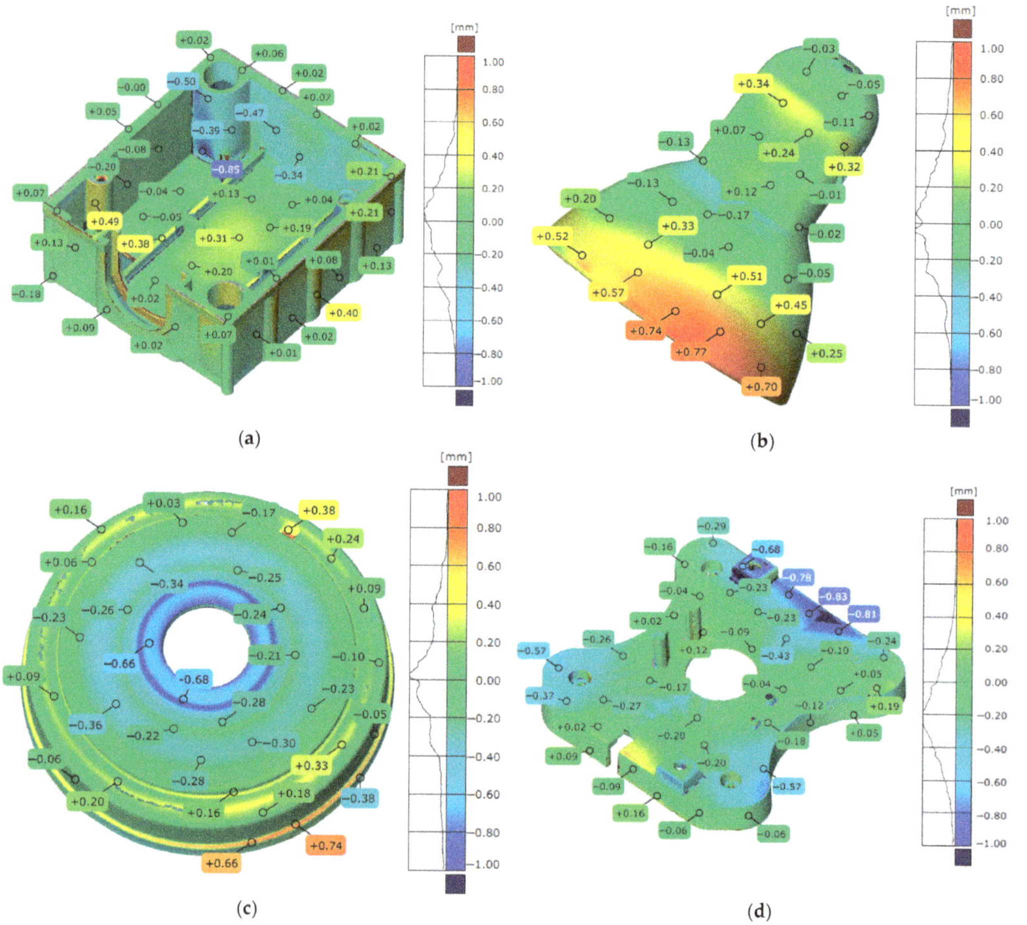

Figure 11. Three-dimensional deviation maps illustrating reconstruction errors for the model: (**a**) electrical box; (**b**) lampshade; (**c**) cover for a vacuum unit; (**d**) battery cover.

Table 4. Statistical parameters representing CAD modeling errors.

Parameters	Electrical Box		Lampshade		Cover for a Vacuum Unit			Battery Cover	
	Detect Primitive	Auto Surfacing	Profile Extraction	Auto Surfacing	Detect Primitive	Profile Extraction	Auto Surfacing	Detect Primitive	Auto Surfacing
Maximum deviation [mm]	1.074	1.381	2.176	2.147	0.874	0.749	1.459	1.669	2.788
Minimum deviation [mm]	−1.762	−1.463	−1.421	−1.602	−0.774	−0.721	−0.828	−1.698	−2.274
Range [mm]	2.835	3.844	3.597	3.749	1.649	1.470	2.287	3.367	5.062
Mean deviation [mm]	−0.108	−0.122	−0.056	−0.196	0.010	−0.012	0.122	0.194	0.306
Standard deviation [mm]	0.291	0.427	0.273	0.428	0.123	0.250	0.461	0.259	0.512

Table 5. Statistical parameters representing additive manufacturing errors.

Parameters	Electrical Box	Lampshade	Cover for a Vacuum Unit	Battery Cover
Maximum deviation [mm]	0.929	0.892	1.373	0.653
Minimum deviation [mm]	−0.385	−0.422	−0.431	−0.848
Range [mm]	1.314	1.314	1.804	1.501
Mean deviation [mm]	−0.050	−0.087	−0.026	−0.137
Standard deviation [mm]	0.091	0.102	0.163	0.109

Table 6. Statistical parameters representing RE + CAD + AM—total errors.

Parameters	Electrical Box	Lampshade	Cover for a Vacuum Unit	Battery Cover
Maximum deviation [mm]	0.834	0.715	1.286	0.754
Minimum deviation [mm]	−0.973	−0.713	−0.937	−0.947
Range [mm]	1.806	1.428	2.223	1.701
Mean deviation [mm]	−0.057	0.125	0.014	−0.102
Standard deviation [mm]	0.382	0.452	0.410	0.409

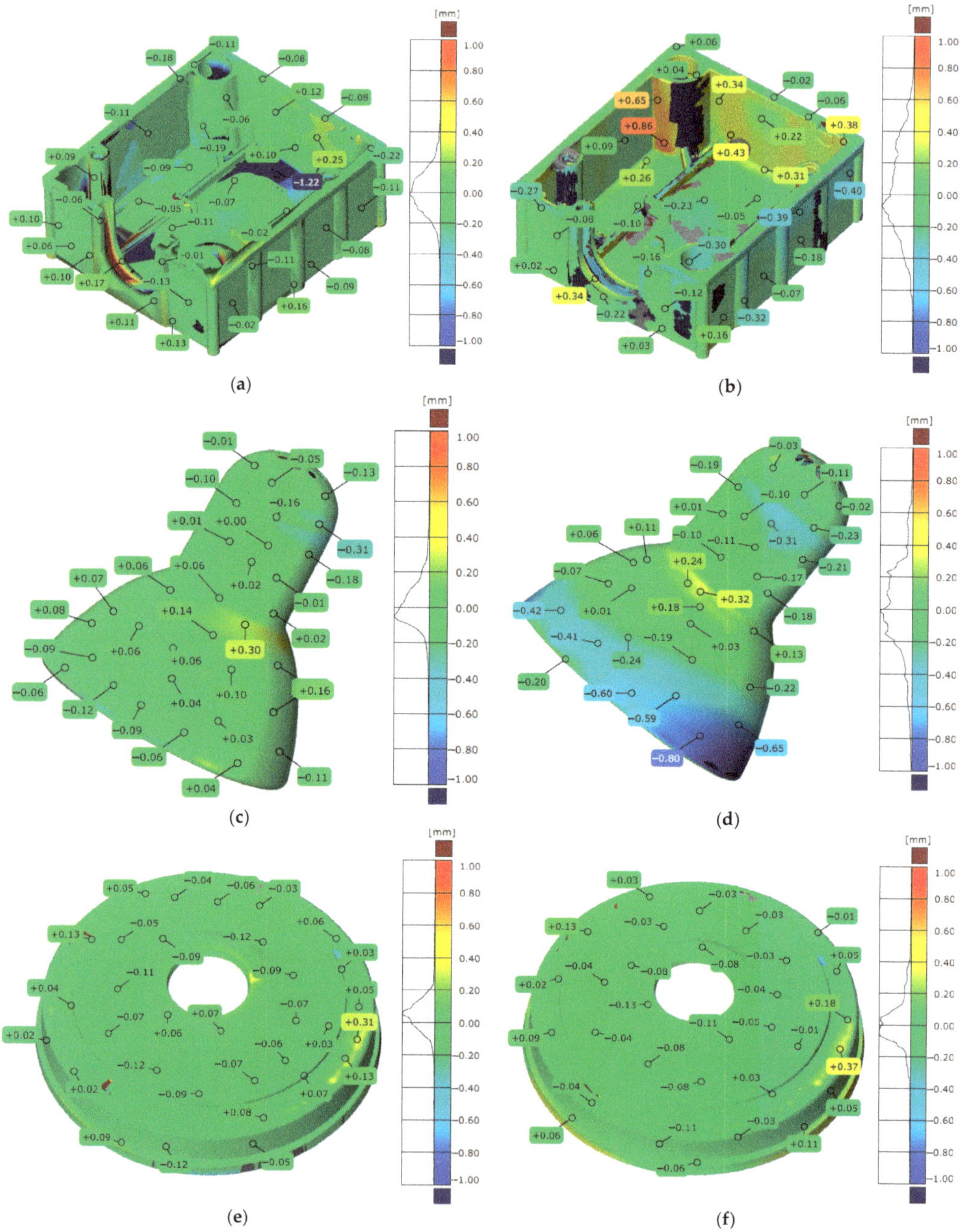

Figure 12. *Cont.*

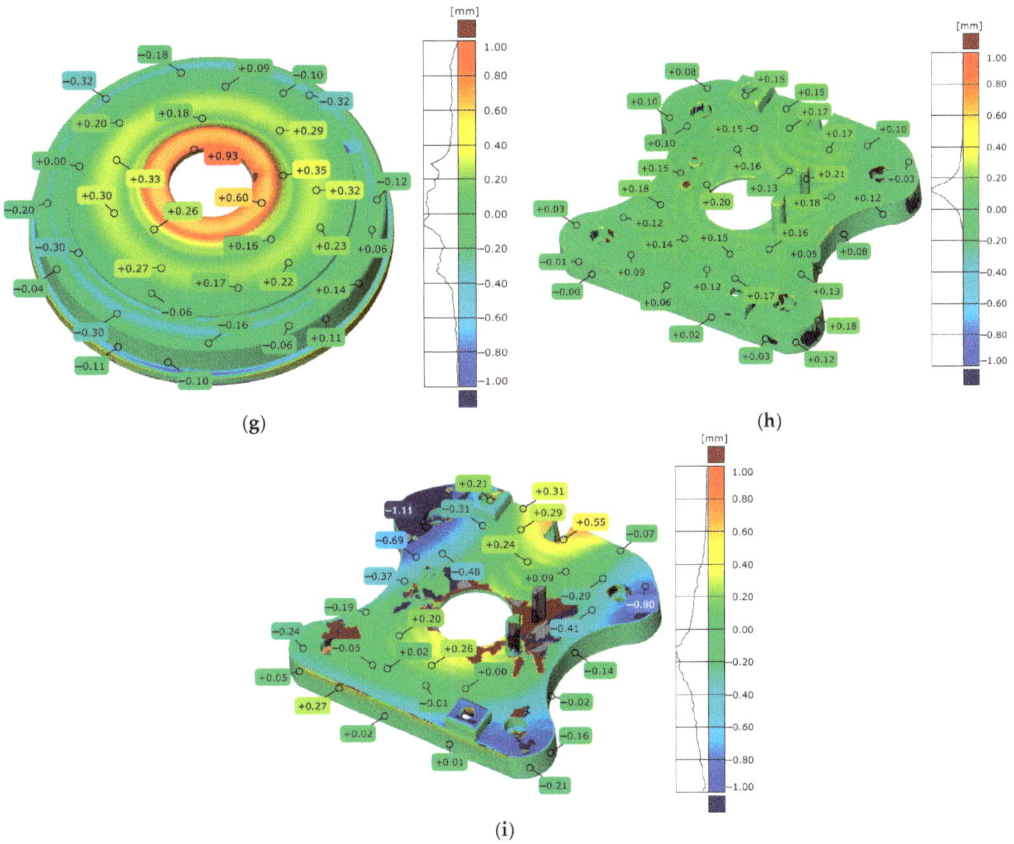

Figure 12. Three-dimensional deviation maps illustrating CAD modeling errors for model: (**a**) electrical box (detect primitives); (**b**) electrical box (auto-surfacing); (**c**) lampshade (profile extraction); (**d**) lampshade (auto-surfacing); (**e**) cover for vacuum unit (detect primitives); (**f**) cover for vacuum unit (profile extraction); (**g**) cover for vacuum unit (auto-surface); (**h**) battery cover (detect primitives); (**i**) battery cover (auto-surface).

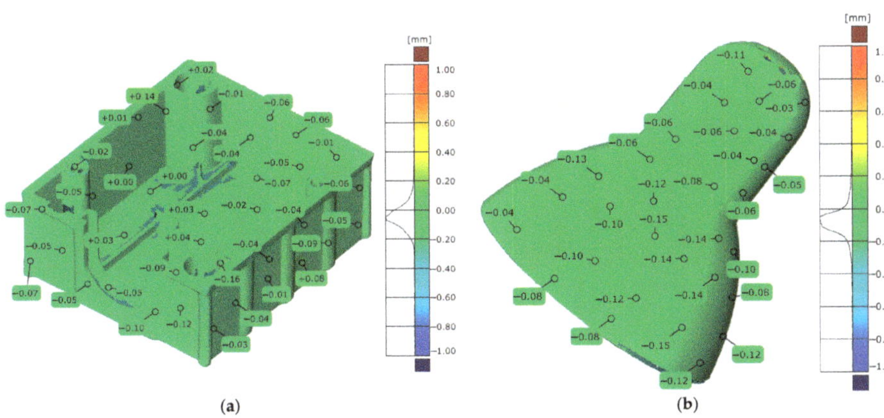

Figure 13. *Cont.*

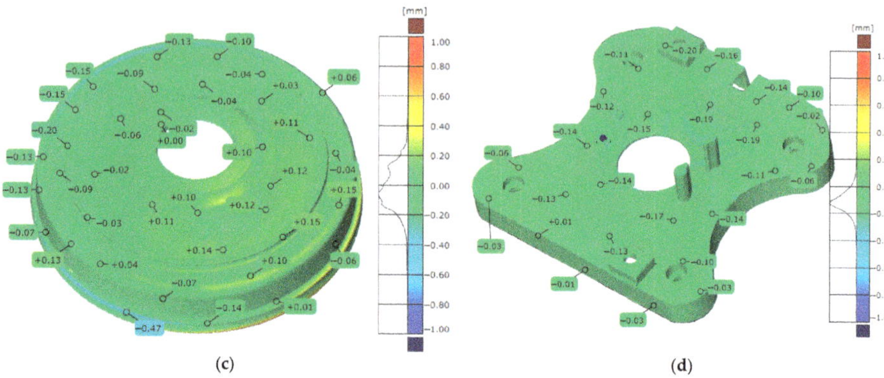

Figure 13. Three-dimensional deviation maps illustrating additive manufacturing errors for the model: (**a**) electrical box; (**b**) lampshade; (**c**) cover for a vacuum unit; (**d**) battery cover.

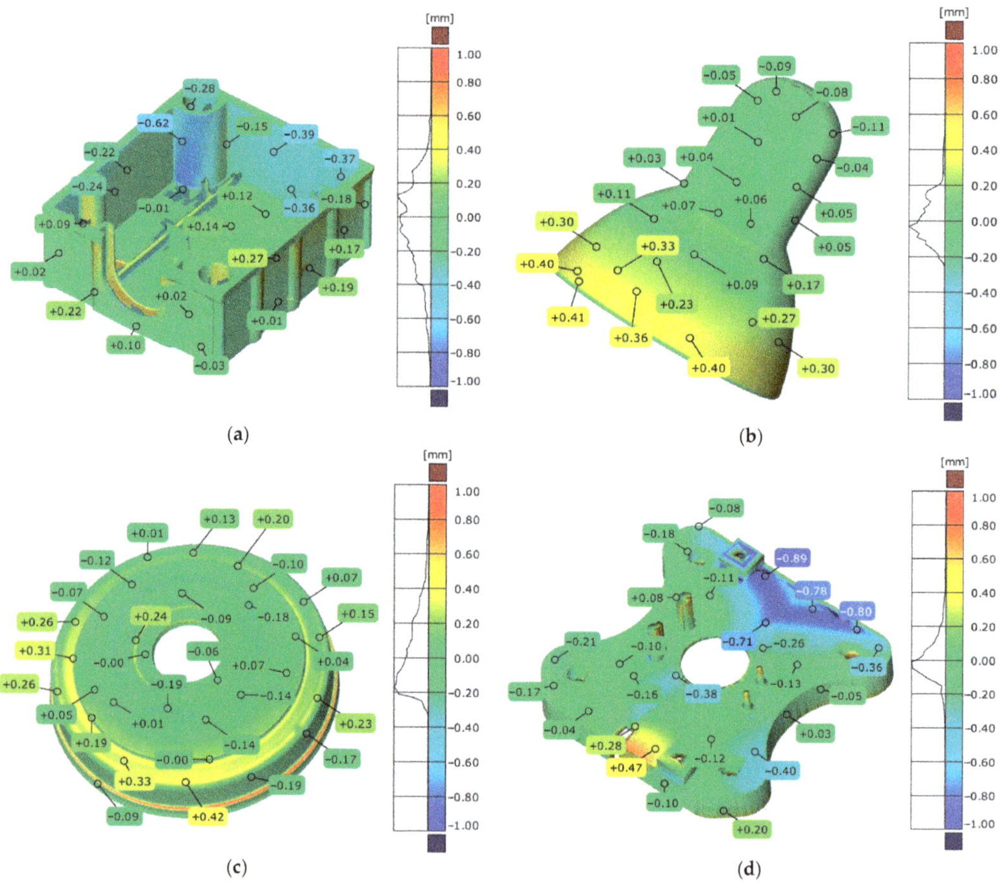

Figure 14. Three-dimensional deviation maps created by directly comparing the nominal model with a scan of the 3D printed model, illustrating total errors for the model: (**a**) electrical box; (**b**) lampshade; (**c**) cover for a vacuum unit; (**d**) battery cover.

Considering the research conducted and the three-dimensional maps of geometric deviations obtained, attention was drawn to the factors affecting the accuracy of the process presented:

Prepare the object for measurements.

- Remove All Contaminants: The model's surface must be free from any dust, dirt, oil, fingerprints, or residue from manufacturing. Even microscopic particles can affect the accuracy of high-precision scanners. Use compressed air, a soft brush, or a lint-free cloth with a suitable solvent (like isopropyl alcohol) that will not damage the part's material. Handle the model with gloves to prevent transferring oils from your hands to the surface.
- Applying Surface Coatings: If the object has highly reflective, transparent, or glossy surfaces, use a matte anti-reflective spray to make them detectable by optical scanners. The coating should be applied as thinly and uniformly as possible to avoid altering the object's true geometry.
- Applying Reference Points: Use high-quality, adhesive reference points that are compatible with your measurement system. Distribute the targets evenly across the model's surface. Place them in non-critical areas to avoid interfering with key geometric features. Ensure there are enough targets to guarantee at least three are visible from any scanning angle.
- Data Acquisition and Reconstruction.
- Securing the Object: The object must be securely fixture to prevent any movement during the measurement process. Vibrations or accidental shifts can cause significant measurement errors. Let the object acclimate to the temperature of the measurement environment to prevent thermal expansion or contraction, which is especially important for polymer parts.
- Scanning Technique: Use a high-precision 3D scanner to capture data. Optimize the scanning process by selecting the appropriate type of measuring head, which determines the size of the measuring area and thus the resolution of the point cloud. In addition, it is necessary to determine the optimal number of measuring steps for the table.
- Alignment Point Clouds: To create a complete and accurate 3D model from multiple scans, it is essential to properly align or "fit" the individual point clouds together. The most reliable method for fitting point clouds together is to use reference points. When reference points are not available, or as a complementary method, a best-fit algorithm can be used. This method works by finding the optimal position and orientation for two overlapping point clouds to minimize the average distance between all corresponding points.
- CAD Modeling
- Clean the 3D Mesh: Before meshing, process the raw data to remove any noise, programming errors of the 3D Mesh, outliers, or duplicate points.
- Parameterization CAD Model: Choose a reconstruction method that is appropriate for the object's geometry.
- Primitive Detection: For models with regular, geometric shapes (planes, cylinders, cones), this method converts the point cloud into a precise, parametric CAD model. Set the appropriate fitting tolerance. Tolerance defines the maximum deviation that points can have from the ideal surface of the primitive in order to be included in it. Too high a tolerance can lead to noisy data being combined into a single primitive, while too low a tolerance will prevent the primitive from being detected at all.
- Profile Extraction: Always select cross-sections at key, representative locations of the object (e.g., at the beginning and end of segments, at locations where the geometry

changes). Use a minimal but sufficient number of profiles to reproduce the shape accurately. Too many profiles can complicate the model, while too few will lead to errors in the final shape. Before extracting profiles, remove noise from the point cloud so that the cross-section lines are as smooth as possible. Ensure that the cross-section planes are perfectly perpendicular to the object's axis so that the profiles are accurate.
- Auto-Surfacing: For organic or complex, free-form shapes, this method automatically generates a surface mesh (NURBS) that is smooth and easy to manipulate in CAD software. Set the appropriate parameters for the algorithm, such as matching tolerance, surface density, and number of patches. Too low a tolerance may cause the model to be noisy, while too high a tolerance will smooth out essential details.
- MEX Additive Manufacturing Process
- Model Orientation: Orient the model to minimize the need for support structures, especially on critical surfaces. Selecting an appropriate orientation also equalizes the stresses along the print layers, which minimizes distortion (e.g., material shrinkage) and increases the strength of the finished part.
- 3D Printing Parameters: Use the thinnest layers possible to increase dimensional accuracy and surface smoothness. You can also optimize the infill pattern to achieve adequate strength while saving material and reducing print time.

Based on the factors determining the accuracy of the reconstruction process, CAD modeling, and additive manufacturing of models using MEX technology, the main guidelines and recommendations were developed in the form of a diagram (Figure 15).

Based on the recommendations provided, paths have been developed to minimize errors in reconstruction, CAD modeling, and additive manufacturing using MEX technology. Table 7 summarizes the average deviations identified during the reconstruction, CAD modeling, and additive manufacturing processes using MEX technology for the analyzed models. Errors in the total process (RE + CAD + AM) were also presented. The obtained ranges of deviations were determined at a confidence level of 0.95.

Additionally, to ensure the quality of the reconstruction, CAD modeling, and additive manufacturing processes, it is crucial to adhere to the standards for geometric accuracy that define the characteristics of the reconstructed models. The geometric accuracy requirements vary significantly for each of the selected components, as each of these parts performs a different function. In the case of a lampshade, there is no single, standardized dimensional tolerance value.

Tolerances are not standardized because lampshades are not precision-engineered parts that require strict mechanical fits. Instead, the acceptable dimensional tolerance is determined by the manufacturer based on the specific material and manufacturing process.

Table 7. The average deviations obtained in the process of reconstruction, CAD modeling, and additive manufacturing using MEX technology for the analyzed models.

Stage	Electrical Box	Lampshade	Cover for a Vacuum Unit	Battery Cover
Reconstruction processs *	±0.6 mm	±0.8 mm	±1 mm	±0.6 mm
CAD modeling	±0.6 mm	±0.6 mm	±0.2 mm	±0.4 mm
Additive manufacturing	±0.2 mm	±0.2 mm	±0.2 mm	±0.2 mm
RE + CAD + AM	±0.5 mm	±0.4 mm	±0.6 mm	±0.4 mm

* Reconstruction errors include measurement and model manufacturing errors.

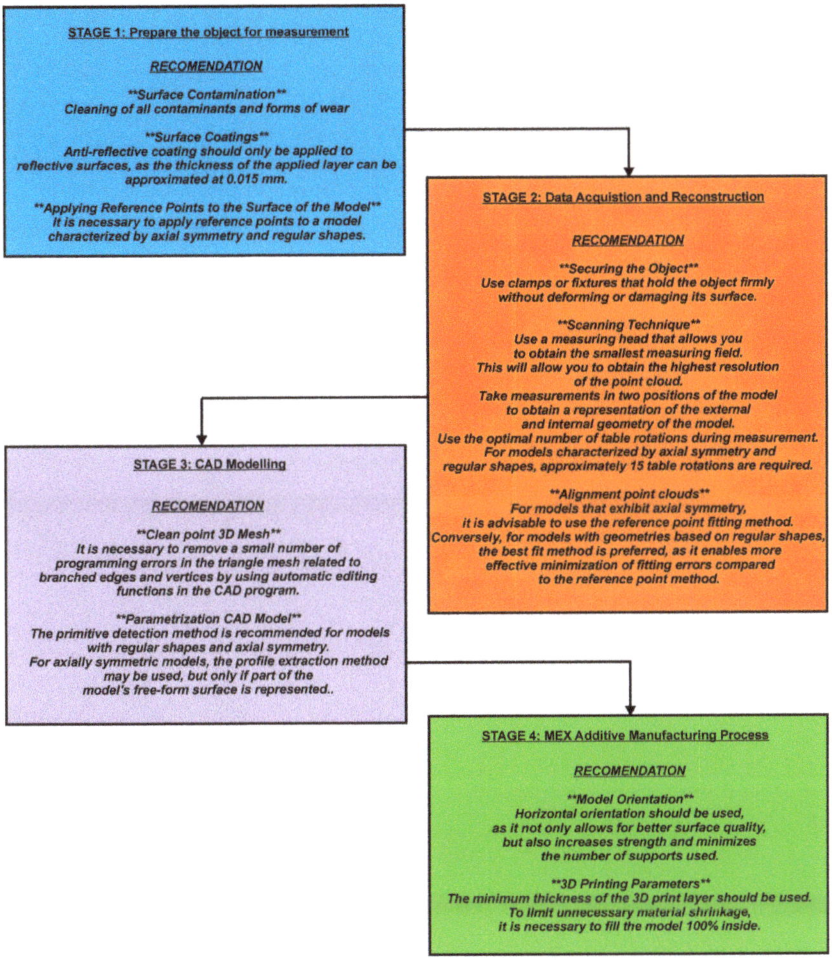

Figure 15. Diagram illustrating strategies for reducing errors in model preparation, geometry reconstruction, CAD modeling, and MEX additive manufacturing for axisymmetric models and regular shape models.

Electrical boxes must comply with the PN-EN IEC 60670 standard [47]. They should accommodate standard switches, sockets, and cables. The main geometric requirements include the following:

- Flatness: Deviations should ensure adhesion to the wall and fittings. The obtained deviation values should be within the range ±0.2 mm.
- Mounting hole spacing: Must be within the range ±0.2 mm to fit the standard socket and switch spacing.
- Depth: Must be uniform and within tolerance, typically ±0.3 mm, to ensure that the hardware fits in the box and does not protrude.

There are no specific standards for the accuracy of vacuum cleaner covers, but guidelines can be provided for critical components:

- Flatness of the sealing surface: The flatness requirements for maintaining tightness should be within the range ±0.2 mm.

- Geometric accuracy at the edges: This is important for the cover to fit the body. The recommended manufacturing tolerance should be within the range of ±0.2 mm.

Moderate geometric accuracy is required for the battery cover. The cover must fit the battery compartment to ensure a secure closure. Key requirements include the following:

- Flatness of contact surfaces: To ensure that the cover does not fall out and protects the interior from dirt, the flatness of the contact surfaces is crucial. The required deviation values should be within the range of ±0.2 mm.
- Dimensional accuracy of the fixing holes: These elements are critical for proper fastening. Tolerances must ensure that the latch will function reliably and that the threads will not loosen. These values should be within the range of ±0.2 mm.

Considering the deviation values observed in critical areas of the models, we examined these values and summarized them in Table 8. However, the tests for the lampshade of a floor lamp were not included due to the absence of specific guidelines. The obtained ranges of deviations were determined at a confidence level of 0.95.

Table 8. The average deviation values in critical areas of the analyzed models.

Type of the Model	Parameters	Stage	Average Deviation Value	Recommended Deviation Values
Electrical box	Flatness	Reconstruction process	±0.1 mm	±0.2 mm
		CAD modeling	±0.08 mm	
		Additive manufacturing	±0.05 mm	
		RE + CAD + AM	±0.1 mm	
	Mounting hole spacing	Reconstruction process	±0.1 mm	±0.2 mm
		CAD modeling	±0.1 mm	
		Additive manufacturing	±0.03 mm	
		RE + CAD + AM	±0.12 mm	
	Depth	Reconstruction process	±0.13 mm	±0.3 mm
		CAD modeling	±0.09 mm	
		Additive manufacturing	±0.04 mm	
		RE + CAD + AM	±0.15 mm	
Vacuum cover	Flatness of the sealing surface	Reconstruction process	±0.2 mm	±0.2 mm
		CAD modeling	±0.09 mm	
		Additive manufacturing	±0.1 mm	
		RE + CAD + AM	±0.15 mm	
	Geometric accuracy at the edges	Reconstruction process	±0.24 mm	±0.2 mm
		CAD modeling	±0.1 mm	
		Additive manufacturing	±0.2 mm	
		RE + CAD + AM	±0.18 mm	
Battery cover	Flatness of contact surfaces	Reconstruction process	±0.2 mm	±0.2 mm
		CAD modeling	±0.15 mm	
		Additive manufacturing	±0.15 mm	
		RE + CAD + AM	±0.11 mm	
	Dimensional accuracy of the fixing holes	Reconstruction process	±0.13 mm	±0.2 mm
		CAD modeling	±0.12 mm	
		Additive manufacturing	±0.1 mm	
		RE + CAD + AM	±0.1 mm	

4. Discussion

4.1. Evaluation of Geometrical Reconstruction Errors

During the geometry reconstruction stage, it is essential to prepare the object for scanning effectively. The quality of the reconstruction heavily depends on the type of geometry being scanned [48]. A crucial aspect of preparing the scanning model involves applying markers to its surface, which helps streamline the measurement process by optimizing the assembly of point clouds obtained from various angular positions of the measurement table. In this research, markers were applied to each selected model. Additionally, special attention was given to the reflectivity of the models' surfaces. Test measurements were conducted to evaluate this aspect. It was determined that only the vacuum cover required a matte layer, which should not exceed 0.015 mm. The measurement process utilized the smallest measurement area available with the GOM Scan 1 system, which is 100 mm × 65 mm × 400 mm. This set up allowed for a maximum point cloud resolution of 0.037 mm, enabling a highly detailed geometry scan in the form of a three-dimensional point cloud. To determine the optimal number of rotations for the measuring table, an empirical approach was used, testing various measurement configurations. It was observed that fewer revolutions of the measuring table made it more challenging to fully digitize the geometry, particularly for the models of the electrical box (Figure 16a) and battery cover. Conversely, with a higher number of scans, an increase in errors was noted during the scan merging and CAD modeling stages [49]. When combining scans to create the final 3D STL model, we observed that the type of geometry influenced the choice of scan combination method. For the electrical box and battery cover models, we used the best-fit method, achieving an accuracy of matching the scans to each other within 0.02 mm. However, due to the axisymmetric nature of the geometries, we encountered issues with the best-fit method, leading us to opt for the feature point-based fitting method instead. The fitting accuracy for the lampshade model reached 0.01 mm, while it was 0.05 mm for the vacuum cover. The higher value for the vacuum cover model may have been affected by the matting layer and wear, corrosion sustained during the operational process [50,51] (Figure 16b).

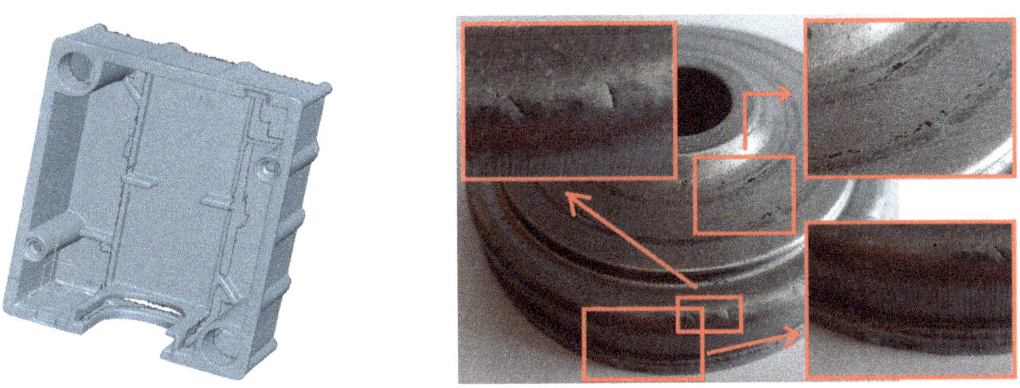

(**a**) (**b**)

Figure 16. Errors in the reconstruction process due to: (**a**) Lack of digitalization of the full geometry; (**b**) Damage to the model surface during exploitation.

Figure 11 and Table 3 present the geometric deviations resulting from the reconstruction process. The results indicated that overall, considering the confidence level of 0.95, the deviation values were typically within the range of ±0.6 mm to ±1 mm. Several factors influence the obtained values:

- The manufacturing process of the models is subject to specific tolerances, which the manufacturer states should remain within ±0.8 mm. This indicates that any discrepancies in the physical dimensions of the models due to production imperfections should not exceed this threshold.
- The measurement system used has its limitations concerning accuracy. For calibration measurements taken on flat surfaces, the maximum allowable error is ±0.02 mm.
- The scanning parameters used during data collection can lead to specific errors. In this case, data is collected at a point cloud resolution of 0.037 mm. This resolution affects the quality and fidelity of the scanned data, which can ultimately influence the overall accuracy of the measurements obtained.
- Scanning reflective surfaces, a matte layer is often applied to mitigate issues caused by glare and reflections. However, the application of this matte finish is not without its limitations; the errors introduced by this process should not exceed 0.015 mm. Proper application is essential to ensure that the scanned data remains as accurate as possible.
- Fitting scans—the process of aligning and merging multiple scan datasets—introduces its own set of potential inaccuracies. The errors associated with this fitting process can vary widely, typically falling within a range of 0.01 mm to 0.05 mm per individual scan. This variance emphasizes the importance of careful alignment and processing to minimize cumulative error in the final output.

It is essential to recognize that the wear and tear on the surface of the scanned model influences the quality of the reconstruction process [52]. The maximum positive and negative deviations, as illustrated in Figure 11, highlight the areas where the most significant surface deformations have occurred. These deviations have an absolute value of nearly 1 mm.

4.2. Evaluation of CAD Modeling Errors

A minimal number of programming errors were encountered during the analysis of the triangle mesh for the scanned models in NX Siemens. These errors primarily involved branching edges and vertices. They were resolved promptly, allowing the parametric modeling process to begin. Various parametric modeling methods were used in the CAD modeling process, with their application depending on the type of geometry involved. For the electrical box model and the battery cover, two CAD modeling techniques were used: one based on the detection of characteristic geometries [18] and the other using the auto-surfacing method [53]. The average fit values for plane or cylindrical surfaces ranged from 0.01 mm to 0.02 mm for both models. However, when attempting global fitting of a parameterized surface to a triangular mesh using the auto-surfacing generation method, it was not possible to achieve such high accuracy. For the electrical box model, the average surface deviations ranged from 0.1 mm to 0.9 mm, while for the battery cover, deviations reached up to 1 mm. Additionally, an increase in the number of surface patches significantly extended the development time of the parametric model. Therefore, it was concluded that this method is not suitable for models with basic geometry types. For the models characterized by axisymmetric, the profile extraction method was utilized during the CAD modeling process. In the cases of the vacuum cover and lamp models, a spline curve was used for curve fitting to define the profile based on a set of points [54]. The average fitting deviations for both models ranged from 0.008 mm to 0.016 mm. However, since the entire geometry of the models was created based on a single profile, the overall deviation values increased significantly. For the lampshade and vacuum cover models, considering the confidence level of 0.95, the deviation values were typically within the range of ±0.4 mm. When the auto-surfacing method was applied to both models, smaller global deviation values were observed compared to the electrical box and battery cover models.

For these models, considering the confidence level of 0.95, the deviation values were typically within the range of ±0.9 mm. For the vacuum cover model, we additionally used a CAD modeling method that involved fitting primitive geometries. This approach helped minimize errors in the developed CAD model. As a result, considering the confidence level of 0.95, the deviation values were typically within the range of ±0.2 mm, demonstrating the effectiveness of this method for CAD modeling. However, this method was not applicable for the lampshade model, as part of its surface featured a free surface rather than a basic surface. Consequently, it was not possible to create the complete CAD geometry using only the primitive surfaces. Considering this, the chosen method for parameterizing and fitting to the acquired measurement data significantly influenced the accuracy of the CAD modeling process. To achieve this, the following errors were identified:

- According to the research findings, using the primitive detection technique allows for the creation of CAD models with an average accuracy tolerance of ±0.2 mm. The error values result from the incorrect matching of the parameterized surface to the obtained point cloud and from connections at the edges and between surfaces.
- According to the research findings, using the profile extraction technique allows for the creation of CAD models with an average accuracy tolerance of ±0.4 mm. The error values obtained result from the incorrect selection of the cross-section against which the profile is extracted, as well as errors in interpolation, curve fitting, and connecting them.
- According to the research findings, using the auto-surfacing technique allows for the creation of CAD models with an average accuracy tolerance of ±0.9 mm. The error values obtained stem from the automatic generation of NURBS surfaces over the entire triangular surface. As a result, to satisfy the fitting conditions, the parameterized surface is often smoothed. This smoothing can complicate the accurate mapping of object edges and the transitions between connected surfaces.

4.3. Evaluation of MEX Additive Manufacturing Errors

The impact of 3D printing parameters on the quality and accuracy of manufactured objects is a critical issue in AM technologies. When creating models using the MEX process, it is essential to consider the appropriate 3D print resolution and the orientation of the model within the 3D printer. Previous research indicates [54,55] that using the highest 3D printing resolution, along with an orientation that maximizes the use of the model's surface along the Z-axis, results in minimized additive manufacturing errors. In examining the manufacturing process of the models created using the MEX additive method, it was observed that the maximum deviation values for most of the models analyzed fall within the range of ±0.2 mm at a confidence level of 0.95. The presented histograms of deviation distribution also confirm this. Only in the case of the vacuum cover model was an increase in deviation values observed, which meant that for a confidence level of 0.95, the maximum deviations were within the range of approximately. ±0.3 mm. This increase may be attributed to the influence of the support material (Figure 17a) and potential errors that occurred during the geometry measurement stage with the GOM Scan 1 scanner (Figure 17b). In addition, the thin-walled construction of the cover may also have contributed to the errors, which could have caused local deformations of the model during the chemical removal of the support material [56].

Despite the geometry errors that occurred, significant efforts were made to achieve the best possible reproduction of the model geometry. To accomplish this, the highest available 3D printing resolution was utilized on the Fortus 360 mc 3D printer. Furthermore, the model was strategically oriented within the 3D printer space to ensure that most surfaces were aligned along the Z-axis. The resulting deviation values are comparable to those reported by the 3D printer manufacturer [57] as well as findings from other scientific publications [58].

(a) (b)

Figure 17. Errors in the MEX additive manufacturing process due to (**a**) occurrence during the manufacture of support material; (**b**) the digitalization of the geometry.

4.4. Evaluation of RE + CAD + AM—Total Errors

The article presents a significant conclusion regarding the discrepancies between the simple summation of deviations and the actual final error. As shown in Tables 7 and 8, relying solely on the summation of tolerances can be overly simplistic and often misleading for quality assessment. The primary reason for this discrepancy lies in error compensation and the statistical nature of how errors accumulate. Notably, an increase in deviations occurs during the geometry reconstruction stage, mainly due to the duplication of model manufacturing errors and measurement inaccuracies. However, this is partially corrected during the CAD modeling and additive manufacturing stages when using MEX technology. A clear example of error compensation can be observed in the lampshade model. During the geometry reconstruction process, the three-dimensional map of deviations predominantly exhibits positive deviations. In contrast, the CAD modeling process reveals an opposite trend. Thus, the errors at these two stages canceled each other out. Ultimately, the deviations observed during the additive manufacturing process do not significantly alter the overall error distribution. A similar situation is observed in Table 8, where critical areas of the model were analyzed based on the deviation values obtained. It is important to note that these deviation values are significantly smaller than their global counterparts. This discrepancy may be attributed to the fact that assessing deviations in a localized area helps minimize various errors, particularly those related to the alignment of the surfaces.

5. Conclusions

This paper presents comprehensive research findings that outline several targeted strategies aimed at significantly reducing measurement errors in various stages of the design and manufacturing process.

- First, it highlights the utilization of an advanced optical scanner, which employs structured light to illuminate the object being measured. This technique not only enhances the clarity of the captured data but also allows for the fine-tuning of scanner settings to improve measurement resolution. Additionally, the research emphasizes the importance of optimizing the number of rotations of the measurement table during the scanning process, as this can lead to more accurate and reliable data capture.

- Furthermore, the paper discusses the selection of an efficient CAD modeling approach. By adopting methods that streamline the modeling process, designers can mitigate the complexities often associated with parametric modeling. This, in turn, reduces potential sources of error that can arise during the conversion from a concept to a digital model.
- The investigation also addresses the challenge of AM errors. It underscores the necessity of carefully selecting the 3D printing layer thickness and determining the optimal model orientation within the 3D printer. These choices play a crucial role in the final output's dimensional accuracy and surface quality, ultimately influencing the integrity of the manufactured part.

It is important to highlight that the research findings in this article demonstrate that evaluating the finished product through direct geometric assessment with three-dimensional deviation maps is a more reliable and realistic measure of quality. This method considers the actual cumulative effects of all errors, including those that are hard to predict, rather than relying solely on theoretical tolerance summation. By implementing these findings, designers can conduct a thorough assessment of accuracy at each stage of the reconstruction process, including scanning, CAD modeling, and additive manufacturing. This comprehensive evaluation allows for a better understanding of the capabilities and limitations of current reverse engineering (RE) methods when preparing model replicas. Such insights are vital for improving the overall quality and precision in manufacturing practices.

Author Contributions: Conceptualization, P.T.; methodology, P.T.; software, P.B., A.C., H.F., T.H. and J.M.; validation, P.T. and J.M.; formal analysis, P.T.; investigation, P.T., P.B., A.C., H.F. and T.H.; writing—original draft preparation, P.B., A.C., H.F. and T.H.; writing—review and editing, P.T.; visualization, P.T. All authors have read and agreed to the published version of the manuscript.

Funding: This research received no external funding.

Data Availability Statement: The raw data supporting the conclusions of this article will be made available by the authors on request.

Conflicts of Interest: The authors declare no conflicts of interest.

References

1. Raja, V. Introduction to reverse engineering. In *Reverse Engineering: An Industrial Perspective*; Springer: London, UK, 2008; pp. 1–9. [CrossRef]
2. She, S.; Lotufo, R.; Berger, T.; Wąsowski, A.; Czarnecki, K. Reverse engineering feature models. In Proceedings of the 33rd International Conference on Software Engineering, Honolulu, HI, USA, 21–28 May 2011; pp. 461–470. [CrossRef]
3. Roithmeier, R. *Measuring Strategies in Tactile Coordinate Metrology*; Carl Zeiss AG: Oberkochen, Germany, 2014.
4. Li, F.; Hiley, J.; Syed, T.M.; Hitchens, C.; Garcia Lopez-Astilleros, M. A region segmentation method to measure multiple features using a tactile scanning probe. *Int. J. Comput. Integr. Manuf.* **2019**, *32*, 569–579. [CrossRef]
5. Wang, X.; Xian, J.; Yang, Y.; Zhang, Y.; Fu, X.; Kang, M. Use of coordinate measuring machine to measure circular aperture complex optical surface. *Measurement* **2017**, *100*, 1–6. [CrossRef]
6. Gapinski, B.; Wieczorowski, M.; Marciniak-Podsadna, L.; Dybala, B.; Ziolkowski, G. Comparison of different method of measurement geometry using CMM, optical scanner and computed tomography 3D. *Procedia Eng.* **2014**, *69*, 255–262. [CrossRef]
7. Li, F.; Longstaff, A.P.; Fletcher, S.; Myers, A. Integrated tactile and optical measuring systems in three dimensional metrology. In Proceedings of the Queen's Diamond Jubilee Computing and Engineering Annual Researchers' Conference 2012: CEARC'12, Huddersfield, UK, 29 February 2012; University of Huddersfield: Huddersfield, UK, 2012.
8. Chen, S.; Chen, H. Research on Triangulation Method of Object Surface with Holes in Reverse Engineering. In Proceedings of the 2009 International Conference on Computational Intelligence and Software Engineering, Wuhan, China, 11–13 December 2009; IEEE: New York, NY, USA, 2012; pp. 1–4. [CrossRef]
9. Liu, J. An adaptive process of reverse engineering from point clouds to CAD models. *Int. J. Comput. Integr. Manuf.* **2020**, *33*, 840–858. [CrossRef]

10. Ali, S.; Durupt, A.; Adragna, P.A.; Bosch-Mauchand, M. A reverse engineering for manufacturing approach. *Comput.-Aided Des. Appl.* **2014**, *11*, 694–703. [CrossRef]
11. Wojnarowska, W.; Kwolek, M.; Miechowicz, S. Selection of a workpiece clamping system for computer-aided subtractive manufacturing of geometrically complex medical models. *Open Eng.* **2021**, *11*, 239–248. [CrossRef]
12. Kumar, A.; Kumar, P.; Singh, H.; Haleem, A.; Mittal, R.K. Integration of reverse engineering with additive manufacturing. In *Advances in Additive Manufacturing*; Elsevier: Amsterdam, The Netherlands, 2023; pp. 43–65. [CrossRef]
13. Vukašinović, N.; Duhovnik, J. Introduction to reverse engineering. In *Advanced CAD Modeling: Explicit, Parametric, Free-Form CAD and Re-Engineering*; Springer International Publishing: Cham, Switzerland, 2018; pp. 165–177. [CrossRef]
14. Mahmood, D.M.; Mahmood, A.Z. Improving Reverse Engineering Processes by using Articulated Arm Coordinate Measuring Machine. *Al-Khwarizmi Eng. J.* **2020**, *16*, 42–54. [CrossRef]
15. Turek, P.; Bezłada, W.; Cierpisz, K.; Dubiel, K.; Frydrych, A.; Misiura, J. Analysis of the accuracy of CAD modeling in engineering and medical industries based on measurement data using reverse engineering methods. *Designs* **2024**, *8*, 50. [CrossRef]
16. Jang, A.; Jeong, S.; Park, M.J.; Ju, Y.K. Structural evaluation by reverse engineering with 3D laser scanner. *ce/papers* **2023**, *6*, 308–314. [CrossRef]
17. Vazzana, A.; Higgins, O.A.; Oxilia, G.; Lugli, F.; Silvestrini, S.; Nava, A.; Benazzi, S. High-accuracy methodology for the integrative restoration of archaeological teeth by using reverse engineering techniques and rapid prototyping. *J. Archaeol. Sci. Rep.* **2022**, *44*, 103511. [CrossRef]
18. Saiga, K.; Ullah, A.S.; Kubo, A. A sustainable reverse engineering process. *Procedia CIRP* **2021**, *98*, 517–522. [CrossRef]
19. Gerbino, S.; Martorelli, M. Reverse Engineering. In *Springer Handbook of Additive Manufacturing*; Springer: Cham, Switzerland, 2023; pp. 253–268. [CrossRef]
20. *ISO/ASTM 52910:2018*; Additive Manufacturing—Design—Requirements, Guidelines and Recommendations. ISO: Geneva, Switzerland, 2018.
21. *ISO/ASTM 52902:2023*; Additive Manufacturing—Test Artefacts—Geometric Capability Assessment of Additive Manufacturing Systems. ISO: Geneva, Switzerland, 2023.
22. *ISO/ASTM 52915:2020*; Specification for Additive Manufacturing File Format (AMF) Version 1.2. ISO: Geneva, Switzerland, 2020.
23. *ISO/ASTM 52950:2021*; Additive Manufacturing—General Principles—Overview of Data Processing. ISO: Geneva, Switzerland, 2021.
24. Zhang, Z.; Zhao, M.; Shen, Z.; Wang, Y.; Jia, X.; Yan, D.M. Interactive reverse engineering of CAD models. *Comput. Aided Geom. Des.* **2024**, *111*, 102339. [CrossRef]
25. Varitis, E. A Reverse Engineering method for 3d parametric modeling of geometries based on processing of laser scanning data and a sweep technique. *IOP Conf. Ser. Mater. Sci. Eng.* **2016**, *161*, 012047. [CrossRef]
26. Turek, P. Evaluation of the auto surfacing methods to create a surface body of the mandible model. *Rep. Mech. Eng.* **2022**, *3*, 46–54. [CrossRef]
27. Kreuz, S.; Zubiri, B.A.; Englisch, S.; Buwen, M.; Kang, S.G.; Ramachandramoorthy, R.; Rolfes, J. Improving reconstructions in nanotomography for homogeneous materials via mathematical optimization. *Nanoscale Adv.* **2024**, *6*, 3934–3947. [CrossRef]
28. Ding, Z.; Lu, Y.; Shao, S.; Qin, Y.; Lu, M.; Song, Z.; Sun, D. Research on 3D Reconstruction Methods for Incomplete Building Point Clouds Using Deep Learning and Geometric Primitives. *Remote Sens.* **2025**, *17*, 399. [CrossRef]
29. Liu, Y.; Obukhov, A.; Wegner, J.D.; Schindler, K. Point2cad: Reverse engineering cad models from 3d point clouds. In Proceedings of the IEEE/CVF Conference on Computer Vision and Pattern Recognition, Seattle, WA, USA, 16–22 June 2024; pp. 3763–3772. [CrossRef]
30. Thengane, V.; Zhu, X.; Bouzerdoum, S.; Phung, S.L.; Li, Y. Foundational Models for 3D Point Clouds: A Survey and Outlook. *arXiv* **2025**, arXiv:2501.18594.
31. Gibson, I.; Rosen, D.; Stucker, B. *Additive Manufacturing Technologies*; Springer: Cham, Switzerland, 2021.
32. Budzik, G.; Dziubek, T.; Kawalec, A.; Turek, P.; Bazan, A.; Dębski, M.; Kuric, I. Geometrical Accuracy of Threaded Elements Manufacture by 3D Printing Process. *Adv. Sci. Technol. Res. J.* **2023**, *17*, 35–45. [CrossRef] [PubMed]
33. Turek, P.; Filip, D.; Przeszłowski, Ł.; Łazorko, A.; Budzik, G.; Snela, S.; Paszkiewicz, A. Manufacturing polymer model of anatomical structures with increased accuracy using cax and am systems for planning orthopedic procedures. *Polymers* **2022**, *14*, 2236. [CrossRef]
34. García Plaza, E.; López, P.J.N.; Torija, M.Á.C.; Muñoz, J.M.C. Analysis of PLA geometric properties processed by FFF additive manufacturing: Effects of process parameters and plate-extruder precision motion. *Polymers* **2019**, *11*, 1581. [CrossRef] [PubMed]
35. Pelzer, L.; Hopmann, C. Additive manufacturing of non-planar layers with variable layer height. *Addit. Manuf.* **2021**, *37*, 101697. [CrossRef]
36. Shen, H.; Guo, S.; Fu, J.; Lin, Z. Building orientation determination based on multi-objective optimization for additive manufacturing. *3D Print. Addit. Manuf.* **2020**, *7*, 186–197. [CrossRef]
37. Matos, M.A.; Rocha, A.M.A.; Pereira, A.I. Improving additive manufacturing performance by build orientation optimization. *Int. J. Adv. Manuf. Technol.* **2020**, *107*, 1993–2005. [CrossRef]

38. Allaire, G.; Bogosel, B. Optimizing supports for additive manufacturing. *Struct. Multidiscip. Optim.* **2018**, *58*, 2493–2515. [CrossRef]
39. Alzyod, H.; Ficzere, P.; Borbas, L. Cost-efficient additive manufacturing: Unraveling the economic dynamics of material Extrusion (MEX) technology. *Int. Rev.* **2024**, *3–4*, 185. [CrossRef]
40. Asami, K.; Kuehne, M.; Röver, T.; Emmelmann, C. Application of Machine Learning in Predicting Quality Parameters in Metal Material Extrusion (MEX/M). *Metals* **2025**, *15*, 505. [CrossRef]
41. *ISO 8015:2011*; Geometrical Product Specifications (GPS)—Fundamentals—Concepts, Principles and Rules. ISO: Geneva, Switzerland, 2011.
42. *ISO 1101:2017*; Geometrical Product Specifications (GPS)—Geometrical Tolerancing—Tolerances of Form, Orientation, Location and Run-Out. ISO: Geneva, Switzerland, 2017.
43. *ISO 286-2:2010*; Geometrical Product Specifications (GPS)—ISO Code System for Tolerances on Linear Sizes—Part 2: Tables of Standard Tolerance Classes and Limit Deviations for Holes and Shafts. ISO: Geneva, Switzerland, 2010.
44. *ISO 22081:2021*; Geometrical Product Specifications (GPS)—Geometrical Tolerancing—General Geometrical Specifications and General Size Specifications. ISO: Geneva, Switzerland, 2021.
45. *ASME Y14.5*; Dimensioning and Tolerancing. ASME: New York, NY, USA, 2018.
46. *VDI/VDE 2634*; Blatt 3 Optical 3D-Measuring Systems—Multiple View Systems Based on Area Scanning. VDI/VDE: Berlin, Germany, 2008.
47. *PN-EN IEC 60670*; Boxes and Enclosures for Accessories for Household and Similar Fixed Electrical Installations. IEC: Geneva, Switzerland, 2021.
48. Buonamici, F.; Carfagni, M.; Furferi, R.; Governi, L.; Lapini, A.; Volpe, Y. Reverse engineering of mechanical parts: A template-based approach. *J. Comput. Des. Eng.* **2018**, *5*, 145–159. [CrossRef]
49. Armstrong, D. Automating Scan-Based Inspection for Repeatable Process Control. *Quality* **2022**, *61*, 17.
50. Milde, J.; Dubnicka, M.; Buransky, I. Impact of powder coating types on dimensional accuracy in optical 3d scanning process. *MM Sci. J.* **2023**, 6800–6806. [CrossRef]
51. Guo, H.; Wang, Y.; Wang, M.; Yang, H.; Gu, B.; Xu, G.; Gao, X. Microstructure, wear, and corrosion resistance of HVOF WC24-Cr3C257-NiCr19 coating. *Tribol. Int.* **2025**, *211*, 110806. [CrossRef]
52. Bagci, E. Reverse engineering applications for recovery of broken or worn parts and re-manufacturing: Three case studies. *Adv. Eng. Softw.* **2009**, *40*, 407–418. [CrossRef]
53. Chang, K.H. A review on shape engineering and design parameterization in reverse engineering. In *Reverse Engineering: Recent Advances and Applications*; BoD–Books on Demand: Norderstedt, Germany, 2012; pp. 161–186. [CrossRef]
54. Turek, P. The Influence of the Layer Thickness Change on the Accuracy of the Zygomatic Bone Geometry Manufactured Using the FDM Technology. *Eng. Proc.* **2022**, *24*, 26. [CrossRef]
55. Turek, P.; Bazan, A.; Bulicz, M. Effect of 3D Printing Orientation on the Accuracy and Surface Roughness of Polycarbonate Samples. *Machines* **2024**, *13*, 9. [CrossRef]
56. Pellizzon, N.; Šeta, B.; Kruse, C.S.; Salajeghe, R.; Spangenberg, J. Investigating thermal strains and chemical shrinkage in tomographic volumetric additive manufacturing. *Addit. Manuf.* **2025**, *105*, 104781. [CrossRef]
57. Bonnie, M. The Accuracy Myth. Don't Make the Mistake of Confusing High Resolution with Accuracy. Available online: https://www.stratasys.com/contentassets/08614c7bcd824cfb99fe37f6af0e0189/wp_fdm_accuracymyth_a4_0116a_web.pdf?v=48fd13 (accessed on 1 August 2025).
58. Turek, P.; Budzik, G. Estimating the Accuracy of Mandible Anatomical Models Manufactured Using Material Extrusion Methods. *Polymers* **2021**, *13*, 2271. [CrossRef]

designs

MDPI

Article

3D-Printed Prosthetic Solutions for Dogs: Integrating Computational Design and Additive Manufacturing

Jeremy Sarpong [1], Khalil Khanafer [1,2,*] and Mohammad Sheikh [1]

[1] College of Innovation and Technology, Mechanical Engineering, University of Michigan, Flint, MI 48502, USA; jsarpong@umich.edu (J.S.); mohammsh@umich.edu (M.S.)
[2] Michigan Institute for Computational Discovery and Engineering, University of Michigan, Ann Arbor, MI 48109, USA
* Correspondence: khanafer@umich.edu

Abstract

This study investigates the mechanical performance of two prosthetic forelimb designs for dogs—one with a solid structure and the other with a perforated structure—using Finite Element Analysis (FEA). Both models were analyzed under static loading conditions representing approximately 60% of a dog's body weight, the typical load borne by the forelimbs. The prosthetics were modeled with ABS plastic, a widely used 3D printing material, and evaluated for Von Mises stress, total deformation, elastic strain, and factor of safety. The analysis showed that both models remained within the elastic limit of the material, indicating that no permanent deformation would occur under the applied loads. The Solid Model demonstrated a significantly higher factor of safety (14) and lower deformation, confirming its structural strength but also highlighting excessive rigidity, increased material use, and higher cost. In contrast, the Perforated Model exhibited slightly higher localized stresses and a lower factor of safety (3.01), yet it still met essential safety requirements while providing greater compliance, flexibility, and material efficiency. These attributes are desirable for comfort, adaptability, and practicality in veterinary applications. Although its long-term durability requires further evaluation, the Perforated Model strikes a more effective balance between safety, comfort, and sustainability. Based on these findings, the perforated design is considered the more suitable option for canine prosthetic development. Future work will extend the analysis to dynamic loading scenarios, such as walking and running, to better simulate real-world performance.

Keywords: finite element analysis (FEA); prosthetics; stress; deformations; strain; mechanical performance; biomechanics; additive manufacturing; 3D-printing; dogs

1. Introduction

Additive manufacturing, or 3D printing, is a process that builds parts layer by layer based on a digital design. This approach is different from traditional methods like machining, which remove material from a solid block to shape a part. One of the key advantages of additive manufacturing is its ability to create complex shapes and internal features that would be difficult—or even impossible—to produce with conventional techniques. Because of this, it is being used more and more in fields like aerospace, where lightweight and high-strength parts are important, and in healthcare, where custom implants and prosthetics can be made to fit individual patients. It is also valuable in the automotive industry for making prototypes quickly and producing parts that need to meet specific performance requirements. Beyond that, it is starting to play a role in renewable energy systems by helping develop components for wind turbines and other equipment. The flexibility and efficiency of additive manufacturing make it a practical and increasingly important tool across many industries [1–3].

Computational design and additive manufacturing have come together in a practical and impactful way to support the advancement of prosthetic devices for both humans and animals. For example, Rahman et al. [4] conducted a finite element analysis (FEA) to evaluate and enhance the performance of prosthetic running blades by exploring alternative composite materials. Using ANSYS (2024R2), they created mechanical models of existing Ossur blade designs and simulated how different materials would impact structural behavior. The study considered two composite materials—polyethylene epoxy and vinylester—as potential replacements for the materials traditionally used in the blades. The results showed that substituting the original materials with these alternatives significantly reduced strain in all three blade designs examined: the Cheetah Blade, Flex-Run Blade, and Flex-Sprint Blade. Specifically, the Cheetah Blade's maximum principal strain decreased from 0.00299 in/in to 0.0026184 in/in—a 12.4% reduction. The Flex-Run and Flex-Sprint blades also showed strain reductions of 11.5% and 12.9%, respectively. These improvements were attributed to the higher Young's modulus of the new materials. Ismail et al. [5] carried out a study focused on designing and developing a below-knee prosthetic leg specifically intended for running. The goal was to create a strong, lightweight, comfortable, and flexible prosthesis that would be more affordable than imported alternatives. To identify the most effective design, the researchers explored six different geometric shapes and three installation concepts using SolidWorks 2024. They applied the finite element method to analyze each design under a static load equivalent to eight times the user's body weight (up to 80 kg). The simulated prosthetic leg was made from carbon fiber composite, which has a tensile strength of 538.83 MPa, a Young's modulus of 76,975.71 MPa, and a Poisson's ratio of 0.14. Among the tested options, the design with 10 mm thickness performed the best, showing a maximum von Mises stress of 414.76 MPa and a safety factor of 1.29. Based on their decision matrix, the most favorable outcome was found in the second installation concept paired with the third geometric design, which received the highest evaluation score of 80.

In veterinary medicine, prosthetic limbs present a promising alternative to traditional approaches like euthanasia or full limb amputation when addressing limb deformities or injuries. Studies by Mich [6] and Carr et al. [7] have shown that animal prosthetics can enhance mobility, prevent joint deterioration, and improve overall quality of life. However, challenges such as skin irritation and mechanical failure of the devices still persist. The final outcome is often influenced by factors such as the level of limb loss and the specific type of prosthetic used [8]. With the emergence of more cost-effective production methods like 3D printing, prosthetic solutions for animals are becoming increasingly accessible [9]. With the growth of additive manufacturing, there has been significant improvement on the design and efficacy of human prosthetics to tackle the problem of limb amputation. Partial limb amputation has become the standard for human patients so that the technology of prosthesis may be used; contrarily, when it comes to animals, total limb amputation has become the standard [10]. It is either this or euthanizing the animal [11].

There is limited research into the design and analysis of prosthetic models for animals as such, this study aims to encourage the use of prosthesis with animals, more specifically, with canines by using additive manufacturing to model different 3D-printed canine prosthetics. It is required that the prosthetic models should be functionally consistent with supporting body weight, provide shock absorption, and flexibility, etc. [11]. There are two types of limb prosthesis for canines: Endo-Exo Prosthesis and Exo-prosthesis [11].

Endo-Prosthesis incorporates the prosthesis into the remaining bone through a process known as osseointegration [12]. This implementation is not very common. Conversely, exo-prosthetics are more conventional and they use external sockets and suspension systems [13]. The models used in this study are exo-prosthetics. The mechanical behavior of the designed models was evaluated using Finite Element Method (FEM). FEM is a powerful computational technique for solving differential and integral equations in various fields of engineering and applied sciences. Regarding this study, this method solves and helps visualize stress, deformation and strain in two different models namely perforated and solid models. To simplify the analysis, the scope of this study was limited to static loading conditions.

2. Materials and Methods

This research compares two 3D-printed dog forelimb prosthetic models: one with perforations at the socket and grooves at the bottom (labeled as Perforated Model) and one completely solid model (labeled as Solid Model) shown in Figure 1. The reason for this comparison is because the perforated and solid prosthetic designs represent two widely considered approaches in the literature, each with distinct mechanical and biological implications. Solid designs are typically favored for their higher structural rigidity, while perforated designs are of interest for their potential to reduce weight, improve fixation, and facilitate tissue integration. However, direct comparisons of these two approaches remain limited, particularly in terms of quantifying their mechanical performance under realistic loading conditions. By addressing this gap, our study provides a clearer rationale for investigating both designs and highlights the importance of understanding their trade-offs for clinical decision-making. The 3D model of the prosthetics was generated using a Computer Aided Design (CAD) software, more specifically, Dassault Systèmes' SOLIDWORKS 2024 software. Various tools and features were used to build the two 3D models. These features include bossed-based extrusion to build the material, extrude cuts to remove material, fillets to round sharp edges and improve strength, patterns and mirroring. High impact Acrylonitrile Butadiene Styrene (ABS) plastic was used as the material for testing. Table 1 shows the material properties of the high impact ABS plastic used for the study while conducting Finite Element Analysis (FEA). ABS was chosen because it is one of the most widely used thermoplastic polymers in additive manufacturing, offering a favorable balance of strength, toughness, and ease of processing. Its mechanical properties (e.g., tensile strength and impact resistance) make it suitable for replicating the structural behavior of prosthetic components under load in an experimental or prototyping context. Furthermore, ABS is cost-effective, readily available, and compatible with standard 3D printing systems, which allows for efficient iteration of design modifications during the research phase. While not intended as the final clinical material, ABS provides a reliable and practical medium for evaluating mechanical performance and validating design concepts prior to transitioning to advanced biocompatible materials. ABS is occasionally used in the fabrication of prosthetic components, particularly in applications where cost-effectiveness, ease of manufacturing, and lightweight properties are prioritized. While not suitable for high-load or permanent prosthetic parts, ABS can be effectively utilized in non-load-bearing elements such as cosmetic covers, temporary sockets, or early-stage prototypes [14]. Its compatibility with 3D printing makes it an attractive material for rapid prototyping and low-resource environments, where design iterations and functional testing can be conducted before transitioning to more durable materials like carbon fiber composites or titanium. However, due to its limited mechanical strength, fatigue resistance, and long-term durability, ABS is typically reserved for applications that do not require high structural performance.

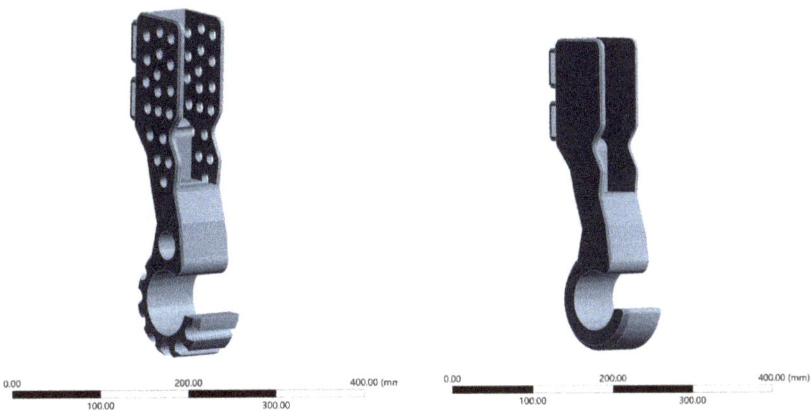

(**a**) (**b**)

Figure 1. (**a**) Perforated Model; (**b**) Solid Model.

Table 1. Mechanical Properties of high-impact ABS used for FEA.

Property	Value
Young's Modulus	2090 MPa
Poisson's Ratio	0.4089
Bulk Modulus	3823.6 MPa
Shear Modulus	741.71 MPa
Tensile Ultimate Strength	36.26 MPa
Tensile Yield Strength	27.11 MPa

Due to the stark difference in designs, the Perforated Model weighs 1.99 lbs while the Solid Model weighs 2.40 lbs.

To ensure consistency in the research, both models were evaluated using Finite Element Analysis (FEA) within the ANSYS Static Structural (ANSYS Mechanical) environment. The analysis involved simulating a standing dog using only one prosthetic (meaning the dog only had one amputation on one of its forelimbs). The analysis also involved studying different weight classes of dogs classified as Light, Medium, Large and Extra-large on both models. These different weight classifications [15] provided the various loads tested on both models. The load was calculated by multiplying the weight of the dog by 0.3, as dogs bear about 60% of their body weight on their forelimbs [16].

Another measure taken to ensure accuracy was verifying that the analysis model in ANSYS Mechanical achieved mesh independence. Various element sizes were tested in the preliminary stages of the research until convergence was reached and there were no discrepancies in the results to ensure the simulation was not mesh dependent on both models. An element size of 1 mm was used for the perforated model to create a fine mesh with a quad dominant sheet body method yielding the following mesh statistics: 53,010 nodes, 28,744 elements, 8675 corner nodes and 44,353 mid nodes. Similarly, the same element size of 1 mm was used for the solid model to create a fine mesh with a quad dominant sheet body method which yielded the following mesh statistics: 47,362 nodes, 26,354 elements, 7572 corner nodes and 39,790 mid nodes. Figure 2 shows a visual representation of the mesh of both models. The perforated model required at least 28,744 elements to achieve mesh independence while the solid model required at least 26,354 elements to achieve convergence.

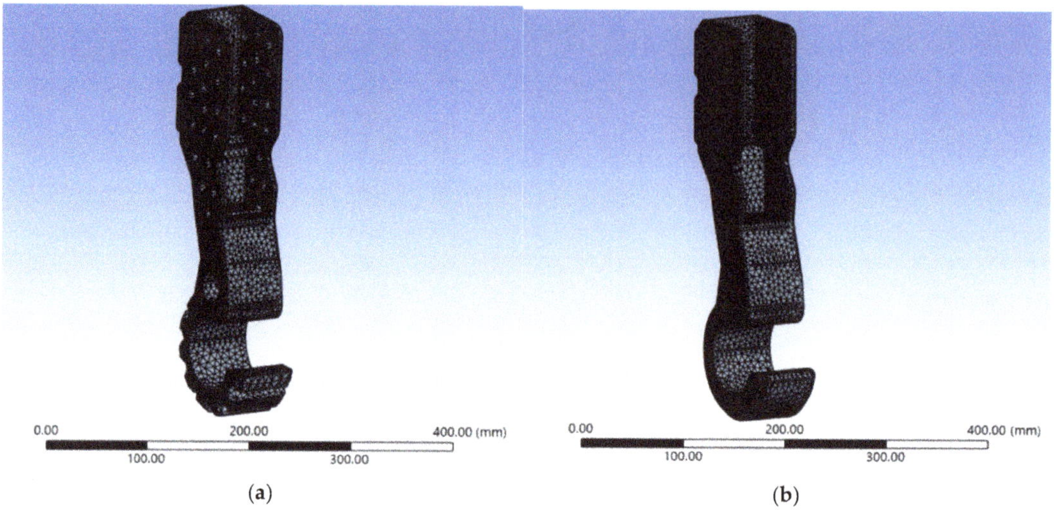

Figure 2. (**a**) Perforated Model Mesh; (**b**) Solid Model Mesh.

Mesh Convergence Verification:

In the preliminary phase of the research, various mesh element sizes were tested in the same conditions (e.g., same loading, boundary conditions) to ensure the FEA model obtained mesh independence. The following element sizes were tested: 20 mm, 10 mm, 5 mm, and 1 mm. Below is an example of testing for mesh convergence of the perforated model with a load of 203.07 N (Table 2). An element size of 1 mm was determined to be sufficient for achieving grid-independent results. Further refinement to 0.5 mm produced a relative error of less than 4%.

Table 2. Mesh Convergence.

Element Size (mm)	Max Stress (MPa)
20	5.88
10	6.61
5	8.33
1	9.10

Boundary Conditions:

The same boundary conditions were used on both models. A fixed support was applied on the inside of the socket of both models, and the load was applied on the point of the footpad which contacted the ground as shown in Figure 3. In a load-bearing static problem such as a dog standing, the load applied by the dog's leg is equal in magnitude to the Ground Reaction Force (GRF) due to the principle of static equilibrium; hence, why the load was placed at the bottom of the footpad.

Lastly, the following results were then solved and studied: Equivalent Stress (Von Mises Stress) in MPa, Total Deformation in mm, and Equivalent Strain in mm/mm. The units were the same for both models. Since the research was solely simulation based, no dogs or animals were tested on.

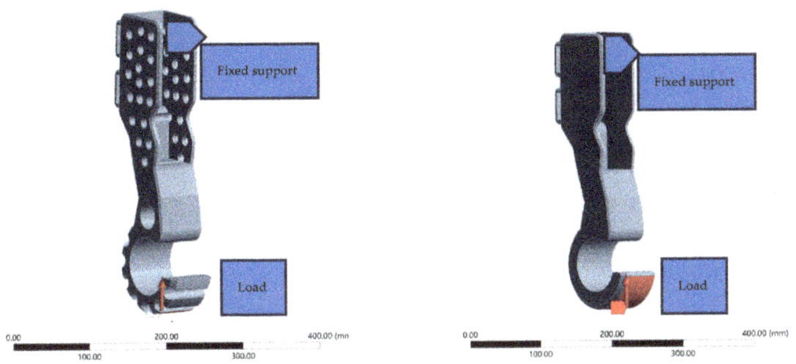

(**a**) (**b**)

Figure 3. (**a**) Perforated Model Boundary Conditions; (**b**) Solid Model Boundary Conditions.

3. Results

3.1. Data Used

Table 3 below shows different dog breeds and their mass in kilograms categorized by their masses:

Table 3. Data used for running the simulation.

Category	Breed	Mass (kg)
Light	Yorkshire Terrier	3
Medium	Cocker Spaniel	14
Large	German Shepherd	34
Extra Large	Newfoundland	69

3.2. Data Obtained for the Perforated Model

Table 4 and Figure 4 below show the results obtained for analysis of the Perforated Model. The following results were obtained: Equivalent Stress (Von Mises Stress), Total Deformation, and Equivalent Strain.

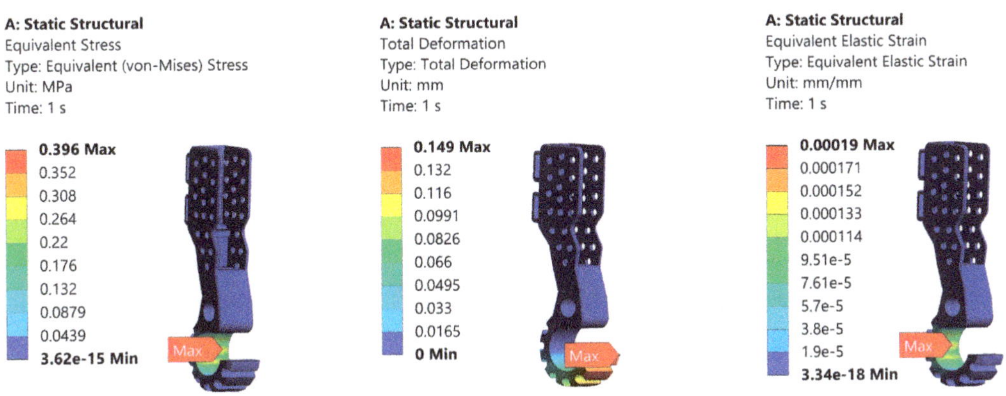

(**a**) (**b**) (**c**)

Figure 4. *Cont.*

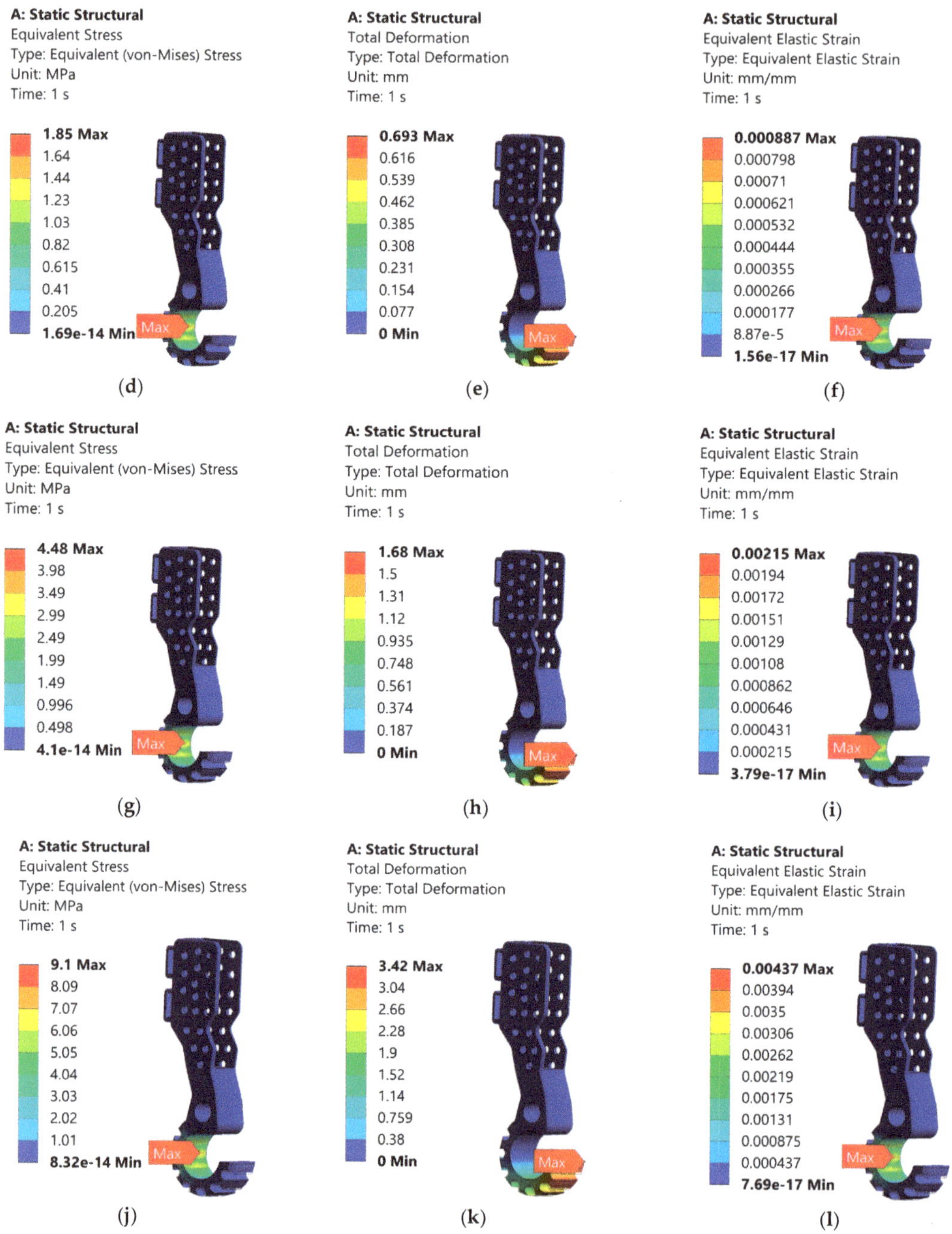

Figure 4. (**a**–**c**) represent the results for Yorkshire Terrier, (**d**–**f**) represent the results for Cocker Spaniel, (**g**–**i**) represent the results for German Shepherd, and (**j**–**l**) represent the results for Newfoundland using perforated model.

Table 4. Mechanical properties of different breeds using perforated model.

Breed	Load (N)	Equivalent Stress (MPa)	Total Deformation (mm)	Equivalent Strain (mm/mm)
Yorkshire Terrier	8.83	0.40	0.15	0.00019011
Cocker Spaniel	41.20	1.85	0.69	0.00088719
German Shepherd	100.06	4.48	1.68	0.00215460
Newfoundland	203.07	9.09	3.42	0.00437260

3.3. Data Obtained for the Solid Model

Table 5 and Figure 5 below show the results obtained for analysis of the Solid Model. The following results were obtained: Equivalent Stress (Von Mises Stress), Total Deformation, and Equivalent Strain.

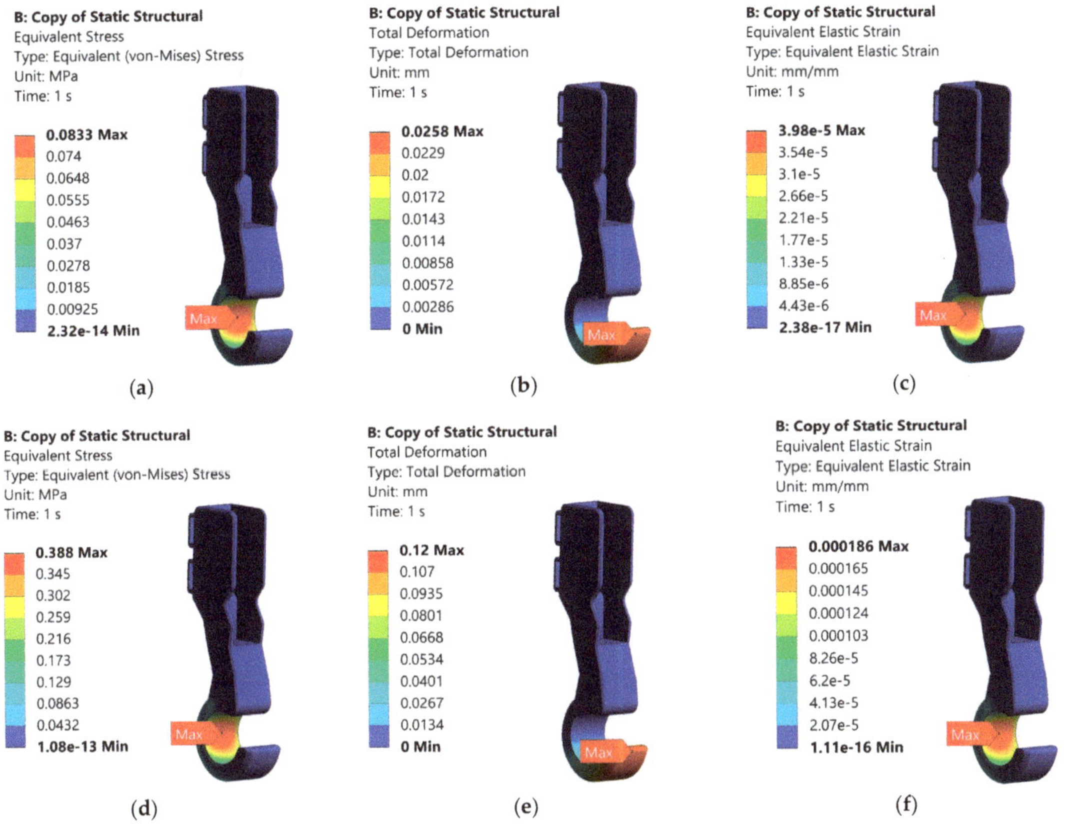

Figure 5. *Cont.*

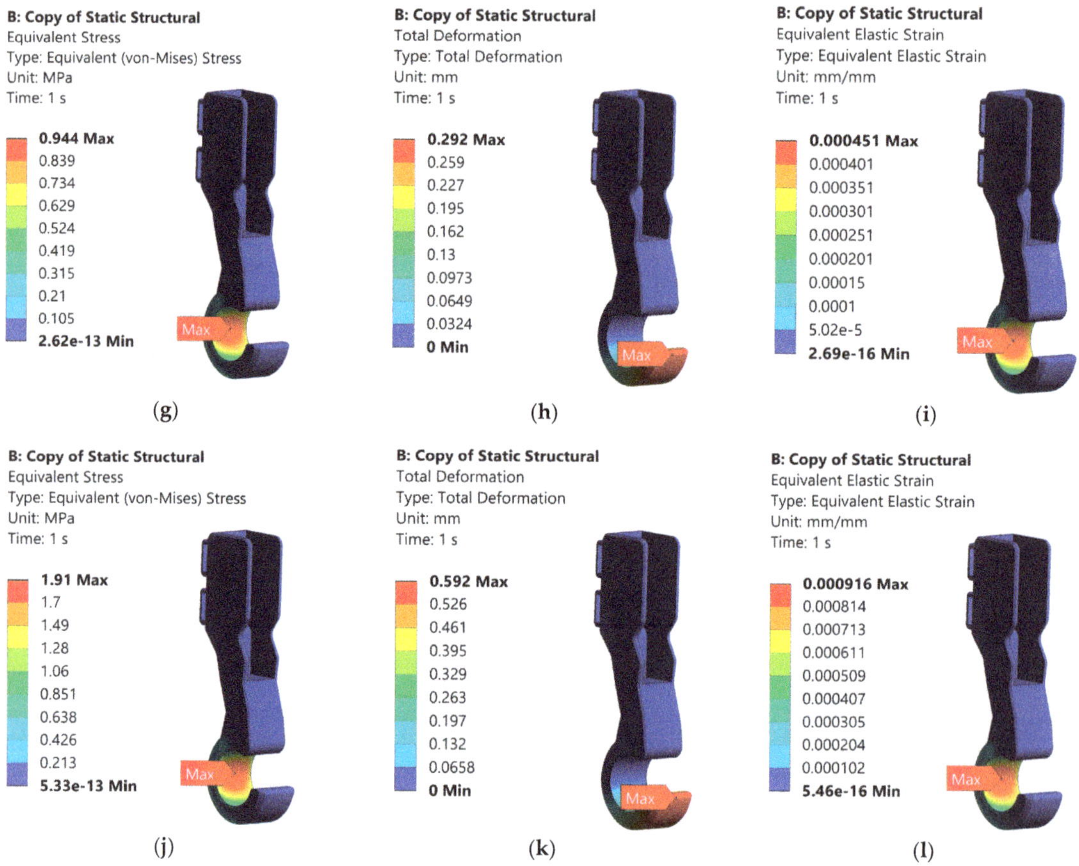

Figure 5. (**a**–**c**) represent the results for Yorkshire Terrier, (**d**–**f**) represent the results for Cocker Spaniel, (**g**–**i**) represent the results for German Shepherd, and (**j**–**l**) represent the results for Newfoundland using solid model.

Table 5. Mechanical properties of different breeds using solid model.

Breed	Load (N)	Equivalent Stress (MPa)	Total Deformation (mm)	Equivalent Strain (mm/mm)
Yorkshire Terrier	8.83	0.08	0.03	0.000039859
Cocker Spaniel	41.20	0.39	0.12	0.000186010
German Shepherd	100.06	0.94	0.29	0.000451730
Newfoundland	203.07	1.92	0.59	0.000916750

3.4. Criteria for Analysis

To proceed ahead with the analysis of the data obtained from the simulations, different criteria need to be set. The first criterion involves the evaluation of the Von Mises Stress obtained:

$$\text{Von Mises Stress} \leq \frac{Yield\ Strength}{Factor\ of\ Safety\ (FoS)} \tag{1}$$

Using the Yield Strength of high impact ABS Plastic of 27.44 MPa (value from ANSYS), and assuming a Factor of Safety of 2, the critical value for Von Mises Stress becomes:

$$\text{Von Mises Stress} \leq 13.72 \text{ MPa} \tag{2}$$

The next criterion is that the Total Deformation should follow [17–19]:

$$1 \text{ mm} \leq \text{Total Deformation} \leq 2\text{–}3 \text{ mm} \tag{3}$$

There are currently no formal industry or veterinary standards that specify an exact deformation range, such as 1–3 mm, for prosthetic limbs in animals. However, this range is commonly referenced in prosthetic design literature based on observations of natural paw pad biomechanics and gait behavior. Research on canine paw pads has shown that during typical walking and standing, the soft tissues compress by up to a few millimeters, helping to absorb ground reaction forces and reduce impact-related stress. Studies such as those by Miao et al. [17] and other biomechanical analyses indicate that paw pads can reduce peak forces by up to 37%, highlighting their important role in shock absorption and load distribution. Based on this, prosthetic designs that produce less than 1 mm of deformation are often considered too stiff to replicate natural behavior, while deformation in the range of 1–2 mm offers moderate compliance. Deformations between 2 and 3 mm are viewed as optimal, as they more closely mimic the mechanical response of the biological paw pad during motion. While this range is not collected in an official guideline, it is supported by empirical evidence and serves as a useful design target in the development of realistic, comfortable, and functional prosthetics for animals. For static load-bearing evaluations in animal prosthetics:

- Ideal deformation: 1–3 mm
- Below 1 mm: too rigid
- Above 3 mm: too soft/unstable

These values should be interpreted in context with stress/strain results, material choice, and gait observations if available. The final criterion involves computing the critical value for the Equivalent Elastic Strain which can be calculated as:

$$\text{Equivalent Elastic Strain} \leq \frac{Yield\ Strength}{Young's\ Modulus} \tag{4}$$

Using the Yield Strength of high impact ABS Plastic of 27.44 MPa (value from ANSYS), and a Young's Modulus value of 2090 MPa (value from ANSYS), the critical value for Equivalent Elastic Strain becomes:

$$\text{Equivalent Elastic Strain} \leq 0.0131 \tag{5}$$

From Equation (1) the minimum Factor of Safety was computed from the highest value of the von mises stress for both models to compare the two models.

4. Discussion

Before analyzing the results, data obtained in Tables 4 and 5 show a proportional increase in the Von Mises stress with increasing load.

Figure 6 illustrates a clear trend: as the applied load increases on both models, the equivalent (von Mises) stress increases correspondingly, demonstrating a direct and nearly linear relationship between load and stress. However, what stands out in the comparison between Figures 2b and 6a is the magnitude of the stress values. The Perforated Model in Figure 6a consistently exhibits much higher equivalent stress levels under the same loading conditions than the Solid Model in Figure 6b. This discrepancy is primarily attributed to differences in geometry and structural integrity. The Perforated Model includes design features such as holes and grooves, which significantly reduce its effective cross-sectional area. With less material available to distribute the applied load, stress is concentrated in the remaining structure, particularly around the geometric discontinuities. In contrast, the Solid Model retains a uniform cross-section, which allows for more even load distribution and better structural stiffness. As a result, it experiences lower stress under identical loading conditions. These findings underscore the impact of geometric design on mechanical performance, especially in components subjected to increasing loads. The presence of perforations weakens the structure's ability to withstand stress, making it more susceptible to failure if not properly accounted for in design. Higher stress concentrations are more likely to occur in the thinner sections of the Perforated Model, as these areas are less capable of distributing the applied load evenly. This phenomenon is clearly visible in Figure 4, where the stress distribution is heavily concentrated around the narrow regions and near the edges of the holes. These localized stresses are a direct result of the geometry, which introduces discontinuities and reduces the effective load-bearing cross-sectional area. In addition to geometric factors, the Perforated Model also exhibits reduced overall structural stiffness compared to a solid counterpart. The presence of openings and grooves in the design increases flexibility, making the structure more prone to deformation when subjected to load. This added flexibility can lead to localized bending, which in turn elevates stress in certain regions, especially near notches or sharp transitions. The combination of reduced stiffness and increased deformation results in more severe stress localization, potentially accelerating the onset of material fatigue or failure. Thus, while the perforated design may offer benefits in weight reduction or design adaptability, it comes at the cost of mechanical performance and stability under stress. Following the criteria listed above in Section 3.4, the following tables were developed:

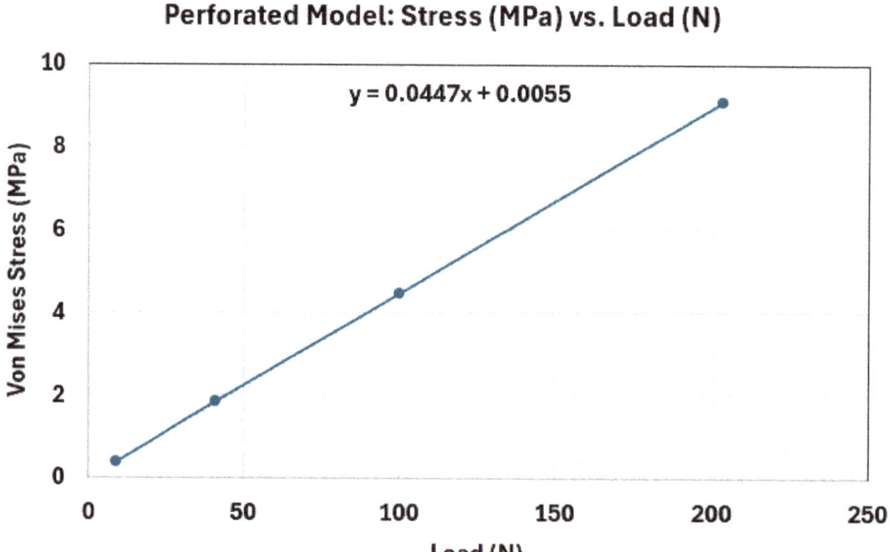

(a)

Figure 6. *Cont.*

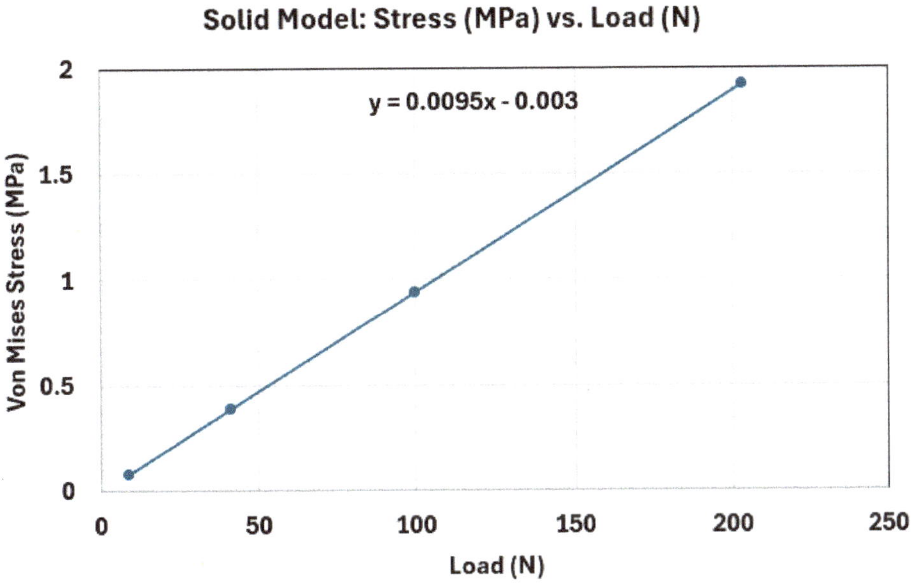

(b)

Figure 6. (**a**) Stress vs. Load for the Perforated Model; (**b**) Stress vs. Load for the Solid Model.

Tables 6 and 7 clearly show that the calculated Von Mises stress values for all the simulated dog breeds remain well below the critical threshold of 13.72 MPa for both the Solid and Perforated prosthetic models. This critical value represents the material's yield strength—essentially the point beyond which permanent deformation would begin to occur. Since none of the stress values approach or exceed this limit, it indicates that the material remains within its elastic range throughout the loading conditions tested. In practical terms, this means that under the simulated loading conditions—representing the range of weights and activities typical for the various dog breeds—the prosthetic limb designs are mechanically sound. The material can withstand the applied forces without any risk of structural failure or irreversible deformation. As a result, both designs are deemed safe and reliable from an engineering standpoint, ensuring the prosthetics will maintain their shape and functionality during regular use. This also highlights the suitability of the chosen material and design approach for veterinary prosthetic applications, where safety, durability, and comfort are critical.

Table 6. Analysis of Von Mises Stress for the Perforated Model.

Breed	Von Mises Stress (MPa)	≤Critical Von Mises Stress
Yorkshire Terrier	0.39549	Yes
Cocker Spaniel	1.84560	Yes
German Shepherd	4.48220	Yes
Newfoundland	9.09620	Yes

Table 7. Analysis of Von Mises Stress for the Solid Model.

Breed	Von Mises Stress (MPa)	≤Critical Von Mises Stress
Yorkshire Terrier	0.083303	Yes

Table 7. *Cont.*

Breed	Von Mises Stress (MPa)	≤Critical Von Mises Stress
Cocker Spaniel	0.388750	Yes
German Shepherd	0.944100	Yes
Newfoundland	1.916000	Yes

Tables 8 and 9 provide insights into the total deformation experienced by the prosthetic footpad models across different dog breeds. While most of the deformation values recorded in both models fall outside the optimal range of 1 mm to 3 mm, the data from Table 8 (Perforated Model) shows values that are noticeably closer to this desired range compared to those in Table 9, which represents the Solid Model. Notably, for the German Shepherd, the Perforated Model exhibited a total deformation that falls within the ideal range. This range is not arbitrary. As supported by prior studies [17–19], a total deformation below 1 mm indicates a structure that is too stiff, offering little to no shock absorption. In other words, the model does not compress or yield under pressure the way a natural dog paw would, potentially resulting in discomfort or long-term stress on the limb. A deformation between 1 and 2 mm is considered acceptable, suggesting moderate cushioning. A value between 2 and 3 mm is ideal—it means the prosthetic footpad mimics the compression behavior of a real paw, offering a balance between support and shock absorption. This biomechanical realism becomes even more important in dynamic activities like walking or running, where effective shock absorption helps prevent stress-related injuries and improves gait comfort. On the other end of the spectrum, deformation values exceeding 3 mm suggest excessive softness. In such cases, the prosthetic may lack the mechanical stability needed to properly support the animal's weight, which could lead to instability or premature failure. As illustrated in Figure 4k, a significant displacement of the model from the wireframe reflects deformation that may compromise structural integrity. In summary, the closer alignment of the Perforated Model's deformation values to the ideal range, particularly in the case of the German Shepherd, highlights its greater potential to deliver a more natural and biomechanically effective solution. This suggests that the Perforated Model may offer a better balance of stiffness and flexibility, allowing the prosthetic to function more like a biological paw and promoting better adaptation by the animal.

Table 8. Analysis of Total Deformation for the Perforated Model.

Breed	Total Deformation (mm)	1 mm ≤ Total Deformation ≤ 2–3 mm
Yorkshire Terrier	0.14858	No
Cocker Spaniel	0.69335	No
German Shepherd	1.68390	Yes
Newfoundland	3.41720	No

Table 9. Analysis of Total Deformation for the Solid Model.

Breed	Total Deformation (mm)	1 mm ≤ Total Deformation ≤ 2–3 mm
Yorkshire Terrier	0.02571	No
Cocker Spaniel	0.11998	No
German Shepherd	0.29138	No
Newfoundland	0.59134	No

Tables 10 and 11 present the Equivalent Strain values for various dog breeds tested using both the Perforated and Solid prosthetic footpad models. In every case, the measured strain is well below the critical threshold of 0.0131 mm/mm. This is an encouraging outcome, as it confirms that all tested configurations remain within the elastic deformation range of the material. In other words, the models are experiencing reversible strain under load—meaning the material can return to its original shape once the load is removed. Remaining far below this strain limit is crucial for ensuring long-term structural integrity and functionality of the prosthetic device. High strain values approaching or exceeding the material's elastic limit could lead to permanent deformation, compromising the prosthetic's performance and safety. However, the results in both models suggest there is a significant margin before such concerns arise, indicating a low risk of material fatigue or failure even under repeated or prolonged use. This is especially important in prosthetic applications where durability and reliability are essential for the comfort and mobility of the animal. By staying well within safe strain limits, both the Perforated and Solid models demonstrate strong potential for real-world use, offering mechanical stability without posing undue stress on the limb or the prosthetic structure itself. Therefore, in terms of equivalent strain performance, both models meet the safety criteria and are suitable for further development or clinical application.

Table 10. Analysis of Equivalent Strain for the Perforated Model.

Breed	Equivalent Strain (mm/mm)	Equivalent Strain $\leq$ 0.0131
Yorkshire Terrier	0.00019011	Yes
Cocker Spaniel	0.00088719	Yes
German Shepherd	0.00215460	Yes
Newfoundland	0.00437260	Yes

Table 11. Analysis of Equivalent Strain for the Solid Model.

Breed	Equivalent Strain (mm/mm)	Equivalent Strain $\leq$ 0.0131
Yorkshire Terrier	0.000039859	Yes
Cocker Spaniel	0.000186010	Yes
German Shepherd	0.000451730	Yes
Newfoundland	0.000916750	Yes

By calculating the minimum Factor of Safety (FoS) for both the Perforated and Solid prosthetic models (Table 12), we gain important insights into their structural reliability under load. In engineering, a FoS greater than 2 is generally considered safe for most applications, and in this case, both models comfortably exceed that threshold. The Solid Model, with an exceptionally high FoS of 14.3, clearly demonstrates that it is structurally sound—but also suggests that it is overengineered for its intended use as a canine forelimb prosthetic. This excessive safety margin means the model uses significantly more material than necessary, leading to an unnecessarily heavy design that could be uncomfortable or impractical for a dog to wear. Furthermore, the added material increases manufacturing costs, making the design less efficient and economically viable. On the other hand, the Perforated Model achieves a more balanced result with a FoS of 3.01. This value is still well above the safety threshold, indicating reliable structural performance, but without the excessive material use found in the Solid Model. Its lighter weight makes it more suitable for animal prosthetics, where comfort, mobility, and realistic biomechanics are crucial. When this safety margin is considered alongside other important performance metrics—such as lower total deformation and adequate strain response—the Perforated Model proves to be not only structurally secure but also a more practical and cost-effective solution. Its design better mimics the natural feel and function of a canine limb, especially under static, load-bearing conditions like standing, making it an ideal candidate for real-world application.

Table 12. Analysis of the Minimum Factor of Safety of Both Models.

Model	Minimum Factor of Safety
Perforated Model	3.01
Solid Model	14.3

5. Conclusions

Based on the findings from the Finite Element Analysis (FEA) simulations, the Solid Model demonstrated the strongest mechanical performance across all tested criteria. It consistently exhibited the lowest values for Von Mises stress, total deformation, and equivalent strain under all applied static loads. Additionally, it achieved a high Factor of Safety (FoS) of 14, clearly indicating exceptional structural reliability and a significant margin before failure. However, despite its superior mechanical performance, the Solid Model is not considered a practical choice for real-world application—especially for canine prosthetics—due to its excessive material use, resulting in a heavier structure. This not only affects user comfort but also presents challenges for large-scale production in terms of cost and material efficiency. In contrast, the Perforated Model offers a more balanced and realistic solution. While it did not outperform the Solid Model in all mechanical metrics, it still fell well within acceptable thresholds. The Von Mises stress values remained comfortably below the critical limit of 13.72 MPa, and its total deformation ranged closer to the ideal 1–3 mm target. This range is considered optimal for mimicking the cushioning effect of a natural paw, which is essential for comfort and shock absorption, particularly under load-bearing conditions such as standing. Furthermore, the Perforated Model maintained a Factor of Safety of 3.01—still well above the acceptable threshold for safety. Its reduced material usage makes it lighter, more cost-effective, and easier to manufacture, without sacrificing necessary structural integrity. It is important to note that this study was limited to static loading scenarios. In real-world settings, animals experience dynamic forces while walking, running, and changing gait. Future work should include testing under cyclic or impact loads to better simulate these conditions. Additionally, experimental validation of the FEA results, improved modeling of limb-socket interaction, and the application of more realistic boundary conditions would significantly enhance the accuracy and applicability of the study. Overall, while the Solid Model sets the benchmark for strength, the Perforated Model strikes a better balance between safety, comfort, and practicality for canine prosthetic applications.

Author Contributions: Conceptualization, K.K. and J.S.; methodology, K.K. and J.S.; software, J.S. and M.S.; validation, K.K. and J.S.; formal analysis, K.K. and J.S.; investigation, K.K. and J.S.; data curation, J.S. and M.S.; writing—original draft preparation, K.K., J.S. and M.S.; writing—review and editing, K.K. and J.S.; visualization, J.S. and M.S.; supervision, K.K.; project administration, K.K. All authors have read and agreed to the published version of the manuscript.

Funding: This research is funded by Institute for Society and Technology—UM Flint and The Hagerman Center for Entrepreneurship & Innovation.

Data Availability Statement: Data are contained within the article.

Conflicts of Interest: The authors declare no conflicts of interest.

References

1. Bhuiyan, M.Z.H.; Khanafer, K.; Rafi, E.I.; Shihab, M.S. Non-Linear Hyperelastic Model Analysis and Numerical Validation of 3D Printed PLA+ Material Incorporating Various Infill Densities. *Machines* **2025**, *13*, 91. [CrossRef]
2. Bhuiyan, Z.H.M.; Khanafer, K. Exploring the Nonlinear Mechanical Characteristics of 3D-Printed ABS with Varying Infill Densities. *J. Manuf. Mater. Process.* **2025**, *9*, 103. [CrossRef]
3. Singh, S.; Ramakrishna, S. Biomedical Applications of Additive Manufacturing: Present and Future. *Curr. Opin. Biomed. Eng.* **2017**, *2*, 105–115. [CrossRef]
4. Rahman, M.; Bennett, T.; Glisson, D.; Beckley, D.; Khan, J. Finite Element Analysis of Prosthetic Running Blades Using Different Composite Materials to Optimize Performance. In Proceedings of the ASME 2014 International Mechanical Engineering Congress and Exposition, Montreal, QC, Canada, 14–20 November 2014; Volume 10: Micro- and Nano-Systems Engineering and Packaging.
5. Ismail, R.; Fitriyana, D.F.; Habibi, M.A.; Sugiyanto; Bayuseno, A. Computational Analysis of Running Prosthesis Design Using Finite Element Method. In Proceedings of the 2020 7th International Conference on Information Technology, Computer, and Electrical Engineering (ICITACEE), Semarang, Indonesia, 24–25 September 2020; pp. 288–293. [CrossRef]
6. Mich, P.M. The Emerging Role of Veterinary Orthotics and Prosthetics (V-OP) in Small Animal Rehabilitation and Pain Management. *Top. Companion Anim. Med.* **2014**, *29*, 10–19. [CrossRef] [PubMed]
7. Carr, B.J.; Canapp, S.; Petrovitch, J.L.; Campana, D.; Canapp, D.; Leasure, C.S. Retrospective Study on External Canine Limb Prosthesis used in 24 patients. *Vet. Evid.* **2018**, *3*, 1–13. [CrossRef]
8. Anderson, D.E.; Jean, G.S. Management of Fractures in Field Settings. *Vet. Clin. N. Am. Food Anim. Pract.* **2008**, *24*, 567–582. [CrossRef] [PubMed]
9. Teixeira, M.; Belinha, J. Review: Prosthetic Limb Replacement in Dogs. In *Advances and Current Trends in Biomechanics*; CRC Press: Boca Raton, FL, USA, 2021. [CrossRef]
10. Wendland, T.M.; Seguin, B.; Duerr, F.M. Prospective Evaluation of Canine Partial Limb Amputation with Socket Prostheses. *Vet. Med. Sci.* **2023**, *9*, 1521–1533. [CrossRef] [PubMed]
11. Arauz, P.G.; Chiriboga, P.; García, M.-G.; Kao, I.; Díaz, E.A. New Technologies Applied to Canine Limb Prostheses: A Review. *Vet. World* **2021**, *14*, 2793–2802. [CrossRef] [PubMed]
12. Popov, V.V.; Muller-Kamskii, G.; Katz-Demyanetz, A.; Kovalevsky, A.; Usov, S.; Trofimcow, D.; Dzhenzhera, G.; Koptyug, A. Additive Manufacturing to Veterinary Practice: Recovery of Bony Defects after the Osteosarcoma Resection in Canines. *Biomed. Eng. Lett.* **2019**, *9*, 97–108. [CrossRef] [PubMed]
13. Kastlunger, T. Design of Prototype Prosthesis for a Canine with a Right Front Limb Deformity as an Alternate Approach to Stabilize Gait and Withstand Gait Forces. Master's Thesis, California Polytechnic State University, San Luis Obispo, CA, USA, 2020. [CrossRef]
14. Motijo-Valenzuel, E.; Ruiz-Navarro, A.; Silva-Conreras, A.; Rubalcava-Aranda, J.; Samano-Hermosillo, E. Design and Validation with Finite Element Analysis of a Veterinary Application Prosthesis for *Cervus nippon*. October 2021. Revista de Investigación Académica Sin Frontera División de Ciencias Econômicas y Sociales. Available online: https://www.researchgate.net/publication/355831761_Design_and_validation_with_finite_element_analysis_of_a_veterinary_application_prosthesis_for_Cervus_nippon (accessed on 7 July 2025).
15. Dog Weight Chart: Weights of Popular Dog Breeds. Available online: https://dogslim.com/articles/dog-weight-chart.html (accessed on 7 July 2025).
16. Fish, F.E.; Sheehan, M.J.; Adams, D.S.; Tennett, K.A.; Gough, W.T. A 60:40 Split: Differential Mass Support in Dogs. *Anat. Rec.* **2021**, *304*, 78–89. [CrossRef] [PubMed]
17. Miao, H.; Fu, J.; Qian, Z.; Ren, L.; Ren, L. How does the canine paw pad attenuate ground impacts? A multi-layer cushion system. *Biol. Open* **2017**, *6*, 1889–1896. [CrossRef] [PubMed]
18. Sierra-Rodríguez, A.; Molines-Barroso, R.J.; Martínez, M.A.; Abenojar-Buendía, J.; Lázaro-Martínez, J.L.; Orejana-García, A.M. Reliability and Validity of the Orthotic Deformation Test. *Appl. Sci.* **2024**, *14*, 8378. [CrossRef]
19. Mei, Y. Investigation of the Microstructure and Mechanical Properties of the Domestic Cat's Paw Pads. *Thai J. Vet. Med.* **2022**, *52*, 359–367. [CrossRef]

MDPI AG
Grosspeteranlage 5
4052 Basel
Switzerland
Tel.: +41 61 683 77 34

Designs Editorial Office
E-mail: designs@mdpi.com
www.mdpi.com/journal/designs

www.ingramcontent.com/pod-product-compliance
Lightning Source LLC
Chambersburg PA
CBHW040813250926
51850CB00007BA/1155
* 9 7 8 3 7 2 5 8 5 2 8 7 1 *